STUDENT SOLUTIONS MANUAL

C TRIMBLE & ASSOCIATES

BASIC COLLEGE MATHEMATICS
SEVENTH EDITION

John Tobey
North Shore Community College

Jeffrey Slater
North Shore Community College

Jamie Blair
Orange Coast College

Jennifer Crawford
Normandale Community College

PEARSON

Boston Columbus Indianapolis New York San Francisco Upper Saddle River
Amsterdam Cape Town Dubai London Madrid Milan Munich Paris Montreal Toronto
Delhi Mexico City Sao Paulo Sydney Hong Kong Seoul Singapore Taipei Tokyo

Reproduced by Pearson from electronic files supplied by the author.

Copyright © 2012, 2009, 2005, 2002 Pearson Education, Inc.
Publishing as Pearson, 75 Arlington Street, Boston, MA 02116.

ISBN-13: 978-0-321-76571-0
ISBN-10: 0-321-76571-0

1 2 3 4 5 6 BRR 16 15 14 13 12

www.pearsonhighered.com

PEARSON

Contents

Chapter 1

1.1 Exercises

1. $6731 = 6000 + 700 + 30 + 1$

3. $108,276 = 100,000 + 8000 + 200 + 70 + 6$

5. $23,761,345$
   $= 20,000,000 + 3,000,000 + 700,000$
   $+ 60,000 + 1000 + 300 + 40 + 5$

7. $103,260,768$
   $= 100,000,000 + 3,000,000 + 200,000$
   $+ 60,000 + 700 + 60 + 8$

9. $600 + 70 + 1 = 671$

11. $9000 + 800 + 60 + 3 = 9863$

13. $40,000 + 800 + 80 + 5 = 40,885$

15. $700,000 + 6000 + 200 = 706,200$

17. a. The digit 7 tells the number of thousands.

    b. The value of the digit 3 is 30,000.

19. a. The digit 2 tells the number of hundred thousands.

    b. The value of the digit 2 is 200,000.

21. $142 =$ one hundred forty-two

23. $9304 =$ nine thousand, three hundred four

25. $36,118 =$ thirty-six thousand, one hundred eighteen

27. $105,261 =$ one hundred five thousand, two hundred sixty-one

29. $14,203,326 =$ fourteen million, two hundred three thousand, three hundred twenty-six

31. $4,302,156,200 =$ four billion, three hundred two million, one hundred fifty-six thousand, two hundred

33. 1561

35. 33,809

37. 100,079,826

39. The word name should be one thousand, nine hundred sixty-five.

41. The estimated population of New York in 1910 was 9 million or 9,000,000.

43. The estimated population of California in 2020 is 42 million or 42,000,000.

45. In 2000, Delta had 946 thousand or 946,000 flights.

47. In 2005, 54,588 thousand or 54,588,000 passengers flew on Northwest flights.

49. a. The digit 5 tells the number of ten thousands.

    b. The digit 2 tells the number of ten billions.

51. a. The digit 2 tells the number of ten thousands.

    b. The digit 1 tells the number of ten millions.

53. 613,001,033,208,003

55. three quintillion, six hundred eighty-two quadrillion, nine hundred sixty-eight trillion, nine billion, nine hundred thirty-one million, nine hundred sixty thousand, seven hundred forty-seven

57. You would obtain 2 E 20. This is 200,000,000,000,000,000,000 in standard notation.

Quick Quiz 1.1

1. $73,952 = 70,000 + 3000 + 900 + 50 + 2$

2. $8,932,475 =$ eight million, nine hundred thirty-two thousand, four hundred seventy-five

3. 964,257

4. Answers may vary. Possible solution: The zeros act as placeholders so the place value of each digit is in the right place: three hundred sixty-eight million, five hundred twenty-two $= 368,000,522$.

1.2 Exercises

1. Answers may vary. Samples are below.

    a. You can change the order of the addends without changing the sum.

    b. You can group the addends in any way without changing the sum.

3.

+	3	5	4	8	0	6	7	2	9	1
2	5	7	6	10	2	8	9	4	11	3
7	10	12	11	15	7	13	14	9	16	8
5	8	10	9	13	5	11	12	7	14	6
3	6	8	7	11	3	9	10	5	12	4
0	3	5	4	8	0	6	7	2	9	1
4	7	9	8	12	4	10	11	6	13	5
1	4	6	5	9	1	7	8	3	10	2
8	11	13	12	16	8	14	15	10	17	9
6	9	11	10	14	6	12	13	8	15	7
9	12	14	13	17	9	15	16	11	18	10

5.
$$\begin{array}{r} 4 \\ 2 \\ 8 \\ +\ 9 \\ \hline 23 \end{array}$$

7.
$$\begin{array}{r} 2 \\ 6 \\ 7 \\ 8 \\ +\ 3 \\ \hline 26 \end{array}$$

9.
$$\begin{array}{r} 18 \\ 36 \\ +\ 3 \\ \hline 57 \end{array}$$

11.
$$\begin{array}{r} 63 \\ 24 \\ +12 \\ \hline 99 \end{array}$$

13.
$$\begin{array}{r} 3315 \\ 726 \\ +\ 84 \\ \hline 4125 \end{array}$$

15.
$$\begin{array}{r} 5631 \\ 2344 \\ +2019 \\ \hline 9994 \end{array}$$

17.
$$\begin{array}{r} 8235 \\ +5626 \\ \hline 13,861 \end{array}$$

19.
$$\begin{array}{r} 62,504 \\ +54,736 \\ \hline 117,240 \end{array}$$

21.
$$\begin{array}{r} 36 \\ 41 \\ 25 \\ 6 \\ +13 \\ \hline 121 \end{array} \qquad \begin{array}{r} 13 \\ 6 \\ 25 \\ 41 \\ +36 \\ \hline 121 \end{array}$$

23.
$$\begin{array}{r} 207 \\ 15 \\ 3 \\ 57 \\ +861 \\ \hline 1143 \end{array} \qquad \begin{array}{r} 861 \\ 57 \\ 3 \\ 15 \\ +207 \\ \hline 1143 \end{array}$$

25.
$$\begin{array}{r} 85 \\ 256 \\ 55 \\ +9734 \\ \hline 10,130 \end{array}$$

27.
$$\begin{array}{r} 1,362,214 \\ 7,002,316 \\ +3,214,896 \\ \hline 11,579,426 \end{array}$$

29.
$$\begin{array}{r} 837,241,000 \\ +298,039,240 \\ \hline 1,135,280,240 \end{array}$$

31.
$$\begin{array}{r} 516,208 \\ 24,317 \\ +1,763,295 \\ \hline 2,303,820 \end{array}$$

33. $25 + 130 + 70 + 75 = 30$

35. $102 + 50 + 98 + 35 + 50 = 335$

37.
$$\begin{array}{r} 455 \\ 186 \\ +\ 82 \\ \hline 723 \end{array}$$
The total amount of money Stephanie spent was $723.

39.
$$\begin{array}{r} 1875 \\ 1930 \\ +1744 \\ \hline 5549 \end{array}$$
The total amount for the three months is $5549.

41.
$$\begin{array}{r} 124 \\ 105 \\ 147 \\ +\ 92 \\ \hline 468 \end{array}$$
468 feet of fencing are needed.

43.
$$\begin{array}{r} 840,000 \\ 88,407 \\ +39,699 \\ \hline 968,106 \end{array}$$
The total area is 968,106 square miles.

45.
$$\begin{array}{r} 7,272,320 \\ 5,104,000 \\ +4,558,400 \\ \hline 16,934,720 \end{array}$$
The total length of these rivers is 16,934,720 yards.

47. a.　　415
　　　　　364
　　　　　159
　　　　+ 196
　　　　‾‾‾‾‾
　　　　1134
1134 students were eligible.

b.　　1134
　　　　　 27
　　　　　 68
　　　　　102
　　　　+　61
　　　　‾‾‾‾‾
　　　　1392
1392 students were considered in all.

49.　　87
　　　　　17
　　　　+ 98
　　　　‾‾‾‾
　　　　202
She drove 202 miles.

51.　　568　　　　　2387
　　　　　682　　　　 −1953
　　　　+ 703　　　　‾‾‾‾‾
　　　　‾‾‾‾‾　　　　 434
　　　　1953
The fourth side is 434 feet.

53. a.　5276
　　　　　2437
　　　　+ 1840
　　　　‾‾‾‾‾
　　　　9553
The total cost for an out-of-state U.S. citizen is $9553.

b.　　3640
　　　　　1926
　　　　+ 1753
　　　　‾‾‾‾‾
　　　　7319
The total cost for an in-state U.S. citizen is $7319.

c.　　8352
　　　　　2855
　　　　+ 1840
　　　　‾‾‾‾‾
　　　　13,047
The total cost for a foreign student is $13,047.

55. $89 + 166 + 23 + 45 + 72 + 190 + 203 + 77 + 18 + 93 + 46 + 73 + 66 = 1161$

57. Answers may vary. A sample is: You could not group the addends in groups that sum to 10s to make column addition easier.

Cumulative Review

58. 76,208,941 = seventy-six million, two hundred eight thousand, nine hundred forty-one

59. 121,000,374 = one hundred twenty-one million, three hundred seventy-four

60. eight million, seven hundred twenty-four thousand, three hundred ninety-six = 8,724,396

61. nine million, fifty-one thousand, seven hundred nineteen = 9,051,719

62. twenty-eight million, three hundred eighty-seven thousand, eighteen = 28,387,018

Quick Quiz 1.2

1.　　56
　　　　　38
　　　　　92
　　　　　17
　　　　+　9
　　　　‾‾‾‾
　　　　212

2.　　831
　　　　　276
　　　　+ 508
　　　　‾‾‾‾‾
　　　　1615

3.　　681,302
　　　　　 5,126
　　　　　18,371
　　　　+ 300,012
　　　　‾‾‾‾‾‾‾
　　　　1,004,811

4. Answers may vary. Possible solution:
　　　 1 2 2
　　　4567
　　　3189
　　+　895
　　‾‾‾‾‾
　　　8651
$7 + 9 + 5 = 21$ (carry the 2)
$2 + 6 + 8 + 9 = 25$ (carry the 2)
$2 + 5 + 1 + 8 = 16$ (carry the 1)
$1 + 4 + 3 = 8$

1.3 Exercises

1. In subtraction, the minuend minus the subtrahend equals the difference. To check, we add the subtrahend and the difference to see if we get the minuend. If we do, the answer is correct.

3. We know that $1683 + 1592 = 32?5$. Therefore, if we add 8 tens and 9 tens, we get 17 tens which is 1 hundred and 7 tens. Then the ? should be replaced by 7.

5.
$$\begin{array}{r} 8 \\ -3 \\ \hline 5 \end{array}$$

7.
$$\begin{array}{r} 15 \\ -9 \\ \hline 6 \end{array}$$

9.
$$\begin{array}{r} 16 \\ -0 \\ \hline 16 \end{array}$$

11.
$$\begin{array}{r} 18 \\ -9 \\ \hline 9 \end{array}$$

13.
$$\begin{array}{r} 11 \\ -4 \\ \hline 7 \end{array}$$

15.
$$\begin{array}{r} 13 \\ -7 \\ \hline 6 \end{array}$$

17.
$$\begin{array}{r} 11 \\ -8 \\ \hline 3 \end{array}$$

19.
$$\begin{array}{r} 15 \\ -6 \\ \hline 9 \end{array}$$

21.
$$\begin{array}{r} 47 \\ -26 \\ \hline 21 \end{array}$$
Check: $\begin{array}{r} 26 \\ +21 \\ \hline 47 \end{array}$

23.
$$\begin{array}{r} 85 \\ -73 \\ \hline 12 \end{array}$$
Check: $\begin{array}{r} 73 \\ +12 \\ \hline 85 \end{array}$

25.
$$\begin{array}{r} 379 \\ -36 \\ \hline 343 \end{array}$$
Check: $\begin{array}{r} 36 \\ +343 \\ \hline 379 \end{array}$

27.
$$\begin{array}{r} 869 \\ -548 \\ \hline 321 \end{array}$$
Check: $\begin{array}{r} 548 \\ +321 \\ \hline 869 \end{array}$

29.
$$\begin{array}{r} 4799 \\ -596 \\ \hline 4203 \end{array}$$
Check: $\begin{array}{r} 596 \\ +4203 \\ \hline 4799 \end{array}$

31.
$$\begin{array}{r} 155,835 \\ -12,600 \\ \hline 143,235 \end{array}$$
Check: $\begin{array}{r} 12,600 \\ +143,235 \\ \hline 155,835 \end{array}$

33.
$$\begin{array}{r} 986,302 \\ -433,201 \\ \hline 553,101 \end{array}$$
Check: $\begin{array}{r} 433,201 \\ +553,101 \\ \hline 986,302 \end{array}$

35.
$$\begin{array}{r} 19 \\ +110 \\ \hline 129 \end{array}$$
Correct

37.
$$\begin{array}{r} 3215 \\ +\ 5781 \\ \hline 8996 \end{array}$$
Incorrect
$$\begin{array}{r} 8596 \\ -\ 3215 \\ \hline 5381 \end{array}$$

39.
$$\begin{array}{r} 5020 \\ +\ 1020 \\ \hline 6040 \end{array}$$
Incorrect
$$\begin{array}{r} 6030 \\ -\ 5020 \\ \hline 1010 \end{array}$$

41.
$$\begin{array}{r} 33,846 \\ +\ 13,023 \\ \hline 46,869 \end{array}$$
Incorrect
$$\begin{array}{r} 47,869 \\ -\ 33,846 \\ \hline 14,023 \end{array}$$

43.
$$\begin{array}{r} 98 \\ -\ 52 \\ \hline 46 \end{array}$$

45.
$$\begin{array}{r} 174 \\ -\ 82 \\ \hline 92 \end{array}$$

47.
$$\begin{array}{r} 647 \\ -\ 263 \\ \hline 384 \end{array}$$

49.
$$\begin{array}{r} 955 \\ -\ 237 \\ \hline 718 \end{array}$$

51.
$$\begin{array}{r} 20,000 \\ -\ 9\ 285 \\ \hline 10,715 \end{array}$$

53.
$$\begin{array}{r} 152,000 \\ -\ 117,908 \\ \hline 34,092 \end{array}$$

55.
$$\begin{array}{r} 45,312 \\ -\ 37,865 \\ \hline 7\ 447 \end{array}$$

57.
$$\begin{array}{r} 2,378,862 \\ -\ 1,469,932 \\ \hline 908,930 \end{array}$$

59. $x + 14 = 19$
$5 + 14 = 19$
$x = 5$

61. $28 = x + 20$
$28 = 8 + 20$
$x = 8$

63. $100 + x = 127$
$100 + 27 = 127$
$x = 27$

65.
$$\begin{array}{r} 1,060,327 \\ -\ 638,017 \\ \hline 422,310 \end{array}$$
Obama received 422,310 votes.

67.
$$\begin{array}{r} 10,707,924 \\ -\ 4,470,510 \\ \hline 6,237,414 \end{array}$$
The population was 6,237,414 less.

69. Total earned = $1280
Total paid out = $318 + $200 = $518
$1280 − $518 = $762 put in checking

71.
$$\begin{array}{r} 5,420,636 \\ -\ 3,806,103 \\ \hline 1,614,533 \end{array}$$
It increased by 1,614,533 people.

73. Illinois = 11,110,285
Indiana + Minnesota = $5,195,392 + 3,806,103$
$= 9,001,495$
Difference $= 11,110,285 − 9,001,495$
$= 2,108,790$ people

75.
$$\begin{array}{r} 11,430,602 \\ -\ 11,110,285 \\ \hline 320,317 \end{array}$$
It increased by 320,317 people.

77.
$$\begin{array}{r} 413,471 \\ -\ 320,317 \\ \hline 93,154 \end{array}$$
It was 93,154 people greater.

79.
$$\begin{array}{r} 225 \\ -\ 132 \\ \hline 93 \end{array}$$
The increase was 93 homes.

81.
$$\begin{array}{r} 96 \\ -\ 83 \\ \hline 13 \end{array}$$
The decrease was 13 homes.

83. change from 2009 to 2010: $150 - 145 = 5$
change from 2010 to 2011: $152 - 145 = 7$
The greatest change occurred between 2010 and 2011.

85. Winchester: 189
Willow Creek: 152
Essex: 83
Manchester: 298
Irving: 59
Harvey: 148
Willow Creek and Harvey were the closest to having the same number of sales in 2011.

87. It is true if a and b represent the same number, for example, if $a = 10$ and $b = 10$.

89.
$$\begin{array}{r} 276 \\ -\ 216 \\ \hline 60 \end{array}$$
Convert 60 feet to the number of 12-foot sections.
Now, $\dfrac{60}{12} = 5$ and each section costs
$60 + $50 = $110.
$$\begin{array}{r} \$110 \\ \times\ \ \ 5 \\ \hline \$550 \end{array}$$

Cumulative Review

91. eight million, four hundred sixty-six thousand, eighty-four = 8,466,084

92. 296,308 = two hundred ninety-six thousand, three hundred eight

93. $25 + 75 + 80 + 20 + 18 = 218$

94.
$$\begin{array}{r} 278,563 \\ +\ 896,187 \\ \hline 1,174,750 \end{array}$$

Quick Quiz 1.3

1.
$$\begin{array}{r} 5392 \\ -\ \ 938 \\ \hline 4454 \end{array}$$

2.
$$\begin{array}{r} 609,240 \\ -\ 386,307 \\ \hline 222,933 \end{array}$$

3.
$$\begin{array}{r} 17,200,300 \\ -\ 11,562,178 \\ \hline 5,638,122 \end{array}$$

4. Answers may vary. Possible solution:
$$\begin{array}{r} {\scriptstyle 12\ \ 13} \\ {\scriptstyle 1\ \ \ 2\ 3\ 15} \\ 12,3\,4\,5 \\ -\ 11,9\,7\ \ 6 \\ \hline 3\ 6\ \ 9 \end{array}$$
Borrow 10 ones from the 4. Change 5 ones to 15 ones. Borrow 10 tens from the 3. Change 3 tens to 13 tens. Borrow 10 hundreds from the 2. Change 2 hundreds to 12 hundreds. Then subtract.

1.4 Exercises

1. Answers may vary. Samples follow.

    a. You can change the order of the factors without changing the product.

    b. You can group the factors in any way without changing the product.

3.

×	6	2	3	8	0	5	7	9	12	4
5	30	10	15	40	0	25	35	45	60	20
7	42	14	21	56	0	35	49	63	84	28
1	6	2	3	8	0	5	7	9	12	4
0	0	0	0	0	0	0	0	0	0	0
6	36	12	18	48	0	30	42	54	72	24
2	12	4	6	16	0	10	14	18	24	8
3	18	6	9	24	0	15	21	27	36	12
8	48	16	24	64	0	40	56	72	96	32
4	24	8	12	32	0	20	28	36	48	16
9	54	18	27	72	0	45	63	81	108	36

5.
$$\begin{array}{r} 32 \\ \times\ 3 \\ \hline 96 \end{array}$$

7.
$$\begin{array}{r} 14 \\ \times 5 \\ \hline 70 \end{array}$$

9.
$$\begin{array}{r} 87 \\ \times\ 6 \\ \hline 522 \end{array}$$

11.
$$\begin{array}{r} 231 \\ \times\ \ 3 \\ \hline 693 \end{array}$$

13.
$$\begin{array}{r} 276 \\ \times\ \ 7 \\ \hline 1932 \end{array}$$

15.
$$\begin{array}{r} 6102 \\ \times\ \ \ \ 3 \\ \hline 18,306 \end{array}$$

17.
$$
\begin{array}{r}
12,203 \\
\times \quad\quad 3 \\
\hline
36,609
\end{array}
$$

19.
$$
\begin{array}{r}
5218 \\
\times \quad\quad 6 \\
\hline
31,308
\end{array}
$$

21.
$$
\begin{array}{r}
12,526 \\
\times \quad\quad 8 \\
\hline
100,208
\end{array}
$$

23.
$$
\begin{array}{r}
344,601 \\
\times \quad\quad 9 \\
\hline
3,101,409
\end{array}
$$

25.
$$
\begin{array}{r}
156 \\
\times \quad 10 \\
\hline
1560
\end{array}
$$

27.
$$
\begin{array}{r}
27,158 \\
\times \quad\quad 100 \\
\hline
2,715,800
\end{array}
$$

29.
$$
\begin{array}{r}
482 \\
\times \quad 1000 \\
\hline
482,000
\end{array}
$$

31.
$$
\begin{array}{r}
37,256 \\
\times \quad 10,000 \\
\hline
372,560,000
\end{array}
$$

33.
$$
\begin{array}{r}
423 \\
\times \quad 20 \\
\hline
8460
\end{array}
$$

35.
$$
\begin{array}{r}
2210 \\
\times \quad 40 \\
\hline
88,400
\end{array}
$$

37.
$$
\begin{array}{r}
14,000 \\
\times \quad 4000 \\
\hline
56,000,000
\end{array}
$$

39.
$$
\begin{array}{r}
514 \\
\times \quad 12 \\
\hline
1028 \\
514 \quad\\
\hline
6168
\end{array}
$$

41.
$$
\begin{array}{r}
146 \\
\times \quad 54 \\
\hline
584 \\
730 \quad\\
\hline
7884
\end{array}
$$

43.
$$
\begin{array}{r}
89 \\
\times \quad 64 \\
\hline
356 \\
534 \quad\\
\hline
5696
\end{array}
$$

45.
$$
\begin{array}{r}
607 \\
\times \quad 25 \\
\hline
3\ 035 \\
12\ 14 \quad\\
\hline
15,175
\end{array}
$$

47.
$$
\begin{array}{r}
544 \\
\times \quad 38 \\
\hline
4352 \\
1\ 632 \quad\\
\hline
20,672
\end{array}
$$

49.
$$
\begin{array}{r}
912 \\
\times \quad 76 \\
\hline
5\ 472 \\
6\ 384 \quad\\
\hline
69,312
\end{array}
$$

51.
$$
\begin{array}{r}
5123 \\
\times \quad 29 \\
\hline
46\ 107 \\
102\ 46 \quad\\
\hline
148,567
\end{array}
$$

53.
$$
\begin{array}{r}
9053 \\
\times \quad 91 \\
\hline
9\ 053 \\
814\ 77 \quad\\
\hline
823,823
\end{array}
$$

55.
$$
\begin{array}{r}
4326 \\
\times \quad 435 \\
\hline
21\ 630 \\
129\ 78 \quad\\
1\ 730\ 4 \quad\quad\\
\hline
1,881,810
\end{array}
$$

57.
```
      458
   ×  314
    1 832
    4 58
  137 4
  143,812
```

59.
```
      2076
   ×   105
    10 380
    00 00
   207 6
   217,980
```

61.
```
       1324
   ×   2004
      5 296
    2 648
    2,653,296
```

63.
```
    12,000
   ×    60
   720,000
```

65.
```
      250
   ×   40
   10,000
```

67.
```
      302
   ×  300
   90,600
```

69. $7 \cdot 2 \cdot 5 = 7 \cdot 10 = 70$

71. $11 \cdot 7 \cdot 4 = 77 \cdot 4 = 308$

73.
```
      412
   ×   33
    1 236
   12 36
   13,596
```

75. $5 \cdot 8 \cdot 4 \cdot 10 = 40 \cdot 4 \cdot 10 = 160 \cdot 10 = 1600$

77. $x = 0$

79. $16 \cdot 24 = 384$
The area of the patio is 384 square feet.

81. $12 \cdot 14 = 168$
$9 \cdot 3 = 27$
$168 + 27 = 195$
They need 195 square feet of carpet.

83.
```
    240
  ×   5
   1200
```
The cost is $1200.

85.
```
    266
  ×  12
    532
    266
   3192
```
The cost for one year is $3192.

87.
```
    34
  × 18
   272
    34
   612
```
He can travel approximately 612 miles.

89.
```
    420
  ×  12
    840
    420
   5040
```
Sylvia earned $5040 during the summer.

91.
```
     32,000,000
  ×         1060
   1 920 000 000
   0 000 000 0
  32 000 000
  33,920,000,000
```
The total yearly income was $33,920,000,000.

93.
```
    18      26      54       36
  × 2     × 0     × 3         0
   36       0     162     + 162
                            198
```
There are 198 black paws in the room.

95.
```
    18      26      54      36
  × 2     × 1     × 0      26
   36      26       0    + 0
                           62
```
There are 62 black ears in the room.

97. $5(x) = 40$
$5(8) = 40$
$x = 8$

99. $72 = 8(x)$
$72 = 8(9)$
$x = 9$

101. No, it would not always be true. In our number system $62 = 60 + 2$. But in roman numerals, $IV \neq I + V$. The digit system in roman numerals involves subtraction. Thus $(XII) \times (IV) \neq (XII \times I) + (XII \times V)$.

Cumulative Review

103.
$$\begin{array}{r} 34,084 \\ -\ 27,328 \\ \hline 6,756 \end{array}$$

104.
$$\begin{array}{r} 263 \\ 27 \\ 891 \\ 5 \\ +\ 63 \\ \hline 1249 \end{array}$$

105. $1278 - (345 + 128) = 1278 - 473 = 805$
There is $805 left in his checking account.

106.
$$\begin{array}{r} 1932 \\ -\ 1772 \\ \hline 160 \end{array}$$
The increase was $160.

107.
$$\begin{array}{r} 45,918 \\ -\ 42,667 \\ \hline 3\ 251 \end{array}$$
The population increased by 3251 people.

108.
$$\begin{array}{r} 1,466,000,000,000 \\ -\ \ 720,800,000,000 \\ \hline 745,200,000,000 \end{array}$$
The increase was $745,200,000,000.

Quick Quiz 1.4

1.
$$\begin{array}{r} 34,986 \\ \times\ \ \ \ \ 5 \\ \hline 174,930 \end{array}$$

2.
$$\begin{array}{r} 79 \\ \times\ 64 \\ \hline 316 \\ 474 \\ \hline 5056 \end{array}$$

3.
$$\begin{array}{r} 698 \\ \times\ \ 297 \\ \hline 4\ 886 \\ 62\ 82 \\ 139\ 6 \\ \hline 207,306 \end{array}$$

4. Answers may vary. Possible solution:
$$\begin{array}{r} 3457 \\ \times\ 2008 \\ \hline 27\ 656 \\ 6\ 914\ 000 \\ \hline 6,941,656 \end{array}$$
The zeros act as placeholders. Think of 3457 multiplied by 2000.

1.5 Exercises

1. a. When you divide a nonzero number by itself, the result is 1.

   b. When you divide a number by 1, the result is that number.

   c. When you divide zero by a nonzero number, the result is zero.

   d. You cannot divide a number by zero. Division by 0 is undefined.

3. $6\overline{)42}$ with quotient 7

5. $8\overline{)24}$ with quotient 3

7. $5\overline{)25}$ with quotient 5

9. $9\overline{)36}$ with quotient 4

11. $7\overline{)21}$ with quotient 3

13. $\dfrac{6}{5\overline{)30}}$

15. $\dfrac{9}{7\overline{)63}}$

17. $\dfrac{9}{8\overline{)72}}$

19. $\dfrac{6}{6\overline{)36}}$

21. $\dfrac{9}{1\overline{)9}}$

23. $\dfrac{0}{10\overline{)0}}$

25. $9 \div 0$
undefined

27. $\dfrac{0}{8} = 0$

29. $6 \div 6 = 1$

31. $\begin{array}{r} 4 \text{ R } 5 \\ 6\overline{)29} \\ \underline{24} \\ 5 \end{array}$

Check: $\begin{array}{r} 6 \\ \times 4 \\ \hline 24 \\ + 5 \\ \hline 29 \end{array}$

33. $\begin{array}{r} 9 \text{ R } 4 \\ 8\overline{)76} \\ \underline{72} \\ 4 \end{array}$

Check: $\begin{array}{r} 8 \\ \times 9 \\ \hline 72 \\ + 4 \\ \hline 76 \end{array}$

35. $\begin{array}{r} 25 \text{ R } 3 \\ 5\overline{)128} \\ \underline{10} \\ 28 \\ \underline{25} \\ 3 \end{array}$

Check: $\begin{array}{r} 25 \\ \times 5 \\ \hline 125 \\ + 3 \\ \hline 128 \end{array}$

37. $\begin{array}{r} 21 \text{ R } 7 \\ 9\overline{)196} \\ \underline{18} \\ 16 \\ \underline{9} \\ 7 \end{array}$

Check: $\begin{array}{r} 21 \\ \times 9 \\ \hline 189 \\ + 7 \\ \hline 196 \end{array}$

39. $\begin{array}{r} 32 \\ 9\overline{)288} \\ \underline{27} \\ 18 \\ \underline{18} \\ 0 \end{array}$

Check: $\begin{array}{r} 32 \\ \times 9 \\ \hline 288 \end{array}$

41. $\begin{array}{r} 37 \\ 5\overline{)185} \\ \underline{15} \\ 35 \\ \underline{35} \\ 0 \end{array}$

Check: $\begin{array}{r} 37 \\ \times 5 \\ \hline 185 \end{array}$

43.
$$
\begin{array}{r}
322 \ R\ 1 \\
4\overline{)1289} \\
\underline{12} \\
8 \\
\underline{8} \\
9 \\
\underline{8} \\
1
\end{array}
$$

45.
$$
\begin{array}{r}
127 \ R\ 1 \\
6\overline{)763} \\
\underline{6} \\
16 \\
\underline{12} \\
43 \\
\underline{42} \\
1
\end{array}
$$

47.
$$
\begin{array}{r}
869 \\
7\overline{)6083} \\
\underline{56} \\
48 \\
\underline{42} \\
63 \\
\underline{63} \\
0
\end{array}
$$

49.
$$
\begin{array}{r}
1238 \ R\ 2 \\
4\overline{)4954} \\
\underline{4} \\
9 \\
\underline{8} \\
15 \\
\underline{12} \\
34 \\
\underline{32} \\
2
\end{array}
$$

51.
$$
\begin{array}{r}
2\ 056 \ R\ 2 \\
8\overline{)16,450} \\
\underline{16} \\
45 \\
\underline{40} \\
50 \\
\underline{48} \\
2
\end{array}
$$

53.
$$
\begin{array}{r}
2\ 562 \ R\ 3 \\
5\overline{)12,813} \\
\underline{10} \\
2\ 8 \\
\underline{2\ 5} \\
31 \\
\underline{30} \\
13 \\
\underline{10} \\
3
\end{array}
$$

55.
$$
\begin{array}{r}
30 \ R\ 5 \\
6\overline{)185} \\
\underline{18} \\
5 \\
\underline{0} \\
5
\end{array}
$$

57.
$$
\begin{array}{r}
5 \ R\ 7 \\
52\overline{)267} \\
\underline{260} \\
7
\end{array}
$$

59.
$$
\begin{array}{r}
8 \\
68\overline{)544} \\
\underline{544} \\
0
\end{array}
$$

61.
$$
\begin{array}{r}
418 \ R\ 8 \\
12\overline{)5024} \\
\underline{48} \\
22 \\
\underline{12} \\
104 \\
\underline{96} \\
8
\end{array}
$$

63.
$$
\begin{array}{r}
48 \ R\ 12 \\
30\overline{)1452} \\
\underline{120} \\
252 \\
\underline{240} \\
12
\end{array}
$$

65.
$$
\begin{array}{r}
845 \\
7\overline{)5915} \\
\underline{56} \\
31 \\
\underline{28} \\
35 \\
\underline{35} \\
0
\end{array}
$$

67.
$$
\begin{array}{r}
210 \text{ R } 8 \\
36\overline{)7568} \\
\underline{72} \\
36 \\
\underline{36} \\
8 \\
\underline{0} \\
8
\end{array}
$$

69.
$$
\begin{array}{r}
14 \text{ R } 2 \\
182\overline{)2550} \\
\underline{182} \\
730 \\
\underline{728} \\
2
\end{array}
$$

71.
$$
\begin{array}{r}
4 \text{ R } 4 \\
174\overline{)700} \\
\underline{696} \\
4
\end{array}
$$

73.
$$
\begin{array}{r}
125 \\
224\overline{)28,000} \\
\underline{22\ 4} \\
5\ 60 \\
\underline{4\ 48} \\
1120 \\
\underline{1120} \\
0
\end{array}
$$

75.
$$
\begin{array}{r}
37 \\
14\overline{)518} \\
\underline{42} \\
98 \\
\underline{98} \\
0
\end{array}
$$
Thus, $518 \div 14 = 37$.
$x = 37$

77.
$$
\begin{array}{r}
61,693 \\
7\overline{)431,851} \\
\underline{42} \\
11 \\
\underline{7} \\
4\ 8 \\
\underline{4\ 2} \\
65 \\
\underline{63} \\
21 \\
\underline{21} \\
0
\end{array}
$$
The daily average is 61,693 runs.

79.
$$
\begin{array}{r}
288 \\
9\overline{)2592} \\
\underline{18} \\
79 \\
\underline{72} \\
72 \\
\underline{72} \\
0
\end{array}
$$
Each pair of skis costs \$288.

81.
$$
\begin{array}{r}
21,053 \\
7\overline{)147,371} \\
\underline{14} \\
7 \\
\underline{7} \\
37 \\
\underline{35} \\
21 \\
\underline{21} \\
0
\end{array}
$$
The cost is \$21,053 per carriage.

83.
$$
\begin{array}{r}
185 \\
56\overline{)10,360} \\
\underline{5\ 6} \\
476 \\
\underline{4\ 48} \\
280 \\
\underline{280} \\
0
\end{array}
$$
Each monitor cost \$185.

85. The smallest number is 330 and
$330 \div 2 = 165$ sandwiches.

87. a. $2 \times 12 = 24$

$$
\begin{array}{r}
1742 \\
\times\ \ 24 \\
\hline
6\,968 \\
34\,84 \\
\hline
41,808
\end{array}
$$

The total distance is 41,808 kilometers.

b.
$$
\begin{array}{r}
50,000 \\
-\ 41,808 \\
\hline
8\,192
\end{array}
$$

The truck can be driven 8192 kilometers more.

89. *a* and *b* must represent the same number. For example, if $a = 12$, then $b = 12$.

Cumulative Review

91.
$$
\begin{array}{r}
108 \\
\times\ 50 \\
\hline
5400
\end{array}
$$

92.
$$
\begin{array}{r}
7162 \\
\times\ \ 145 \\
\hline
35\,810 \\
286\,48 \\
716\,2 \\
\hline
1,038,490
\end{array}
$$

93.
$$
\begin{array}{r}
316,214 \\
+\ 89,981 \\
\hline
406,195
\end{array}
$$

94.
$$
\begin{array}{r}
1,360,000 \\
-\ 1,293,156 \\
\hline
66,844
\end{array}
$$

Quick Quiz 1.5

1.
$$
\begin{array}{r}
467 \\
9\overline{)4203} \\
36 \\
\hline
60 \\
54 \\
\hline
63 \\
63 \\
\hline
0
\end{array}
$$

2.
$$
\begin{array}{r}
3287\ \text{R }3 \\
8\overline{)26,299} \\
24 \\
\hline
2\,2 \\
1\,6 \\
\hline
69 \\
64 \\
\hline
59 \\
56 \\
\hline
3
\end{array}
$$

3.
$$
\begin{array}{r}
328 \\
76\overline{)24,928} \\
22\,8 \\
\hline
2\,12 \\
1\,52 \\
\hline
608 \\
608 \\
\hline
0
\end{array}
$$

4. Answers may vary. Possible solution: Consider how many times the first digit of the divisor, 4, can be divided into the first two digits of the dividend, 29. Try 7 first. You will find this to be too large, so try 6. This works.

How Am I Doing? Sections 1.1–1.5

1. 78,310,436 = seventy-eight million, three hundred ten thousand, four hundred thirty-six

2. 38,247 = 30,000 + 8000 + 200 + 40 + 7

3. 5,064,122

4. 17,487 thousand or 17,487,000 students were enrolled in 2005.

5. 20,080 thousand or 20,080,000 students are expected to be enrolled in 2017.

6.
$$
\begin{array}{r}
13 \\
31 \\
88 \\
43 \\
+\ 69 \\
\hline
244
\end{array}
$$

7.
$$
\begin{array}{r}
28,318 \\
5,039 \\
+\ 17,213 \\
\hline
50,570
\end{array}
$$

8. $\begin{array}{r} 833{,}576 \\ +\ 517{,}885 \\ \hline 1{,}351{,}461 \end{array}$

9. $\begin{array}{r} 5728 \\ -1735 \\ \hline 3993 \end{array}$

10. $\begin{array}{r} 100{,}450 \\ -\ 24{,}139 \\ \hline 76{,}311 \end{array}$

11. $\begin{array}{r} 45{,}861{,}413 \\ -\ 43{,}879{,}761 \\ \hline 1{,}981{,}652 \end{array}$

12. $9 \times 6 \times 1 \times 2 = 54 \times 1 \times 2 = 54 \times 2 = 108$

13. $50 \times 10 \times 200 = 500 \times 200 = 100{,}000$

14. $\begin{array}{r} 2658 \\ \times\ \ \ \ 7 \\ \hline 18{,}606 \end{array}$

15. $\begin{array}{r} 68 \\ \times\ 55 \\ \hline 340 \\ 340\ \ \\ \hline 3740 \end{array}$

16. $\begin{array}{r} 365 \\ \times\ \ \ 908 \\ \hline 2\,920 \\ 328\,50\ \ \\ \hline 331{,}420 \end{array}$

17. $\begin{array}{r} 10{,}605\ \ \ \ \ \\ 8\overline{)84{,}840} \\ \underline{8}\ \ \ \ \ \ \ \ \ \\ 4\,8\ \ \ \ \ \\ 4\,8\ \ \ \ \ \\ \hline 40\ \ \\ \underline{40}\ \ \\ \hline 0 \end{array}$

18. $\begin{array}{r} 7{,}376\ \text{R 1} \\ 7\overline{)51{,}633} \\ \underline{49}\ \ \ \ \ \ \ \\ 2\,6\ \ \ \ \\ \underline{2\,1}\ \ \ \ \\ 53\ \ \\ \underline{49}\ \ \\ 43 \\ \underline{42} \\ 1 \end{array}$

19. $\begin{array}{r} 26\ \text{R 8} \\ 76\overline{)1984} \\ \underline{152}\ \ \\ 464 \\ \underline{456} \\ 8 \end{array}$

20. $\begin{array}{r} 139 \\ 42\overline{)5838} \\ \underline{42}\ \ \ \\ 163 \\ \underline{126} \\ 378 \\ \underline{378} \\ 0 \end{array}$

1.6 Exercises

1. 5^3 means $5 \times 5 \times 5 = 125$.

3. In exponent form, the <u>base</u> is the number that is multiplied.

5. To ensure consistency we
 1. perform operations inside parentheses
 2. simplify any expressions with exponents
 3. multiply or divide from left to right
 4. add or subtract from left to right

7. $6 \times 6 \times 6 \times 6 = 6^4$

9. $4 \times 4 \times 4 \times 4 \times 4 \times 4 \times 4 = 4^7$

11. $9 \times 9 \times 9 \times 9 = 9^4$

13. $7 = 7^1$

15. $2^4 = 2 \times 2 \times 2 \times 2 = 16$

17. $4^3 = 4 \times 4 \times 4 = 64$

19. $6^2 = 6 \times 6 = 36$

21. $10^4 = 10 \times 10 \times 10 \times 10 = 10,000$

23. $1^{17} = 1$

25. $2^6 = 2 \times 2 \times 2 \times 2 \times 2 \times 2 = 64$

27. $3^5 = 3 \times 3 \times 3 \times 3 \times 3 = 243$

29. $15^2 = 15 \times 15 = 225$

31. $7^3 = 7 \times 7 \times 7 = 343$

33. $4^4 = 4 \times 4 \times 4 \times 4 = 256$

35. $9^0 = 1$

37. $25^2 = 25 \times 25 = 625$

39. $10^6 = 10 \times 10 \times 10 \times 10 \times 10 \times 10 = 1,000,000$

41. $13^2 = 13 \times 13 = 169$

43. $9^1 = 9$

45. $8^2 = 8 \times 8 = 64$

47. $3^2 + 1^2 = 9 + 1 = 10$

49. $2^3 + 10^2 = 8 + 100 = 108$

51. $8^3 + 8 = 512 + 8 = 520$

53. $8 \times 7 - 20 = 56 - 20 = 36$

55. $3 \times 9 - 10 \div 2 = 27 - 10 \div 2 = 27 - 5 = 22$

57. $48 \div 2^3 + 4 = 48 \div 8 + 4 = 6 + 4 = 10$

59. $3 \times 6^2 - 50 = 3 \times 36 - 50 = 108 - 50 = 58$

61. $\begin{aligned} 10^2 + 3 \times (8 - 3) &= 10^2 + 3(5) \\ &= 100 + 3(5) \\ &= 100 + 15 \\ &= 115 \end{aligned}$

63. $(400 \div 20) \div 20 = 20 \div 20 = 1$

65. $950 \div (25 \div 5) = 950 \div 5 = 190$

67. $(14)(4) - (14 + 4) = (14)(4) - 18 = 56 - 18 = 38$

69. $3^2 + 4^2 \div 2^2 = 9 + 16 \div 4 = 9 + 4 = 13$

71. $\begin{aligned} (6)(7) - (12 - 8) \div 4 &= (6)(7) - 4 \div 4 \\ &= 42 - 4 \div 4 \\ &= 42 - 1 \\ &= 41 \end{aligned}$

73. $100 - 3^2 \times 4 = 100 - 9 \times 4 = 100 - 36 = 64$

75. $5^2 + 2^2 + 3^3 = 25 + 4 + 27 = 56$

77. $\begin{aligned} 72 \div 9 \times 3 \times 1 \div 2 &= 8 \times 3 \times 1 \div 2 \\ &= 24 \times 1 \div 2 \\ &= 24 \div 2 \\ &= 12 \end{aligned}$

79. $12^2 - 2 \times 0 \times 5 \times 6 = 144 - 0 = 144$

81. $4^2 \times 6 \div 3 = 16 \times 6 \div 3 = 96 \div 3 = 32$

83. $60 - 2 \times 4 \times 5 + 10 = 60 - 40 + 10 = 20 + 10 = 30$

85. $3 + 3^2 \times 6 + 4 = 3 + 9 \times 6 + 4 = 3 + 54 + 4 = 61$

87. $\begin{aligned} 32 \div 2 \times (3 - 1)^4 &= 32 \div 2 \times 2^4 \\ &= 32 \div 2 \times 16 \\ &= 16 \times 16 \\ &= 256 \end{aligned}$

89. $\begin{aligned} 3^2 \times 6 \div 9 + 4 \times 3 &= 9 \times 6 \div 9 + 4 \times 3 \\ &= 54 \div 9 + 4 \times 3 \\ &= 6 + 4 \times 3 \\ &= 6 + 12 \\ &= 18 \end{aligned}$

91. $4^0 + 5^3 + 9^1 = 1 + 125 + 9 = 135$

93. $\begin{aligned} 1200 - 2^3(3) \div 6 &= 1200 - 8(3) \div 6 \\ &= 1200 - 24 \div 6 \\ &= 1200 - 4 \\ &= 1196 \end{aligned}$

95. $120 \div (30 + 10) - 1 = 120 \div 40 - 1 = 3 - 1 = 2$

97. $120 \div 30 + 10 - 1 = 4 + 10 - 1 = 14 - 1 = 13$

99. $7 \times 3 - (9-7)^3 + 3^0 = 7 \times 3 - 2^3 + 3^0$
$$= 7 \times 3 - 8 + 1$$
$$= 21 - 8 + 1$$
$$= 13 + 1$$
$$= 14$$

101. $23(60)(60) + 56(60) + 4$
$$= 82,800 + 3360 + 4$$
$$= 86,164$$
It is 86,164 seconds.

Cumulative Review

103. a. 3

　　　b. 2,000,000

104. 200,765,909

105. 261,763,002 = two hundred sixty-one million, seven hundred sixty-three thousand, two

106. Perimeter: $2 \times 250 + 2 \times 480 = 500 + 960$
$$= 1460 \text{ feet}$$
1460 feet of fencing is needed to surround field.
Area: $250 \times 480 = 120,000$ square feet
120,000 square feet of grass must be planted.

Quick Quiz 1.6

1. $12 \times 12 \times 12 \times 12 \times 12 = 12^5$

2. $6^4 = 6 \times 6 \times 6 \times 6 = 1296$

3. $42 - 2^5 + 3 \times (9-6)^3 = 42 - 2^5 + 3 \times 3^3$
$$= 42 - 32 + 3 \times 27$$
$$= 42 - 32 + 81$$
$$= 10 + 81$$
$$= 91$$

4. Answers may vary. Possible solution:
$7 \times 6 \div 3 \times 4^2 - 2$

$= 7 \times 6 \div 3 \times 16 - 2$	Exponents
$= 42 \div 3 \times 16 - 2$	Multiply.
$= 14 \times 16 - 2$	Divide.
$= 224 - 2$	Multiply.
$= 222$	Subtract.

1.7 Exercises

1. Locate the rounding place. If the digit to the right of the rounding place is 5 or greater than 5, round up. If the digit to the right of the rounding place is less than 5, round down. Examples will vary.

3. 8<u>3</u> rounds to 80 since 3 is less than 5.

5. 6<u>5</u> rounds to 70 since 5 is 5 or more.

7. 16<u>8</u> rounds to 170 since 8 is 5 or more.

9. 743<u>8</u> rounds to 7440 since 8 is 5 or more.

11. 296<u>1</u> rounds to 2960 since 1 is less than 5.

13. 2<u>4</u>7 rounds to 200 since 4 is less than 5.

15. 27<u>8</u>1 rounds to 2800 since 8 is 5 or more.

17. 7<u>6</u>92 rounds to 7700 since 9 is 5 or more.

19. 7<u>6</u>21 rounds to 8000 since 6 is 5 or more.

21. 1<u>4</u>89 rounds to 1000 since 4 is less than 5.

23. 27,<u>8</u>63 rounds to 28,000 since 8 is 5 or more.

25. 8<u>3</u>2,400 rounds to 800,000 since 3 is less than 5.

27. 15,<u>1</u>69,873 stars rounds to 15,000,000 stars since 1 is less than 5.

29. a. 373,50<u>4</u>,000 rounds to 373,500,000 since 0 is less than 5.

　　　b. 3<u>7</u>3,504,000 rounds to 400,000,000 since 7 is 5 or more.

31. a. 3,7<u>0</u>5,392 square miles rounds to 3,700,000 square miles since 0 is less than 5.
9,5<u>9</u>6,960 square kilometers rounds to 9,600,000 square kilometers since 9 is 5 or more.

　　　b. 3,70<u>5</u>,392 square miles rounds to 3,710,000 square miles since 5 is 5 is 5 or more.
9,59<u>6</u>,960 square kilometers rounds to 9,600,000 square kilometers since 6 is 5 or more.

33.
$$
\begin{array}{r}
800 \\
300 \\
+\ 200 \\
\hline
1300
\end{array}
$$

35.
$$
\begin{array}{r}
40 \\
70 \\
100 \\
+\ 20 \\
\hline
230
\end{array}
$$

37. $200,000 + 50,000 + 9000 = 259,000$

39.
$$\begin{array}{r} 300,000 \\ -\ 70,000 \\ \hline 230,000 \end{array}$$

41.
$$\begin{array}{r} 800,000 \\ -\ 80,000 \\ \hline 720,000 \end{array}$$

43. $30,000,000 - 20,000,000 = 10,000,000$

45. $50 \times 60 = 3000$

47. $1000 \times 8 = 8000$

49. $600,000 \times 300 = 180,000,000$

51.
$$\begin{array}{r} 150 \\ 40\overline{)6000} \\ \underline{40} \\ 200 \\ \underline{200} \\ 0 \end{array}$$

53.
$$\begin{array}{r} 10,000 \\ 40\overline{)400,000} \\ \underline{40} \\ 0 \end{array}$$

55. $4,000,000 \div 800 = 5000$

57.
$$\begin{array}{r} 400 \\ 500 \\ 900 \\ +\ 200 \\ \hline 2000 \end{array}$$
Incorrect

59.
$$\begin{array}{r} 100,000 \\ 50,000 \\ +\ 40,000 \\ \hline 190,000 \end{array}$$
Incorrect

61.
$$\begin{array}{r} 300,000 \\ -\ 90,000 \\ \hline 210,000 \end{array}$$
Correct

63.
$$\begin{array}{r} 80,000,000 \\ -\ 50,000,000 \\ \hline 30,000,000 \end{array}$$
Incorrect

65.
$$\begin{array}{r} 400 \\ \times\ 30 \\ \hline 12,000 \end{array}$$
Incorrect

67.
$$\begin{array}{r} 6000 \\ \times\ 70 \\ \hline 420,000 \end{array}$$
Correct

69.
$$\begin{array}{r} 2000 \\ 40\overline{)80,000} \end{array}$$
Correct

71.
$$\begin{array}{r} 500 \\ 400\overline{)200,000} \end{array}$$
Correct

73. $20 \times 20 = 400$
The estimated square footage of the garage is 400 square feet.

75.
$$\begin{array}{r} 6,000,000 \\ 4,000,000 \\ +\ 2,000,000 \\ \hline 12,000,000 \end{array}$$
The estimated total population is 12,000,000 people.

77. $300 \times 100 = 30,000$
The estimate is a total of 30,000 pizzas.

79.
$$\begin{array}{r} 90,000,000 \\ -\ 69,400,000 \\ \hline 20,600,000 \end{array}$$
The estimated difference is 20,600,000 passengers.

81. $590,000 - 270,000 = 320,000$
The estimated difference in area is 320,000 square miles.

83. a. $8,000,000,000 \div 20,000 = 400,000$
It will take approximately 400,000 hours.

    b. $400,000 \div 20 = 20,000$
It will take approximately 20,000 days.

Cumulative Review

85. $26 \times 3 + 20 \div 4 = 78 + 20 \div 4 = 78 + 5 = 83$

86. $5^2 + 3^2 - (17 - 10) = 5^2 + 3^2 - 7$
$$= 25 + 9 - 7$$
$$= 34 - 7$$
$$= 27$$

87. $3 \times (16 \div 4) + 8 \times 2 = 3 \times 4 + 8 \times 2$
$$= 12 + 8 \times 2$$
$$= 12 + 16$$
$$= 28$$

88. $126 + 4 - (20 \div 5)^3 = 126 + 4 - 4^3$
$$= 126 + 4 - 64$$
$$= 130 - 64$$
$$= 66$$

89.
```
      5489
   ×    67
    38 423
   329 34
   367,763
```

90.
```
        87
  52)4524
      416
      364
      364
        0
```

Quick Quiz 1.7

1. 92,3<u>5</u>4 rounds to 92,400 since 5 is 5 or more.

2. 2,34<u>2</u>,786 rounds to 2,340,000 since 2 is less than 5.

3.
```
          8000
   ×    300,000
   2,400,000,000
```

4. Answers may vary. Possible solution:
   Since the digit to the right of the millions place is less than 5 (it is 4), round down to 682,000,000.

1.8 Exercises

1.
```
     40,300
   − 31,500
      8 800
```
The repairs will cost $8800.

3.
```
      120
   ×   13
      360
      120
     1560
```
There are 1560 bagels in Paula's order.

5.
```
        7
   12)84
       84
        0
```
The unit cost for the tomato sauce is 7¢ per ounce.

7.
```
        64
   13)832
       78
       52
       52
        0
```
Each pair cost $64 on average.

9.
```
    400,000,000
  − 309,163,000
     90,837,000
```
The increase in population from 2010 to 2043 is expected to be 90,837,000.

11.
```
     7356
     3257
     4777
   + 4992
    20,382
```
The gross revenue was $20,382.

13.
```
   24,111        793
      327      − 327
   +  793        466
   25,231
```
The total for the three groups is 25,231. There are 466 more volunteers than full-time staff.

15. $480 \div 60 = 8$
$100{,}000 \times 8 = 800{,}000$
The population will increase by 80,000 people.

17.
$$\begin{array}{ccc} 15 & 9 & 5 \\ \times\ 6 & \times 8 & \times 6 \\ \hline 90 & 72 & 30 \end{array}$$
$90 + 72 + 30 = 192$
She made $192.

19. Deposits: $132 + (4 \times 715) = 132 + 2860 = 2992$
Withdrawals: $(2 \times 575) + 482 = 1150 + 482$
$\qquad\qquad\qquad = 1632$
$$\begin{array}{r} 2992 \\ -\ 1632 \\ \hline 1360 \end{array}$$
The balance in her checking account will be
$1360.

21.
$$\begin{array}{ccc} 250 & 57 & 21{,}250 \\ \times\ 85 & \times\ 85 & -\ 4\,845 \\ \hline 1\,250 & 285 & 16{,}405 \\ 20\,00 & 456 & \\ \hline 21{,}250 & 4845 & \end{array}$$
Her profit is $16,405.

23.
$$\begin{array}{r} 15{,}276 \\ -\ 14{,}926 \\ \hline 350 \end{array}$$
$350 \div 14 = 25$
Her car gets 25 miles per gallon.

25. oaks: $3 \times 18 = 54$
maples: $2 \times 54 = 108$
pines: $7 \times 108 = 756$
birches = 18
$18 + 54 + 108 + 756 = 936$
There are 54 oaks, 108 maple trees, and 756 pine trees. In total there are 936 trees.

27.
$$\begin{array}{r} 626 \\ 407 \\ 207 \\ +\ 161 \\ \hline 1401 \end{array}$$
1401 stations play some kind of adult contemporary music.

29.
$$\begin{array}{cc} 479 & 1997 \\ +\ 367 & -\ 846 \\ \hline 846 & 1151 \end{array}$$
There are 1151 more country stations than classic rock or classic hits stations.

31.
$$\begin{array}{r} 90 \\ -\ 33 \\ \hline 57 \end{array}$$
57,000,000 more households had cable television in 2004.

33.
$$\begin{array}{cc} 101 & 101 \\ -\ 90 & +\ 11 \\ \hline 11 & 112 \end{array}$$
112,000,000 households will have cable television in 2014.

Cumulative Review

35. $7^3 = 7 \times 7 \times 7 = 343$

36. $3 \times 2^3 + 15 \div 3 - 4 \times 2 = 3 \times 8 + 15 \div 3 - 4 \times 2$
$\qquad\qquad\qquad\qquad = 24 + 15 \div 3 - 4 \times 2$
$\qquad\qquad\qquad\qquad = 24 + 5 - 4 \times 2$
$\qquad\qquad\qquad\qquad = 24 + 5 - 8$
$\qquad\qquad\qquad\qquad = 29 - 8$
$\qquad\qquad\qquad\qquad = 21$

37.
$$\begin{array}{r} 126 \\ \times\ 38 \\ \hline 1008 \\ 378 \\ \hline 4788 \end{array}$$

38.
$$\begin{array}{r} 258 \\ 12\overline{)3096} \\ \underline{24} \\ 69 \\ \underline{60} \\ 96 \\ \underline{96} \\ 0 \end{array}$$

39.
$$\begin{array}{r} 96 \\ 123 \\ 57 \\ +\ 526 \\ \hline 802 \end{array}$$

40. $509,263$
         $\underline{-\ 485,978}$
            $23,285$

41. $526,195,\underline{7}26$ rounds to $526,196,000$ because 7 is 5 or more.

42. $3,400,603,025$

Quick Quiz 1.8

1.

$$
\begin{array}{r}
269 \\
16\overline{)4304} \\
\underline{32} \\
110 \\
\underline{96} \\
144 \\
\underline{144} \\
0
\end{array}
$$

Each club member will pay $269.

2. Deposits: $471 + 198 + 276 + 347 = 1292$
Checks: $49 + 227 + 158 = 434$
       1292
      $\underline{-\ 434}$
        858
Her new balance will be $858.

3.
$$
\begin{array}{ccc}
11 & 14 & 5 \\
\underline{\times\ 2} & \underline{\times\ 6} & \underline{\times\ 4} \\
22 & 84 & 20
\end{array}
$$
$22 + 84 + 20 = 126$
The total cost for the Tobey family was $126.

4. Answers may vary. Possible solution:
Round each number so there is one nonzero digit. Then divide the total cost by the number of cars.
$$
\begin{array}{r}
20,000 \\
40\overline{)800,000}
\end{array}
$$
The estimate is $20,000 each.

Use Math To Save Money

1. Their three credit cards are maxed out at $8000, they have hospital debt of $12,000, they owe $2000 on their car, and they owe friends $100 and $300.
$100, $300, $2000, $8000, $8000, $8000, $12,000

2. $3 \times \$25 + \$50 + \$200 + 2 \times \$20$
$= \$75 + \$50 + \$200 + \$40$
$= \$365$

3. Their three smallest debts are the $100 and $300 loans from friends and their $2000 car loan.

4. The total amount of the minimum monthly payments for the smallest two debts is $2 \times \$20 = \$40$.

5. After 2 months, the balance on the car loan is $\$2000 - \$400 = \$1600$. Then they begin paying $240 per month.
$$
\begin{array}{r}
6\ \text{R}\ 160 \\
240\overline{)1600} \\
\underline{1440} \\
160
\end{array}
$$
It takes them about 7 months to pay off their car loan.

6. Answers will vary.

You Try It

1. The digit 7 is in the hundred thousands place.

2. $132,259$
$= 100,000 + 30,000 + 2000 + 200 + 50 + 9$

3. $58,872,150 =$ fifty-eight million, eight hundred seventy-two thousand, one hundred fifty

4. 478
         134
         260
        $\underline{+\ 73}$
         945

5. $23,495$
        $\underline{-\ 19,297}$
          $4\ 198$

6. $5 \times 2 \times 4 \times 6 \times 8 = 10 \times 4 \times 6 \times 8$
$= 40 \times 6 \times 8$
$= 240 \times 8$
$= 1920$

7.
$$
\begin{array}{r}
532 \\
\times\ 167 \\
\hline
3\ 724 \\
31\ 92 \\
53\ 2 \\
\hline
88,844
\end{array}
$$

8.
$$
\begin{array}{r}
628 \\
135\overline{)84,780} \\
81\ 0 \\
\hline
3\ 78 \\
2\ 70 \\
\hline
1\ 080 \\
1\ 080 \\
\hline
0
\end{array}
$$

9. a. $9 \times 9 \times 9 \times 9 \times 9 = 9^5$

  b. $7^4 = 7 \times 7 \times 7 \times 7 = 2401$

10. $6^2 \div 2 \times 3 - (10 - 8)^2 = 6^2 \div 2 \times 3 - 2^2$
$$
\begin{aligned}
&= 36 \div 2 \times 3 - 4 \\
&= 18 \times 3 - 4 \\
&= 54 - 4 \\
&= 50
\end{aligned}
$$

11. a. 338,912 rounds to 339,000 since 9 is 5 or more.

  b. 745,830 rounds to 700,000 since 4 is less than 5.

12. $90,000 \times 2000 = 180,000,000$

Chapter 1 Review Problems

1. 892 = eight hundred ninety-two

2. 109,276 = one hundred nine thousand, two hundred seventy-six

3. 4364 = 4000 + 300 + 60 + 4

4. 42,166,037 = 40,000,000 + 2,000,000 + 100,000 + 60,000 + 6000 + 30 + 7

5. 5302

6. 1,328,828

7.
$$
\begin{array}{r}
76 \\
+\ 39 \\
\hline
115
\end{array}
$$

8.
$$
\begin{array}{r}
235 \\
+\ 165 \\
\hline
400
\end{array}
$$

9.
$$
\begin{array}{r}
12 \\
28 \\
34 \\
+\ 76 \\
\hline
150
\end{array}
$$

10.
$$
\begin{array}{r}
123 \\
61 \\
9 \\
84 \\
+\ 123 \\
\hline
400
\end{array}
$$

11.
$$
\begin{array}{r}
226 \\
134 \\
+\ 647 \\
\hline
1007
\end{array}
$$

12.
$$
\begin{array}{r}
52,134 \\
+\ 7\ 966 \\
\hline
60,100
\end{array}
$$

13.
$$
\begin{array}{r}
1\ 356 \\
2\ 892 \\
561 \\
89 \\
+\ 9\ 805 \\
\hline
14,703
\end{array}
$$

14.
$$
\begin{array}{r}
36 \\
-\ 19 \\
\hline
17
\end{array}
$$

15.
$$
\begin{array}{r}
126 \\
-\ 99 \\
\hline
27
\end{array}
$$

16.
$$
\begin{array}{r}
543 \\
-\ 372 \\
\hline
171
\end{array}
$$

17. $$\begin{array}{r} 7000 \\ -\ 845 \\ \hline 6155 \end{array}$$

18. $$\begin{array}{r} 201,340 \\ -\ 120,618 \\ \hline 80,722 \end{array}$$

19. $$\begin{array}{r} 6,325,034 \\ -\ \ 89,023 \\ \hline 6,236,011 \end{array}$$

20. $$\begin{array}{r} 5,412,022 \\ -\ \ 79,031 \\ \hline 5,332,991 \end{array}$$

21. $8 \times 1 \times 9 \times 2 = 8 \times 9 \times 2 = 72 \times 2 = 144$

22. $7 \times 6 \times 0 \times 4 = 42 \times 0 \times 4 = 0 \times 4 = 0$

23. $2 \cdot 5 \cdot 10 \cdot 8 = 10 \cdot 10 \cdot 8 = 100 \cdot 8 = 800$

24. $621 \times 100 = 62,100$

25. $84,312 \times 1000 = 84,312,000$

26. $78 \times 10,000 = 780,000$

27. $$\begin{array}{r} 3492 \\ \times\ \ \ \ 7 \\ \hline 24,444 \end{array}$$

28. $$\begin{array}{r} 6257 \\ \times\ \ \ \ 8 \\ \hline 50,056 \end{array}$$

29. $$\begin{array}{r} 58 \\ \times\ 32 \\ \hline 1856 \end{array}$$

30. $$\begin{array}{r} 73 \\ \times\ 24 \\ \hline 292 \\ 146\ \ \\ \hline 1752 \end{array}$$

31. $$\begin{array}{r} 709 \\ \times\ \ 36 \\ \hline 4\ 254 \\ 21\ 27\ \ \\ \hline 25,524 \end{array}$$

32. $$\begin{array}{r} 123 \\ \times\ 714 \\ \hline 492 \\ 1\ 23\ \ \\ 86\ 1\ \ \ \ \\ \hline 87,822 \end{array}$$

33. $$\begin{array}{r} 431 \\ \times\ 623 \\ \hline 1\ 293 \\ 8\ 62\ \ \\ 258\ 6\ \ \ \ \\ \hline 268,513 \end{array}$$

34. $$\begin{array}{r} 1782 \\ \times\ \ 305 \\ \hline 8910 \\ 534\ 60\ \ \ \\ \hline 543,510 \end{array}$$

35. $$\begin{array}{r} 2057 \\ \times\ \ 124 \\ \hline 8\ 228 \\ 41\ 14\ \ \\ 205\ 7\ \ \ \ \\ \hline 255,068 \end{array}$$

36. $$\begin{array}{r} 3182 \\ \times\ \ \ 35 \\ \hline 15\ 910 \\ 95\ 46\ \ \\ \hline 111,370 \end{array}$$

37. $$\begin{array}{r} 1200 \\ \times\ \ 6000 \\ \hline 7,200,000 \end{array}$$

38. $$\begin{array}{r} 100,000 \\ \times\ \ \ \ 20,000 \\ \hline 2,000,000,000 \end{array}$$

39. $20 \div 10 = 2$

40. $40 \div 8 = 5$

41. $0 \div 8 = 0$

42. $7 \div 1 = 7$

43. $\dfrac{81}{9} = 9$

44. $\dfrac{42}{6} = 7$

45. $\dfrac{5}{0}$ undefined

46. $\dfrac{24}{6} = 4$

47.
$$
\begin{array}{r}
125 \\
6\overline{)750} \\
\underline{6} \\
15 \\
\underline{12} \\
30 \\
\underline{30} \\
0
\end{array}
$$

48.
$$
\begin{array}{r}
207 \\
9\overline{)1863} \\
\underline{18} \\
063 \\
\underline{63} \\
0
\end{array}
$$

49.
$$
\begin{array}{r}
2\,504 \\
6\overline{)15{,}024} \\
\underline{12} \\
3\,0 \\
\underline{3\,0} \\
024 \\
\underline{24} \\
0
\end{array}
$$

50.
$$
\begin{array}{r}
3\,064 \\
8\overline{)24{,}512} \\
\underline{24} \\
51 \\
\underline{48} \\
32 \\
\underline{32} \\
0
\end{array}
$$

51.
$$
\begin{array}{r}
36{,}958 \\
6\overline{)221{,}748} \\
\underline{18} \\
41 \\
\underline{36} \\
5\,7 \\
\underline{5\,4} \\
34 \\
\underline{30} \\
48 \\
\underline{48} \\
0
\end{array}
$$

52.
$$
\begin{array}{r}
36{,}921 \\
5\overline{)184{,}605} \\
\underline{15} \\
34 \\
\underline{30} \\
4\,6 \\
\underline{4\,5} \\
10 \\
\underline{10} \\
5 \\
\underline{5} \\
0
\end{array}
$$

53.
$$
\begin{array}{r}
15{,}046 \text{ R } 3 \\
8\overline{)120{,}371} \\
\underline{8} \\
40 \\
\underline{40} \\
0\,37 \\
\underline{32} \\
51 \\
\underline{48} \\
3
\end{array}
$$

54.
$$
\begin{array}{r}
7 \text{ R } 21 \\
67\overline{)490} \\
\underline{469} \\
21
\end{array}
$$

55.
$$
\begin{array}{r}
31 \text{ R } 15 \\
21\overline{)666} \\
\underline{63} \\
36 \\
\underline{21} \\
15
\end{array}
$$

56.
$$\begin{array}{r} 60 \text{ R } 22 \\ 53\overline{)3202} \\ \underline{318} \\ 22 \end{array}$$

57.
$$\begin{array}{r} 195 \\ 45\overline{)8775} \\ \underline{45} \\ 427 \\ \underline{405} \\ 225 \\ \underline{225} \\ 0 \end{array}$$

58.
$$\begin{array}{r} 54 \\ 132\overline{)7128} \\ \underline{660} \\ 528 \\ \underline{528} \\ 0 \end{array}$$

59.
$$\begin{array}{r} 19 \\ 204\overline{)3876} \\ \underline{204} \\ 1836 \\ \underline{1836} \\ 0 \end{array}$$

60. $21 \times 21 \times 21 = 21^3$

61. $8 \times 8 \times 8 \times 8 \times 8 = 8^5$

62. $10 \times 10 \times 10 \times 10 \times 10 \times 10 = 10^6$

63. $2^6 = 2 \times 2 \times 2 \times 2 \times 2 \times 2 = 64$

64. $3^4 = 3 \times 3 \times 3 \times 3 = 81$

65. $5^3 = 5 \times 5 \times 5 = 125$

66. $7^2 = 7 \times 7 = 49$

67. $9^2 = 9 \times 9 = 81$

68. $6^3 = 6 \times 6 \times 6 = 216$

69. $7 + 2 \times 3 - 5 = 7 + 6 - 5 = 13 - 5 = 8$

70. $\begin{aligned} 2^5 + 4 - (5 + 3^2) &= 32 + 4 - (5 + 9) \\ &= 32 + 4 - 14 \\ &= 36 - 14 \\ &= 22 \end{aligned}$

71. $34 - 9 \div 9 \times 12 = 34 - 1 \times 12 = 34 - 12 = 22$

72. $\begin{aligned} 2^3 \times 5 \div 8 + 3 \times 4 &= 8 \times 5 \div 8 + 3 \times 4 \\ &= 40 \div 8 + 3 \times 4 \\ &= 5 + 3 \times 4 \\ &= 5 + 12 \\ &= 17 \end{aligned}$

73. $\begin{aligned} 2^3 + 4 \times 5 - 32 \div (1 + 3)^2 &= 2^3 + 4 \times 5 - 32 \div 4^2 \\ &= 8 + 4 \times 5 - 32 \div 16 \\ &= 8 + 20 - 2 \\ &= 26 \end{aligned}$

74. $\begin{aligned} 6 \times 3 &+ 3 \times 5^2 - 63 \div (5 - 2)^2 \\ &= 6 \times 3 + 3 \times 5^2 - 63 \div 3^2 \\ &= 6 \times 3 + 3 \times 25 - 63 \div 9 \\ &= 18 + 75 - 7 \\ &= 86 \end{aligned}$

75. 336<u>4</u> rounds to 3360 since 4 is less than 5.

76. 589<u>5</u> rounds to 5900 since 5 is 5 or more.

77. 42,64<u>4</u> rounds to 42,640 since 4 is less than 5.

78. 12,<u>3</u>50 rounds to 12,000 since 3 is less than 5.

79. 22,<u>9</u>86 rounds to 23,000 since 9 is 5 or more.

80. 202,<u>4</u>98 rounds to 202,000 since 4 is less than 5.

81. 4,6<u>4</u>9,320 rounds to 4,600,000 since 4 is less than 5.

82. 9,99<u>5</u>,312 rounds to 10,000,000 since 5 is 5 or more.

83.
$$\begin{array}{r} 20,000 \\ 8\,000 \\ + 40,000 \\ \hline 68,000 \end{array}$$

84.
$$\begin{array}{r} 30,000 \\ - 20,000 \\ \hline 10,000 \end{array}$$

85.
$$\begin{array}{r} 3,000,000 \\ \times \qquad 900 \\ \hline 2,700,000,000 \end{array}$$

86.
$$\begin{array}{r} 4000 \\ 20\overline{)80,000} \\ \underline{80} \\ 0 \end{array}$$

87.
$$\begin{array}{r} 25 \\ \times \quad 7 \\ \hline 175 \end{array}$$
He typed 175 words.

88.
$$\begin{array}{r} 2462 \\ 1997 \\ + 2561 \\ \hline 7020 \end{array}$$
7020 people visited the festival during these three months.

89.
$$\begin{array}{r} 14,630 \\ - \quad 4\,329 \\ \hline 10,301 \end{array}$$
There was 10,301 feet between them.

90.
$$\begin{array}{r} 4330 \\ + \quad 268 \\ \hline 4598 \end{array} \qquad \begin{array}{r} 4598 \\ - 1250 \\ \hline 3348 \end{array}$$
Gerardo will have to pay $3348 for tuition and books.

91.
$$\begin{array}{r} 74 \\ 112\overline{)8288} \\ \underline{784} \\ 448 \\ \underline{448} \\ 0 \end{array}$$
The cost was $74 per bed.

92.

Deposits	Withdrawals
24	18
105	145
36	250
+ 177	+ 461
342	874

$810 + 342 - 874 = 278$
Her balance will be $278.

93.
$$\begin{array}{r} 56,720 \\ - 56,320 \\ \hline 400 \text{ miles} \end{array}$$

$$\begin{array}{r} 25 \\ 16\overline{)400} \\ \underline{32} \\ 80 \\ \underline{80} \\ 0 \end{array}$$
He got 25 miles per gallon.

94.
$$\begin{array}{r} 15 \\ \times 65 \\ \hline 975 \end{array} \quad \begin{array}{r} 60 \\ \times 12 \\ \hline 720 \end{array} \quad \begin{array}{r} 42 \\ \times 8 \\ \hline 336 \end{array} \quad \begin{array}{r} 975 \\ 720 \\ + \quad 336 \\ \hline 2031 \end{array}$$
The total price is $2031.

95.
$$\begin{array}{r} 63,500 \\ - 21,800 \\ \hline 41,700 \end{array}$$
41,700 thousand tons or 41,700,000 tons more were recovered in 2000 than in 1985.

96.
$$\begin{array}{r} 55,000,000 \\ - 33,600,000 \\ \hline 21,400,000 \end{array}$$
The greatest increase was 21,400,000 tons between 1990 and 1995.

97. From 2000 to 2010:
$$\begin{array}{r} 84,500 \\ - 63,500 \\ \hline 21,000 \end{array}$$

$$\begin{array}{r} 84,500 \\ + 21,000 \\ \hline 105,500 \end{array}$$
There would be 105,500 thousand tons or 105,500,000 tons recovered in 2020.

98.
$$\begin{array}{r} 205 \\ 36 \\ 1983 \\ + \quad 60 \\ \hline 2284 \end{array}$$

99.
$$\begin{array}{r} 56,793 \\ - 48,926 \\ \hline 7\,867 \end{array}$$

100.
$$\begin{array}{r} 396 \\ \times\ \ 28 \\ \hline 3\ 168 \\ 7\ 92\ \ \ \\ \hline 11{,}088 \end{array}$$

101.
$$\begin{array}{r} 129 \\ 37\overline{)4773} \\ \underline{37}\ \ \ \ \\ 107\ \ \\ \underline{74}\ \ \\ 333 \\ \underline{333} \\ 0 \end{array}$$

102. $4\times12-(12+9)+2^3\div4 = 4\times12-21+2^3\div4$
$$= 4\times12-21+8\div4$$
$$= 48-21+8\div4$$
$$= 48-21+2$$
$$= 29$$

103.
$$\begin{array}{cccc} 699 & 78 & 2097 & 3000 \\ \times\ \ 3 & \times\ \ 2 & +\ 156 & -2253 \\ \hline 2097 & 156 & 2253 & 747 \end{array}$$
He has $747 in his account.

104. a.
$$\begin{array}{r} 22 \\ \times\ 15 \\ \hline 110 \\ 22\ \ \\ \hline 330 \end{array}$$
The patio is 330 square feet.

b. $2(22) + 2(15) = 44 + 30 = 74$
He would need 74 feet of fence.

How Am I Doing? Chapter 1 Test

1. $44{,}007{,}635 = $ forty-four million, seven thousand, six hundred thirty-five

2. $26{,}859 = 20{,}000 + 6000 + 800 + 50 + 9$

3. three million, five hundred eighty-one thousand, seventy-six $= 3{,}581{,}076$

4.
$$\begin{array}{r} 189 \\ 26 \\ 12 \\ 528 \\ +\ 76 \\ \hline 831 \end{array}$$

5.
$$\begin{array}{r} 763 \\ 220 \\ +\ 508 \\ \hline 1491 \end{array}$$

6.
$$\begin{array}{r} 135{,}484 \\ 2{,}376 \\ 81{,}004 \\ +\ 100{,}113 \\ \hline 318{,}977 \end{array}$$

7.
$$\begin{array}{r} 8961 \\ -\ 894 \\ \hline 8067 \end{array}$$

8.
$$\begin{array}{r} 501{,}760 \\ -\ 328{,}902 \\ \hline 172{,}858 \end{array}$$

9.
$$\begin{array}{r} 18{,}400{,}100 \\ -\ 13{,}174{,}332 \\ \hline 5{,}225{,}768 \end{array}$$

10. $1 \times 6 \times 9 \times 7 = 6 \times 9 \times 7 = 54 \times 7 = 378$

11.
$$\begin{array}{r} 45 \\ \times\ 96 \\ \hline 270 \\ 405\ \ \\ \hline 4320 \end{array}$$

12.
$$\begin{array}{r} 326 \\ \times\ \ 592 \\ \hline 652 \\ 29\ 34\ \ \\ 163\ 0\ \ \ \\ \hline 192{,}992 \end{array}$$

13.
$$\begin{array}{r} 18{,}491 \\ \times\ \ \ \ \ 7 \\ \hline 129{,}437 \end{array}$$

14.

$$
\begin{array}{r}
3\ 014\ \text{R}\ 1 \\
5\overline{)15{,}071} \\
\underline{15} \\
0 \\
\underline{\ 0} \\
7 \\
\underline{5} \\
21 \\
\underline{20} \\
1
\end{array}
$$

15.

$$
\begin{array}{r}
2\ 358 \\
6\overline{)14{,}148} \\
\underline{12} \\
21 \\
\underline{18} \\
34 \\
\underline{30} \\
48 \\
\underline{48} \\
0
\end{array}
$$

16.

$$
\begin{array}{r}
352 \\
37\overline{)13{,}024} \\
\underline{11\ 1} \\
1\ 92 \\
\underline{1\ 85} \\
74 \\
\underline{74} \\
0
\end{array}
$$

17. $14 \times 14 \times 14 = 14^3$

18. $2^6 = 2 \times 2 \times 2 \times 2 \times 2 \times 2 = 64$

19. $5 + 6^2 - 2 \times (9-6)^2 = 5 + 6^2 - 2 \times 3^2$
$$
\begin{aligned}
&= 5 + 36 - 2 \times 9 \\
&= 5 + 36 - 18 \\
&= 41 - 18 \\
&= 23
\end{aligned}
$$

20. $2^4 + 3^3 + 28 \div 4 = 16 + 27 + 28 \div 4$
$$
\begin{aligned}
&= 16 + 27 + 7 \\
&= 43 + 7 \\
&= 50
\end{aligned}
$$

21. $4 \times 6 + 3^3 \times 2 + 23 \div 23 = 4 \times 6 + 27 \times 2 + 23 \div 23$
$$
\begin{aligned}
&= 24 + 27 \times 2 + 23 \div 23 \\
&= 24 + 54 + 23 \div 23 \\
&= 24 + 54 + 1 \\
&= 78 + 1 \\
&= 79
\end{aligned}
$$

22. 94,7<u>6</u>8 rounds to 94,800 since 6 is 5 or more.

23. 6,46<u>2</u>,431 rounds to 6,460,000 since 2 is less than 5.

24. 5,2<u>7</u>8,963 rounds to 5,300,000 since 7 is 5 or more.

25. $5{,}000{,}000 \times 30{,}000 = 150{,}000{,}000{,}000$

26. $1000 + 3000 + 4000 + 8000 = 16{,}000$

27.

$$
\begin{array}{r}
2\ 148 \\
15\overline{)32{,}220} \\
\underline{30} \\
2\ 2 \\
\underline{1\ 5} \\
72 \\
\underline{60} \\
120 \\
\underline{120} \\
0
\end{array}
$$

Each person paid $2148.

28.

$$
\begin{array}{r}
602 \\
-\ 135 \\
\hline
467
\end{array}
$$

The boy is 467 feet from the other side of the river.

29. $3 \times 2 + 1 \times 45 + 2 \times 21 + 2 \times 17 = 6 + 45 + 42 + 34$
$$
= 127
$$
His total bill was $127.

30.

$$
\begin{array}{rr}
31 & 885 \\
902 & 103 \\
+\ 399 & 26 \\
\hline
1332 & 17 \\
& +\quad 9 \\
\cline{2-2}
& 1040
\end{array}
$$

Her balance is $1332 - $1040 = $292.

31.
$$
\begin{array}{r}
6800 \\
\times\ \ 110 \\
\hline
0000 \\
68\ 00 \\
6800 \\
\hline
748,000
\end{array}
$$

The area of the runway is 748,000 square feet.

32. $2 \times 8 + 2 \times 15 = 16 + 30 = 46$
The perimeter is 46 feet. He should purchase 46 feet of fence.

Chapter 2

2.1 Exercises

1. A <u>fraction</u> can be used to represent part of a whole or part of a group.

3. In a fraction, the <u>denominator</u> tells the total number of parts in the whole or in the group.

5. The number on the top, 3, is the numerator, and the number on the bottom, 5, is the denominator.

7. The number on the top, 7, is the numerator, and the number on the bottom, 8, is the denominator.

9. The number on the top, 1, is the numerator, and the number on the bottom, 17, is the denominator.

11. One out of three equal parts is shaded. The fraction is $\frac{1}{3}$.

13. Seven out of nine equal parts are shaded. The fraction is $\frac{7}{9}$.

15. Three out of four equal parts are shaded. The fraction is $\frac{3}{4}$.

17. Three out of seven equal parts are shaded. The fraction is $\frac{3}{7}$.

19. Two out of five equal parts are shaded. The fraction is $\frac{2}{5}$.

21. Seven out of ten equal parts are shaded. The fraction is $\frac{7}{10}$.

23. Five out of eight equal parts are shaded. The fraction is $\frac{5}{8}$.

25. Four out of seven circles are shaded. The fraction is $\frac{4}{7}$.

27. Seven out of eight triangles are shaded. The fraction is $\frac{7}{8}$.

29. Nine out of fifteen circles are shaded. The fraction is $\frac{9}{15}$.

31. $\frac{1}{5}$; divide a rectangular bar into 5 equal parts. Then shade 1 part.

33. $\frac{3}{8}$; divide a rectangular bar into 8 equal parts. Then shade 3 parts.

35. $\frac{7}{10}$; divide a rectangular bar into 10 equal parts. Then shade 7 parts.

37. $\dfrac{\text{number of women}}{\text{total number of students}} = \dfrac{51}{95}$

39. $\dfrac{\text{construction earnings}}{\text{pool table price}} = \dfrac{329}{950}$

41. $\dfrac{\text{roast}}{\text{total}} = \dfrac{89}{122+89} = \dfrac{89}{211}$

43. $\dfrac{\text{number of balsam firs}}{\text{total number of trees}} = \dfrac{9}{9+12+5} = \dfrac{9}{26}$

45. $\dfrac{\text{number of comedies and romances}}{\text{total number of items}}$
$= \dfrac{12+9}{12+15+8+9}$
$= \dfrac{21}{44}$

47. a. $\dfrac{50+40}{94+101} = \dfrac{90}{195}$

    b. $\dfrac{3+19}{94+101} = \dfrac{22}{195}$

49. $\dfrac{0}{6}$ is the amount of money each of 6 business

owners get if the business has a profit of $0.

Cumulative Review

51.
```
    18
    27
    34
    16
   125
 +  21
  ‾‾‾‾
   241
```

52.
```
   56,203
 − 42,987
  ‾‾‾‾‾‾‾
   13,216
```

53.
```
      3178
   ×    46
  ‾‾‾‾‾‾‾‾
    19 068
  127 12
  ‾‾‾‾‾‾‾‾
   146,188
```

54.
```
         1258 R 4
   24)30,196
      24
      ‾‾
       6 1
       4 8
       ‾‾‾
       1 39
       1 20
       ‾‾‾‾
         196
         192
         ‾‾‾
           4
```

Quick Quiz 2.1

1. Four out of seven equal parts are shaded. The

fraction is $\dfrac{4}{7}$.

2. $\dfrac{\text{number of students who drive a car}}{\text{total number of students}} = \dfrac{204}{371}$

3. $\dfrac{\text{number lifting weights}}{\text{total number of people}} = \dfrac{8+5}{8+5+7+13} = \dfrac{13}{33}$

4. Answers may vary. Possible solution: There were 120 new businesses. If 30 of the new restaurants went out of business and 25 of the new businesses that were not restaurants went out of business, then
$120 - (30 + 25) = 120 - 55 = 65$ new businesses did not go out of business:
$$\dfrac{\text{not go out of business}}{\text{new businesses}} = \dfrac{65}{120}$$

2.2 Exercises

1. By comparing the given list of whole numbers to the list of the first 15 prime numbers in Section 2.2, we see that 11, 19, 41, and 5 are prime.

3. A <u>composite</u> <u>number</u> is a whole number greater than 1 that can be divided by whole numbers other than itself and 1.

5. $56 = 2 \times 2 \times 2 \times 7$ or $2^3 \times 7$; answers may vary.

7. $15 = 3 \times 5$

9. $35 = 5 \times 7$

11. $49 = 7 \times 7 = 7^2$

13. $16 = 4 \times 4 = 2 \times 2 \times 2 \times 2 = 2^4$

15. $55 = 5 \times 11$

17. $63 = 7 \times 9 = 7 \times 3 \times 3 = 7 \times 3^2$ or $3^2 \times 7$

19. $84 = 4 \times 21 = 2 \times 2 \times 3 \times 7 = 2^2 \times 3 \times 7$

21. $54 = 6 \times 9 = 2 \times 3 \times 3 \times 3 = 2 \times 3^3$

23. $120 = 10 \times 12 = 2 \times 5 \times 2 \times 2 \times 3 = 2^3 \times 3 \times 5$

25. $184 = 4 \times 46 = 2 \times 2 \times 2 \times 23 = 2^3 \times 23$

27. 47 is prime.

29. $57 = 3 \times 19$

31. 67 is prime.

33. $62 = 2 \times 31$

35. 89 is prime.

37. 127 is prime.

Copyright © 2012 Pearson Education, Inc.

39. $121 = 11 \times 11$ or 11^2

41. $145 = 5 \times 29$

43. $\dfrac{18}{27} = \dfrac{18 \div 9}{27 \div 9} = \dfrac{2}{3}$

45. $\dfrac{36}{48} = \dfrac{36 \div 12}{48 \div 12} = \dfrac{3}{4}$

47. $\dfrac{54}{84} = \dfrac{54 \div 6}{84 \div 6} = \dfrac{9}{14}$

49. $\dfrac{260}{290} = \dfrac{260 \div 10}{290 \div 10} = \dfrac{26}{29}$

51. $\dfrac{5}{30} = \dfrac{5 \times 1}{5 \times 2 \times 3} = \dfrac{1}{2 \times 3} = \dfrac{1}{6}$

53. $\dfrac{66}{68} = \dfrac{2 \times 3 \times 11}{2 \times 2 \times 2 \times 11} = \dfrac{3}{4}$

55. $\dfrac{30}{45} = \dfrac{2 \times 3 \times 5}{3 \times 3 \times 5} = \dfrac{2}{3}$

57. $\dfrac{60}{75} = \dfrac{2 \times 2 \times 3 \times 5}{3 \times 5 \times 5} = \dfrac{4}{5}$

59. $\dfrac{48}{66} = \dfrac{6 \times 8}{6 \times 11} = \dfrac{8}{11}$

61. $\dfrac{63}{108} = \dfrac{3 \times 3 \times 7}{2 \times 2 \times 3 \times 3 \times 3} = \dfrac{7}{12}$

63. $\dfrac{88}{121} = \dfrac{11 \times 8}{11 \times 11} = \dfrac{8}{11}$

65. $\dfrac{120}{200} = \dfrac{40 \times 3}{40 \times 5} = \dfrac{3}{5}$

67. $\dfrac{220}{260} = \dfrac{11 \times 20}{13 \times 20} = \dfrac{11}{13}$

69. $\dfrac{4}{16} \overset{?}{=} \dfrac{7}{28}$
$4 \times 28 \overset{?}{=} 16 \times 7$
$112 = 112$
Yes

71. $\dfrac{12}{40} \overset{?}{=} \dfrac{3}{13}$
$12 \times 13 \overset{?}{=} 40 \times 3$
$156 \neq 120$
No

73. $\dfrac{23}{27} \overset{?}{=} \dfrac{92}{107}$
$23 \times 107 \overset{?}{=} 27 \times 92$
$2461 \neq 2484$
No

75. $\dfrac{27}{57} \overset{?}{=} \dfrac{45}{95}$
$27 \times 95 \overset{?}{=} 57 \times 45$
$2565 = 2565$
Yes

77. $\dfrac{60}{95} \overset{?}{=} \dfrac{12}{19}$
$60 \times 19 \overset{?}{=} 95 \times 12$
$1140 = 1140$
Yes

79. $\begin{array}{r} 128 \\ -\ 32 \\ \hline 96 \end{array}$

$\dfrac{96}{128} = \dfrac{3 \times 32}{4 \times 32} = \dfrac{3}{4}$

$\dfrac{3}{4}$ of the deliveries were just one pizza.

81. $\dfrac{\text{number who failed}}{\text{total number}} = \dfrac{12}{96} = \dfrac{1 \times 12}{8 \times 12} = \dfrac{1}{8}$
Since 12 failed, $96 - 12 = 84$ passed.
$\dfrac{\text{number who passed}}{\text{total number}} = \dfrac{84}{96} = \dfrac{7 \times 12}{8 \times 12} = \dfrac{7}{8}$

$\dfrac{1}{8}$ of the class failed; $\dfrac{7}{8}$ of the class passed.

83. $\dfrac{\text{amount saved}}{\text{amount earned}} = \dfrac{6000}{8400} = \dfrac{5 \times 1200}{7 \times 1200} = \dfrac{5}{7}$

Amelia saved $\dfrac{5}{7}$ of her earnings.

85. Total student body is:
$1100 + 1700 + 900 + 500 + 300 = 4500$
Short commute is: $\dfrac{1700}{4500} = \dfrac{17 \times 100}{45 \times 100} = \dfrac{17}{45}$

87. $\dfrac{500+300}{4500} = \dfrac{800}{4500} = \dfrac{8\times100}{45\times100} = \dfrac{8}{45}$

$\dfrac{8}{45}$ have a long or very long commute.

Cumulative Review

89.
$$
\begin{array}{r}
386 \\
\times\ \ 425 \\
\hline
1\,930 \\
7\,72 \\
154\,4 \\
\hline
164{,}050
\end{array}
$$

90.
$$
\begin{array}{r}
1296 \\
12\overline{)15{,}552} \\
\underline{12} \\
3\,5 \\
\underline{2\,4} \\
1\,15 \\
\underline{1\,08} \\
72 \\
\underline{72} \\
0
\end{array}
$$

91.
$$
\begin{array}{r}
3200 \\
\times\ \ \ 300 \\
\hline
960{,}000
\end{array}
$$

92.
$$
\begin{array}{r}
2{,}734{,}603{,}864 \\
-1{,}835{,}300{,}000 \\
\hline
899{,}303{,}864
\end{array}
$$

Avatar generated \$899,303,864 more than *Titanic*.

Quick Quiz 2.2

1. $\dfrac{25}{35} = \dfrac{5\times5}{5\times7} = \dfrac{5}{7}$

2. $\dfrac{14}{84} = \dfrac{2\times7}{2\times2\times3\times7} = \dfrac{1}{6}$

3. $\dfrac{40}{105} = \dfrac{2\times2\times2\times5}{3\times5\times7} = \dfrac{8}{21}$

4. Answers may vary. Possible solution:
Consider factors of 195.
$195 = 3 \times 5 \times 13$
Now find whether 231 has a factor of 3 or 5 or

13. It has a factor of 3.
$\dfrac{195}{231} = \dfrac{3\times5\times13}{3\times7\times11} = \dfrac{65}{77}$

2.3 Exercises

1. a. Multiply the whole number by the denominator of the fraction.

   b. Add the numerator of the fraction to the product formed in step (a).

   c. Write the sum found in step (b) over the denominator of the fraction.

3. $2\dfrac{1}{3} = \dfrac{2\times3+1}{3} = \dfrac{7}{3}$

5. $2\dfrac{3}{7} = \dfrac{2\times7+3}{7} = \dfrac{17}{7}$

7. $9\dfrac{2}{9} = \dfrac{9\times9+2}{9} = \dfrac{83}{9}$

9. $10\dfrac{2}{3} = \dfrac{10\times3+2}{3} = \dfrac{32}{3}$

11. $11\dfrac{3}{5} = \dfrac{11\times5+3}{5} = \dfrac{58}{5}$

13. $7\dfrac{2}{7} = \dfrac{7\times7+2}{7} = \dfrac{51}{7}$

15. $20\dfrac{1}{6} = \dfrac{20\times6+1}{6} = \dfrac{121}{6}$

17. $10\dfrac{11}{12} = \dfrac{10\times12+11}{12} = \dfrac{131}{12}$

19. $7\dfrac{9}{10} = \dfrac{7\times10+9}{10} = \dfrac{79}{10}$

21. $8\dfrac{1}{25} = \dfrac{8\times25+1}{25} = \dfrac{201}{25}$

23. $5\dfrac{5}{12} = \dfrac{5\times12+5}{12} = \dfrac{65}{12}$

25. $164\dfrac{2}{3} = \dfrac{164\times3+2}{3} = \dfrac{494}{3}$

27. $8\frac{11}{15} = \frac{8 \times 15 + 11}{15} = \frac{131}{15}$

29. $6\frac{19}{30} = \frac{6 \times 30 + 19}{30} = \frac{199}{30}$

31.
$$3\overline{)4}$$
$$\underline{3}$$
$$1$$

$\frac{4}{3} = 1\frac{1}{3}$

33.
$$4\overline{)11}$$
$$\underline{8}$$
$$3$$

$\frac{11}{4} = 2\frac{3}{4}$

35.
$$7\overline{)15}$$
$$\underline{14}$$
$$1$$

$\frac{15}{7} = 2\frac{1}{7}$

37.
$$8\overline{)27}$$
$$\underline{24}$$
$$3$$

$\frac{27}{8} = 3\frac{3}{8}$

39.
$$4\overline{)100}$$
$$\underline{8}$$
$$20$$
$$\underline{20}$$
$$0$$

$\frac{100}{4} = 25$

41.
$$9\overline{)86}$$
$$\underline{81}$$
$$5$$

$\frac{86}{9} = 9\frac{5}{9}$

43.
$$3\overline{)70}$$
$$\underline{6}$$
$$10$$
$$\underline{9}$$
$$1$$

$\frac{70}{3} = 23\frac{1}{3}$

45.
$$4\overline{)25}$$
$$\underline{24}$$
$$1$$

$\frac{25}{4} = 6\frac{1}{4}$

47.
$$10\overline{)57}$$
$$\underline{50}$$
$$7$$

$\frac{57}{10} = 5\frac{7}{10}$

49.
$$2\overline{)35}$$
$$\underline{2}$$
$$15$$
$$\underline{14}$$
$$1$$

$\frac{35}{2} = 17\frac{1}{2}$

51.
$$7\overline{)91}$$
$$\underline{7}$$
$$21$$
$$\underline{21}$$
$$0$$

$\frac{91}{7} = 13$

53. $15\overline{)210}$ with quotient 14

$$\frac{14}{15\overline{)210}}$$
$$\underline{15}$$
$$60$$
$$\underline{60}$$
$$0$$

$$\frac{210}{15} = 14$$

55. $17\overline{)102}$ with quotient 6

$$\frac{6}{17\overline{)102}}$$
$$\underline{102}$$
$$0$$

$$\frac{102}{17} = 6$$

57. $25\overline{)180}$ with quotient 7

$$\frac{7}{25\overline{)180}}$$
$$\underline{175}$$
$$5$$

$$\frac{180}{25} = 7\frac{5}{25} = 7\frac{1}{5}$$

59. $\dfrac{3}{6} = \dfrac{1\times3}{2\times3} = \dfrac{1}{2}$

$5\dfrac{3}{6} = 5\dfrac{1}{2}$

61. $\dfrac{11}{66} = \dfrac{11\times1}{11\times6} = \dfrac{1}{6}$

$4\dfrac{11}{66} = 4\dfrac{1}{6}$

63. $\dfrac{18}{72} = \dfrac{1\times18}{4\times18} = \dfrac{1}{4}$

$15\dfrac{18}{72} = 15\dfrac{1}{4}$

65. $\dfrac{24}{6} = \dfrac{6\times4}{6\times1} = 4$

67. $\dfrac{36}{15} = \dfrac{12\times3}{5\times3} = \dfrac{12}{5}$

69. $\dfrac{105}{28} = \dfrac{3\times5\times7}{2\times2\times7} = \dfrac{15}{4}$

71. $126\overline{)340}$ with quotient 2

$$\frac{2}{126\overline{)340}}$$
$$\underline{242}$$
$$88$$

$$\frac{340}{126} = 2\frac{88}{126}$$
$$\frac{88}{126} = \frac{2\times44}{2\times63} = \frac{44}{63}$$
$$\frac{340}{126} = 2\frac{88}{126} = 2\frac{44}{63}$$

73. $280\overline{)580}$ with quotient 2

$$\frac{2}{280\overline{)580}}$$
$$\underline{560}$$
$$20$$

$$\frac{580}{280} = 2\frac{20}{280}$$
$$\frac{20}{280} = \frac{1\times20}{14\times20} = \frac{1}{14}$$
$$\frac{580}{280} = 2\frac{20}{280} = 2\frac{1}{14}$$

75. $296\overline{)508}$ with quotient 1

$$\frac{1}{296\overline{)508}}$$
$$\underline{296}$$
$$212$$

$$\frac{508}{296} = 1\frac{212}{296}$$
$$\frac{212}{296} = \frac{4\times53}{4\times74} = \frac{53}{74}$$
$$\frac{508}{296} = 1\frac{212}{296} = 1\frac{53}{74}$$

77. $360\dfrac{2}{3} = \dfrac{360\times3+2}{3} = \dfrac{1082}{3}$

The art department is using $\dfrac{1082}{3}$ yards.

79.

$$3\overline{)151}$$ with quotient 50

$$\frac{15}{1}$$
$$\frac{0}{1}$$

$$\frac{151}{3} = 50\frac{1}{3}$$

Damage was done to $50\frac{1}{3}$ acres.

81.

$$8\overline{)1131}$$ with quotient 141

$$\frac{8}{33}$$
$$\frac{32}{11}$$
$$\frac{8}{3}$$

$$\frac{1131}{8} = 141\frac{3}{8}$$

The cafeteria used $141\frac{3}{8}$ pounds of white flour.

83. No; 101 is prime and is not a factor of 5687.

Cumulative Review

85.
$$\begin{array}{r} 1,398,210 \\ -\,1,137,963 \\ \hline 260,247 \end{array}$$

86. $20,000 \times 100,000 = 2,000,000,000$

87. $300,000 \div 1000 = 300$

88. $\dfrac{156-98}{156} = \dfrac{58}{156} = \dfrac{2\times29}{2\times78} = \dfrac{29}{78}$

$\dfrac{29}{78}$ of his new e-mails were not spam.

Quick Quiz 2.3

1. $4\dfrac{7}{13} = \dfrac{4\times13+7}{13} = \dfrac{59}{13}$

2.

$$12\overline{)89}$$ with quotient 7

$$\frac{84}{5}$$

$$\frac{89}{12} = 7\frac{5}{12}$$

3. $\dfrac{42}{14} = \dfrac{3\times14}{1\times14} = \dfrac{3}{1} = 3$

4. Answers may vary. Possible solution: Multiply the whole number, 5, by the denominator of the fraction, 13. Add the result to the numerator of the fraction. Write this sum over the denominator of the fraction.

$$5\frac{6}{13} = \frac{5\times13+6}{13} = \frac{65+6}{13} = \frac{71}{13}$$

2.4 Exercises

1. $\dfrac{3}{5} \times \dfrac{7}{11} = \dfrac{3\times7}{5\times11} = \dfrac{21}{55}$

3. $\dfrac{3}{4} \times \dfrac{5}{13} = \dfrac{3\times5}{4\times13} = \dfrac{15}{52}$

5. $\dfrac{6}{5} \times \dfrac{10}{12} = \dfrac{6\times10}{5\times12} = \dfrac{60}{60} = 1$

7. $\dfrac{5}{36} \times \dfrac{9}{20} = \dfrac{5}{\overset{}{\underset{4}{36}}} \times \dfrac{\overset{1}{\cancel{9}}}{\overset{1}{\underset{4}{\cancel{20}}}} = \dfrac{1}{16}$

9. $\dfrac{12}{25} \times \dfrac{5}{11} = \dfrac{12}{\underset{5}{\cancel{25}}} \times \dfrac{\overset{1}{\cancel{5}}}{11} = \dfrac{12}{55}$

11. $\dfrac{9}{10} \times \dfrac{35}{12} = \dfrac{\overset{3}{\cancel{9}}}{\underset{2}{\cancel{10}}} \times \dfrac{\overset{7}{\cancel{35}}}{\underset{4}{\cancel{12}}} = \dfrac{21}{8}$ or $2\dfrac{5}{8}$

13. $8 \times \dfrac{3}{7} = \dfrac{8}{1} \times \dfrac{3}{7} = \dfrac{24}{7}$ or $3\dfrac{3}{7}$

15. $\dfrac{15}{12} \times 8 = \dfrac{5}{\underset{3}{\cancel{12}}} \times \dfrac{\overset{2}{\cancel{8}}}{1} = \dfrac{10}{3}$ or $3\dfrac{1}{3}$

17. $\dfrac{4}{9} \times \dfrac{3}{7} \times \dfrac{7}{8} = \dfrac{\overset{1}{\cancel{4}}}{\underset{3}{\cancel{9}}} \times \dfrac{\overset{1}{\cancel{3}}}{\underset{1}{\cancel{7}}} \times \dfrac{\overset{1}{\cancel{7}}}{\underset{2}{\cancel{8}}} = \dfrac{1}{6}$

19. $\dfrac{5}{4} \times \dfrac{9}{10} \times \dfrac{8}{3} = \dfrac{5}{4} \times \dfrac{3 \times 3}{2 \times 5} \times \dfrac{2 \times 4}{3} = 3$

21. $2\dfrac{5}{6} \times \dfrac{3}{17} = \dfrac{17}{6} \times \dfrac{3}{17} = \dfrac{1}{2} \times \dfrac{1}{1} = \dfrac{1}{2}$

23. $10 \times 3\dfrac{1}{10} = \dfrac{10}{1} \times \dfrac{31}{10} = 31$

25. $1\dfrac{3}{16} \times 0 = 0$

27. $3\dfrac{7}{8} \times 1 = 3\dfrac{7}{8}$

29. $1\dfrac{1}{4} \times 3\dfrac{2}{3} = \dfrac{5}{4} \times \dfrac{11}{3} = \dfrac{55}{12}$ or $4\dfrac{7}{12}$

31. $2\dfrac{3}{10} \times \dfrac{3}{5} = \dfrac{23}{10} \times \dfrac{3}{5} = \dfrac{69}{50}$ or $1\dfrac{19}{50}$

33. $4\dfrac{1}{5} \times 8\dfrac{1}{3} = \dfrac{21}{5} \times \dfrac{25}{3} = \dfrac{7}{1} \times \dfrac{5}{1} = 35$

35. $6\dfrac{2}{5} \times \dfrac{1}{4} = \dfrac{32}{5} \times \dfrac{1}{4} = \dfrac{8}{5} \times \dfrac{1}{1} = \dfrac{8}{5}$ or $1\dfrac{3}{5}$

37. $\dfrac{11}{15} \times \dfrac{35}{33} = \dfrac{11}{3 \times 5} \times \dfrac{5 \times 7}{3 \times 11} = \dfrac{1}{3} \times \dfrac{7}{3} = \dfrac{7}{9}$

39. $\begin{aligned} 2\dfrac{3}{8} \times 5\dfrac{1}{3} &= \dfrac{19}{8} \times \dfrac{16}{3} \\ &= \dfrac{19}{8} \times \dfrac{2 \times 8}{3} \\ &= \dfrac{19}{1} \times \dfrac{2}{3} \\ &= \dfrac{38}{3} \text{ or } 12\dfrac{2}{3} \end{aligned}$

41. $\dfrac{4}{9} \cdot x = \dfrac{28}{81}$

Since $4 \cdot 7 = 28$ and $9 \cdot 9 = 81$,

$\dfrac{4}{9} \cdot \dfrac{7}{9} = \dfrac{28}{81}$.

Therefore, $x = \dfrac{7}{9}$.

43. $\dfrac{7}{13} \cdot x = \dfrac{56}{117}$

Since $7 \cdot 8 = 56$ and $13 \cdot 9 = 117$,

$\dfrac{7}{13} \cdot \dfrac{8}{9} = \dfrac{56}{117}$.

Therefore, $x = \dfrac{8}{9}$.

45. $\begin{aligned} 8\dfrac{3}{4} \times 4\dfrac{1}{3} &= \dfrac{8 \times 4 + 3}{4} \times \dfrac{4 \times 3 + 1}{3} \\ &= \dfrac{35}{4} \times \dfrac{13}{3} \\ &= \dfrac{455}{12} \\ &= 37\dfrac{11}{12} \end{aligned}$

The area of the forest where he is hiding is

$37\dfrac{11}{12}$ square miles.

47. $360 \times 4\dfrac{1}{3} = \dfrac{360}{1} \times \dfrac{13}{3} = 120 \times 13 = 1560$

The plane can go 1560 miles.

49. $90\dfrac{1}{2} \times 18 = \dfrac{181}{2} \times \dfrac{18}{1} = 181 \times 9 = 1629$

She would need 1629 grams of cheese.

51. $\dfrac{2}{3} \times 7998 = \dfrac{2}{3} \times \dfrac{3 \times 2666}{1} = 5332$

5332 students are under 25 years of age.

53. $12,064 \times \dfrac{1}{32} = \dfrac{12,064}{32} = 377$

He will have heard from 377 companies.

55. $4\dfrac{1}{4} \times 1\dfrac{1}{3} = \dfrac{17}{4} \times \dfrac{4}{3} = \dfrac{17}{3}$

$\dfrac{17}{3} \times \dfrac{1}{3} = \dfrac{17}{9} = 1\dfrac{8}{9}$

She jogged in the rain $1\dfrac{8}{9}$ miles.

57. The step of dividing the numerator and denominator by the same number allows us to work with smaller numbers when we do the multiplication. Also, this allows us to avoid the step of having to simplify the fraction in the final answer.

Cumulative Review

59.

$$\begin{array}{r} 529 \\ 31\overline{)16,399} \\ \underline{15\ 5} \\ 89 \\ \underline{62} \\ 279 \\ \underline{279} \\ 0 \end{array}$$

The average number of cars using the bridge in one day is 529 cars.

60.

$$\begin{array}{r} 368 \\ 42\overline{)15,456} \\ \underline{12\ 6} \\ 2\ 85 \\ \underline{2\ 52} \\ 336 \\ \underline{336} \\ 0 \end{array}$$

The average number of calls made per month by one salesperson is 368 calls.

61. $\dfrac{78-41}{78} = \dfrac{37}{78}$

$\dfrac{37}{78}$ of the cars were made in the United States.

62. $\dfrac{96-15}{96} = \dfrac{81}{96} = \dfrac{3\times 27}{3\times 32} = \dfrac{27}{32}$

$\dfrac{27}{32}$ of the class passed the first exam.

Quick Quiz 2.4

1. $32 \times \dfrac{5}{16} = \dfrac{32}{1} \times \dfrac{5}{16} = \dfrac{2\times 16}{1} \times \dfrac{5}{16} = \dfrac{2}{1} \times \dfrac{5}{1} = 10$

2. $\dfrac{11}{13} \times \dfrac{4}{5} = \dfrac{11\times 4}{13\times 5} = \dfrac{44}{65}$

3. $4\dfrac{1}{3} \times 2\dfrac{3}{4} = \dfrac{13}{3} \times \dfrac{11}{4} = \dfrac{143}{12}$ or $11\dfrac{11}{12}$

4. Answers may vary. Possible solution: Write 6 as $\dfrac{6}{1}$. Change the mixed number to an improper fraction, $4\dfrac{3}{5} = \dfrac{23}{5}$. Then multiply and simplify.

$\dfrac{6}{1} \cdot \dfrac{23}{5} = \dfrac{138}{5}$ or $27\dfrac{3}{5}$

2.5 Exercises

1. Think of a simple problem like $3 \div \dfrac{1}{2}$. One way to think of it is, how many $\dfrac{1}{2}$'s can be placed in 3? For example, how many $\dfrac{1}{2}$-pound rocks could be put in a bag that holds 3 pounds of rocks? The answer is 6. If we inverted the first fraction by mistake, we would have $\dfrac{1}{3} \times \dfrac{1}{2} = \dfrac{1}{6}$. We know that is wrong since there are obviously several $\dfrac{1}{2}$-pound rocks in a bag that holds 3 pounds of rocks. The answer $\dfrac{1}{6}$ would make no sense.

3. $\dfrac{7}{16} \div \dfrac{3}{4} = \dfrac{7}{16} \times \dfrac{4}{3} = \dfrac{7}{4} \times \dfrac{1}{3} = \dfrac{7}{12}$

5. $\dfrac{2}{3} \div \dfrac{4}{27} = \dfrac{2}{3} \times \dfrac{27}{4} = \dfrac{1}{1} \times \dfrac{9}{2} = \dfrac{9}{2}$ or $4\dfrac{1}{2}$

7. $\dfrac{7}{18} \div \dfrac{21}{6} = \dfrac{7}{18} \times \dfrac{6}{21} = \dfrac{1}{3} \times \dfrac{1}{3} = \dfrac{1}{9}$

9. $\dfrac{5}{9} \div \dfrac{1}{5} = \dfrac{5}{9} \times \dfrac{5}{1} = \dfrac{25}{9}$ or $2\dfrac{7}{9}$

11. $\dfrac{4}{15} \div \dfrac{4}{15} = \dfrac{4}{15} \times \dfrac{15}{4} = 1$

13. $\dfrac{3}{7} \div \dfrac{7}{3} = \dfrac{3}{7} \times \dfrac{3}{7} = \dfrac{9}{49}$

15. $\dfrac{4}{5} \div 1 = \dfrac{4}{5} \times \dfrac{1}{1} = \dfrac{4}{5}$

17. $\dfrac{3}{11} \div 4 = \dfrac{3}{11} \times \dfrac{1}{4} = \dfrac{3}{44}$

19. $1 \div \dfrac{7}{27} = \dfrac{1}{1} \times \dfrac{27}{7} = \dfrac{27}{7}$ or $3\dfrac{6}{7}$

21. $0 \div \dfrac{3}{17} = 0 \times \dfrac{17}{3} = 0$

23. $\dfrac{18}{19} \div 0$

Division by 0 is undefined.

25. $8 \div \dfrac{4}{5} = \dfrac{8}{1} \times \dfrac{5}{4} = \dfrac{2}{1} \times \dfrac{5}{1} = \dfrac{10}{1} = 10$

27. $\dfrac{7}{8} \div 4 = \dfrac{7}{8} \times \dfrac{1}{4} = \dfrac{7}{32}$

29. $\dfrac{9}{16} \div \dfrac{3}{4} = \dfrac{9}{16} \times \dfrac{4}{3} = \dfrac{3}{4} \times \dfrac{1}{1} = \dfrac{3}{4}$

31. $3\dfrac{1}{4} \div 2\dfrac{1}{4} = \dfrac{13}{4} \div \dfrac{9}{4} = \dfrac{13}{4} \times \dfrac{4}{9} = \dfrac{13}{1} \times \dfrac{1}{9} = \dfrac{13}{9}$ or $1\dfrac{4}{9}$

33. $6\dfrac{2}{5} \div 3\dfrac{1}{5} = \dfrac{32}{5} \div \dfrac{16}{5} = \dfrac{32}{5} \times \dfrac{5}{16} = \dfrac{2}{1} \times \dfrac{1}{1} = 2$

35. $6000 \div \dfrac{6}{5} = \dfrac{6000}{1} \times \dfrac{5}{6} = \dfrac{1000}{1} \times \dfrac{5}{1} = 5000$

37. $\dfrac{\frac{4}{5}}{200} = \dfrac{4}{5} \times \dfrac{1}{200} = \dfrac{1}{5} \times \dfrac{1}{50} = \dfrac{1}{250}$

39. $\dfrac{\frac{5}{8}}{\frac{25}{7}} = \dfrac{5}{8} \times \dfrac{7}{25} = \dfrac{1}{8} \times \dfrac{7}{5} = \dfrac{7}{40}$

41. $3\dfrac{1}{5} \div \dfrac{1}{5} = \dfrac{16}{5} \div \dfrac{1}{5} = \dfrac{16}{5} \times \dfrac{5}{1} = \dfrac{16}{1} \times \dfrac{1}{1} = 16$

43. $2\dfrac{1}{3} \times \dfrac{1}{6} = \dfrac{7}{3} \times \dfrac{1}{6} = \dfrac{7}{18}$

45. $5\dfrac{1}{4} \div 2\dfrac{5}{8} = \dfrac{21}{4} \div \dfrac{21}{8} = \dfrac{21}{4} \times \dfrac{8}{21} = \dfrac{1}{1} \times \dfrac{2}{1} = 2$

47. $5 \div 1\dfrac{1}{4} = \dfrac{5}{1} \div \dfrac{5}{4} = \dfrac{5}{1} \times \dfrac{4}{5} = \dfrac{1}{1} \times \dfrac{4}{1} = 4$

49. $5\dfrac{2}{3} \div 2\dfrac{1}{4} = \dfrac{17}{3} \div \dfrac{9}{4} = \dfrac{17}{3} \cdot \dfrac{4}{9} = \dfrac{68}{27}$ or $2\dfrac{14}{27}$

51. $\dfrac{7}{2} \div 3\dfrac{1}{2} = \dfrac{7}{2} \div \dfrac{7}{2} = \dfrac{7}{2} \times \dfrac{2}{7} = 1$

53. $\dfrac{13}{25} \times 2\dfrac{1}{3} = \dfrac{13}{25} \times \dfrac{7}{3} = \dfrac{91}{75}$ or $1\dfrac{16}{75}$

55. $3\dfrac{3}{4} \div 9 = \dfrac{15}{4} \times \dfrac{1}{9} = \dfrac{5}{4} \times \dfrac{1}{3} = \dfrac{5}{12}$

57. $\dfrac{5}{3\frac{1}{6}} = \dfrac{5}{\frac{19}{6}} = \dfrac{5}{1} \times \dfrac{6}{19} = \dfrac{30}{19}$ or $1\dfrac{11}{19}$

59. $\dfrac{0}{4\frac{3}{8}} = \dfrac{0}{\frac{35}{8}} = 0 \times \dfrac{8}{35} = 0$

61. $\dfrac{\frac{7}{12}}{3\frac{2}{3}} = \dfrac{7}{12} \div \dfrac{11}{3} = \dfrac{7}{12} \times \dfrac{3}{11} = \dfrac{7}{4} \times \dfrac{1}{11} = \dfrac{7}{44}$

63. $4\dfrac{2}{5} \times 2\dfrac{8}{11} = \dfrac{22}{5} \times \dfrac{30}{11} = \dfrac{2}{1} \times \dfrac{6}{1} = 12$

65. $x \div \dfrac{4}{3} = \dfrac{21}{20}$

$x \cdot \dfrac{3}{4} = \dfrac{21}{20}$

$\dfrac{7}{5} \cdot \dfrac{3}{4} = \dfrac{21}{20}$

$x = \dfrac{7}{5}$

67. $x \div \dfrac{10}{7} = \dfrac{21}{100}$

$x \cdot \dfrac{7}{10} = \dfrac{21}{100}$

$\dfrac{3}{10} \cdot \dfrac{7}{10} = \dfrac{21}{100}$

$x = \dfrac{3}{10}$

69. $20\dfrac{1}{4} \div 9 = \dfrac{81}{4} \div \dfrac{9}{1}$

$\qquad = \dfrac{81}{4} \times \dfrac{1}{9}$

$\qquad = \dfrac{9 \times 9 \times 1}{4 \times 9}$

$\qquad = \dfrac{9}{4}$

$\qquad = 2\dfrac{1}{4}$

Each vat will hold $2\dfrac{1}{4}$ gallons.

71. $125 \div 3\dfrac{1}{3} = \dfrac{125}{1} \div \dfrac{10}{3}$

$\qquad = \dfrac{125}{1} \times \dfrac{3}{10}$

$\qquad = \dfrac{5 \times 25 \times 3}{5 \times 2}$

$\qquad = \dfrac{75}{2}$

$\qquad = 37\dfrac{1}{2}$

His average speed was $37\dfrac{1}{2}$ miles per hour.

73. $38\dfrac{2}{3} \div \dfrac{2}{3} = \dfrac{116}{3} \div \dfrac{2}{3}$

$\qquad = \dfrac{116}{3} \times \dfrac{3}{2}$

$\qquad = \dfrac{58 \times 2 \times 3}{3 \times 2}$

$\qquad = 58$

58 students will be fed.

75. $\dfrac{150}{1\frac{1}{2}} = \dfrac{150}{\frac{3}{2}} = \dfrac{150}{1} \times \dfrac{2}{3} = 100$

100 large Styrofoam cups can be filled.

77. $4\dfrac{3}{4} \div \dfrac{5}{6} = \dfrac{19}{4} \times \dfrac{6}{5}$

$\qquad = \dfrac{19 \times 2 \times 3}{2 \times 2 \times 5}$

$\qquad = \dfrac{57}{10}$

$\qquad = 5\dfrac{7}{10}$ attempts

It took six drill attempts.

79. $15 \div 5 = 3$

Exact $= 14\dfrac{2}{3} \div 5\dfrac{1}{6} = \dfrac{44}{3} \times \dfrac{6}{31} = \dfrac{88}{31} = 2\dfrac{26}{31}$

It is off by only $\dfrac{5}{31}$.

Cumulative Review

81. 39,576,304 = thirty-nine million, five hundred seventy-six thousand, three hundred four

82. 509,270 = 500,000 + 9000 + 200 + 70

83. 126 + 34 + 9 + 891 + 12 + 27 = 1099

84. 87,595,631

85. $\dfrac{9}{10} + \dfrac{3}{5} = \dfrac{9}{10} + \dfrac{3 \times 2}{5 \times 2} = \dfrac{9}{10} + \dfrac{6}{10} = \dfrac{15}{10} = \dfrac{3}{2}$ or $1\dfrac{1}{2}$

86. $\dfrac{17}{20} - \dfrac{3}{4} = \dfrac{17}{20} - \dfrac{3 \times 5}{4 \times 5} = \dfrac{17}{20} - \dfrac{15}{20} = \dfrac{2}{20} = \dfrac{1}{10}$

Quick Quiz 2.5

1. $\dfrac{15}{24} \div \dfrac{5}{6} = \dfrac{15}{24} \times \dfrac{6}{5} = \dfrac{\overset{3}{\cancel{15}}}{\underset{4}{\cancel{24}}} \times \dfrac{\overset{1}{\cancel{6}}}{\underset{1}{\cancel{5}}} = \dfrac{3}{4}$

2. $6\dfrac{1}{3} \div 2\dfrac{5}{12} = \dfrac{19}{3} \div \dfrac{29}{12}$

$\qquad = \dfrac{19}{3} \times \dfrac{12}{29}$

$\qquad = \dfrac{19}{1} \times \dfrac{4}{29}$

$\qquad = \dfrac{76}{29}$ or $2\dfrac{18}{29}$

3. $7\dfrac{3}{4} \div 4 = \dfrac{31}{4} \times \dfrac{1}{4} = \dfrac{31}{16}$ or $1\dfrac{15}{16}$

4. Answers may vary. Possible solution: Write 7 as $\dfrac{7}{1}$. Change the mixed number to an improper fraction, $3\dfrac{3}{5} = \dfrac{18}{5}$. Then write the division problem as a multiplication problem and simplify.

$7 \div 3\dfrac{3}{5} = \dfrac{7}{1} \div \dfrac{18}{5} = \dfrac{7}{1} \times \dfrac{5}{18} = \dfrac{35}{18}$ or $1\dfrac{17}{18}$

How Am I Doing? Sections 2.1–2.5

1. Three out of eight equal parts are shaded. The fraction is $\dfrac{3}{8}$.

2. $\dfrac{\text{number from outside the country}}{\text{total number}}$
$= \dfrac{800}{3500 + 2600 + 800}$
$= \dfrac{800}{6900}$
$= \dfrac{8 \times 100}{69 \times 100}$
$= \dfrac{8}{69}$

3. $\dfrac{\text{number defective}}{\text{total number}} = \dfrac{10}{224} = \dfrac{2 \times 5}{2 \times 112} = \dfrac{5}{112}$

4. $\dfrac{4}{28} = \dfrac{4 \div 4}{28 \div 4} = \dfrac{1}{7}$

5. $\dfrac{13}{39} = \dfrac{13 \div 13}{39 \div 13} = \dfrac{1}{3}$

6. $\dfrac{16}{112} = \dfrac{16 \div 16}{112 \div 16} = \dfrac{1}{7}$

7. $\dfrac{175}{200} = \dfrac{175 \div 25}{200 \div 25} = \dfrac{7}{8}$

8. $\dfrac{44}{121} = \dfrac{44 \div 11}{121 \div 11} = \dfrac{4}{11}$

9. $3\dfrac{2}{3} = \dfrac{3 \times 3 + 2}{3} = \dfrac{11}{3}$

10. $15\dfrac{1}{3} = \dfrac{15 \times 3 + 1}{3} = \dfrac{46}{3}$

11.
$$
\begin{array}{r}
20 \\
4\overline{)81} \\
8 \\
\hline
01 \\
0 \\
\hline
1
\end{array}
$$
$\dfrac{81}{4} = 20\dfrac{1}{4}$

12.
$$
\begin{array}{r}
5 \\
5\overline{)29} \\
25 \\
\hline
4
\end{array}
$$
$\dfrac{29}{5} = 5\dfrac{4}{5}$

13.
$$
\begin{array}{r}
2 \\
17\overline{)36} \\
34 \\
\hline
2
\end{array}
$$
$\dfrac{36}{17} = 2\dfrac{2}{17}$

14. $\dfrac{5}{11} \times \dfrac{1}{4} = \dfrac{5 \times 1}{11 \times 4} = \dfrac{5}{44}$

15. $\dfrac{3}{7} \times \dfrac{14}{9} = \dfrac{3 \times 2 \times 7}{7 \times 3 \times 3} = \dfrac{2}{3}$

16. $3\dfrac{1}{3} \times 5\dfrac{1}{3} = \dfrac{10}{3} \times \dfrac{16}{3} = \dfrac{160}{9}$ or $17\dfrac{7}{9}$

17. $\dfrac{3}{7} \div \dfrac{3}{7} = \dfrac{3}{7} \times \dfrac{7}{3} = 1$

18. $\dfrac{7}{16} \div \dfrac{7}{8} = \dfrac{7}{16} \times \dfrac{8}{7} = \dfrac{7 \times 8}{2 \times 8 \times 7} = \dfrac{1}{2}$

19. $6\dfrac{4}{7} \div 1\dfrac{5}{21} = \dfrac{46}{7} \div \dfrac{26}{21}$
$= \dfrac{46}{7} \times \dfrac{21}{26}$
$= \dfrac{2 \times 23 \times 3 \times 7}{7 \times 2 \times 13}$
$= \dfrac{69}{13}$ or $5\dfrac{4}{13}$

20. $12 \div \dfrac{4}{7} = \dfrac{12}{1} \times \dfrac{7}{4} = \dfrac{3}{1} \times \dfrac{7}{1} = 21$

How Am I Doing? Test on Sections 2.1–2.5

1. $\dfrac{\text{number answered correctly}}{\text{total number of questions}} = \dfrac{33}{40}$

2. $\dfrac{\text{number of correct weight}}{\text{total number}} = \dfrac{340}{340+112}$
$$= \dfrac{340}{452}$$
$$= \dfrac{340 \div 4}{452 \div 4}$$
$$= \dfrac{85}{113}$$

3. $\dfrac{19}{38} = \dfrac{19 \div 19}{38 \div 19} = \dfrac{1}{2}$

4. $\dfrac{40}{56} = \dfrac{40 \div 8}{56 \div 8} = \dfrac{5}{7}$

5. $\dfrac{24}{66} = \dfrac{24 \div 6}{66 \div 6} = \dfrac{4}{11}$

6. $\dfrac{125}{155} = \dfrac{125 \div 5}{155 \div 5} = \dfrac{25}{31}$

7. $\dfrac{50}{140} = \dfrac{50 \div 10}{140 \div 10} = \dfrac{5}{14}$

8. $\dfrac{84}{36} = \dfrac{84 \div 12}{36 \div 12} = \dfrac{7}{3}$ or $2\dfrac{1}{3}$

9. $12\dfrac{2}{3} = \dfrac{12 \times 3 + 2}{3} = \dfrac{38}{3}$

10. $4\dfrac{1}{8} = \dfrac{4 \times 8 + 1}{8} = \dfrac{33}{8}$

11. $\begin{array}{r} 6 \\ 7\overline{)45} \\ \underline{42} \\ 3 \end{array}$

   $\dfrac{45}{7} = 6\dfrac{3}{7}$

12. $\begin{array}{r} 8 \\ 9\overline{)75} \\ \underline{72} \\ 3 \end{array}$

   $\dfrac{75}{9} = 8\dfrac{3}{9} = 8\dfrac{1}{3}$

13. $\dfrac{3}{8} \times \dfrac{7}{11} = \dfrac{3 \times 7}{8 \times 11} = \dfrac{21}{88}$

14. $\dfrac{35}{16} \times \dfrac{4}{5} = \dfrac{7}{4} \times \dfrac{1}{1} = \dfrac{7}{4}$ or $1\dfrac{3}{4}$

15. $18 \times \dfrac{5}{6} = \dfrac{18}{1} \times \dfrac{5}{6} = \dfrac{3 \times 6 \times 5}{6} = 3 \times 5 = 15$

16. $\dfrac{3}{8} \times 44 = \dfrac{3}{8} \times \dfrac{44}{1} = \dfrac{3 \times 4 \times 11}{2 \times 4} = \dfrac{3 \times 11}{2} = \dfrac{3}{2}$ or $16\dfrac{1}{2}$

17. $2\dfrac{1}{3} \times 5\dfrac{3}{4} = \dfrac{7}{3} \times \dfrac{23}{4} = \dfrac{161}{12}$ or $13\dfrac{5}{12}$

18. $24 \times 3\dfrac{1}{3} = \dfrac{24}{1} \times \dfrac{10}{3} = \dfrac{8}{1} \times \dfrac{10}{1} = 80$

19. $\dfrac{4}{7} \div \dfrac{3}{4} = \dfrac{4}{7} \times \dfrac{4}{3} = \dfrac{4 \times 4}{7 \times 3} = \dfrac{16}{21}$

20. $\dfrac{8}{9} \div \dfrac{1}{6} = \dfrac{8}{9} \times \dfrac{6}{1} = \dfrac{8 \times 3 \times 2}{3 \times 3} = \dfrac{8 \times 2}{3} = \dfrac{16}{3}$ or $5\dfrac{1}{3}$

21. $5\dfrac{1}{4} \div \dfrac{3}{4} = \dfrac{21}{4} \times \dfrac{4}{3} = \dfrac{7 \times 3 \times 4}{4 \times 3} = 7$

22. $5\dfrac{3}{5} \div 2\dfrac{1}{3} = \dfrac{28}{5} \div \dfrac{7}{3} = \dfrac{28}{5} \times \dfrac{3}{7} = \dfrac{4}{5} \times \dfrac{3}{1} = \dfrac{12}{5}$ or $2\dfrac{2}{5}$

23. $2\dfrac{1}{4} \times 3\dfrac{1}{2} = \dfrac{9}{4} \times \dfrac{7}{2} = \dfrac{63}{8}$ or $7\dfrac{7}{8}$

24. $6 \times 2\dfrac{1}{3} = \dfrac{6}{1} \times \dfrac{7}{3} = \dfrac{2 \times 3 \times 7}{3} = 2 \times 7 = 14$

25. $12 \div 2\dfrac{1}{4} = \dfrac{12}{1} \div \dfrac{9}{4} = \dfrac{12}{1} \times \dfrac{4}{9} = \dfrac{3 \times 4 \times 4}{3 \times 3} = \dfrac{16}{3}$ or $5\dfrac{1}{3}$

26. $5\dfrac{3}{4} \div 2 = \dfrac{23}{4} \div 2 = \dfrac{23}{4} \times \dfrac{1}{2} = \dfrac{23}{4 \times 2} = \dfrac{23}{8}$ or $2\dfrac{7}{8}$

27. $\dfrac{13}{20} \div \dfrac{4}{5} = \dfrac{13}{20} \times \dfrac{5}{4} = \dfrac{13 \times 5}{5 \times 4 \times 4} = \dfrac{13}{16}$

28. $\dfrac{5}{3} \div 5 = \dfrac{5}{3} \times \dfrac{1}{5} = \dfrac{5 \times 1}{3 \times 5} = \dfrac{1}{3}$

29. $\dfrac{7}{10} \times \dfrac{20}{23} = \dfrac{7 \times 2 \times 10}{10 \times 23} = \dfrac{14}{23}$

30. $\dfrac{14}{25} \times \dfrac{65}{42} = \dfrac{7 \times 2 \times 13 \times 5}{5 \times 5 \times 2 \times 3 \times 7} = \dfrac{13}{5 \times 3} = \dfrac{13}{15}$

31. $5\dfrac{1}{4} \times 8\dfrac{3}{4} = \dfrac{21}{4} \times \dfrac{35}{4} = \dfrac{735}{16} = 45\dfrac{15}{16}$

The area of the garden is $45\dfrac{15}{16}$ square feet.

32. $2\dfrac{2}{3} \times 1\dfrac{1}{2} = \dfrac{8}{3} \times \dfrac{3}{2} = \dfrac{4 \times 2 \times 3}{3 \times 2} = 4$

She will need 4 cups of flour.

33. $62\dfrac{1}{2} \times \dfrac{3}{4} = \dfrac{125}{2} \times \dfrac{3}{4} = \dfrac{375}{8}$ or $46\dfrac{7}{8}$

She drove $46\dfrac{7}{8}$ miles on the highway.

34. $12\dfrac{3}{8} \div \dfrac{3}{4} = \dfrac{99}{8} \times \dfrac{4}{3}$

$= \dfrac{3 \times 33 \times 4}{4 \times 2 \times 3}$

$= \dfrac{33}{2}$

$= 16\dfrac{1}{2}$ packages

He had 16 full packages with $\dfrac{3}{4} \times \dfrac{1}{2} = \dfrac{3}{8}$ pound left over.

35. $136 \times \dfrac{3}{8} = \dfrac{136}{1} \times \dfrac{3}{8}$

$= \dfrac{17 \times 8 \times 3}{8}$

$= 17 \times 3$

$= 51$

51 computers have Windows 7 installed on them.

36. $\dfrac{3}{10} \times 82,000 = \dfrac{3}{10} \times \dfrac{82,000}{1}$

$= \dfrac{3}{1} \times \dfrac{8200}{1}$

$= 24,600$

The average household uses 24,600 gallons of water per year for showers and baths.

37. $102 \div 12\dfrac{3}{4} = \dfrac{102}{1} \div \dfrac{51}{4} = \dfrac{102}{1} \times \dfrac{4}{51} = \dfrac{2 \times 51 \times 4}{1 \times 51} = 8$

Yung Kim's kitchen is 8 feet wide.

38. $56\dfrac{1}{2} \div 8\dfrac{1}{4} = \dfrac{113}{2} \div \dfrac{33}{4}$

$= \dfrac{113}{2} \times \dfrac{4}{33}$

$= \dfrac{113 \times 2 \times 2}{2 \times 33}$

$= \dfrac{226}{33}$

$= 6\dfrac{28}{33}$

He can make 6 full tents, with

$8\dfrac{1}{4} \times \dfrac{28}{33} = \dfrac{33}{4} \times \dfrac{28}{33} = 7$ yards left over.

39. $32\dfrac{4}{5} \div \dfrac{4}{5} = \dfrac{164}{5} \times \dfrac{5}{4} = 41$

He can use it for 41 days.

2.6 Exercises

1. 8 and 12
Multiples of 8: 8, 16, 24, 32, 48, ...
Multiples of 12: 12, 24, 36, 48, 60, ...
The least common multiple is 24.

3. 20 and 50
Multiples of 20: 20, 40, 60, 80, 100, ...
Multiplies of 50: 50, 100, 150, 200, 250, ...
The least common multiple is 100.

5. 12 and 15
Multiples of 12: 12, 24, 36, 48, 60, ...
Multiples of 15: 15, 30, 45, 60, 75, ...
The least common multiple is 60.

7. 10 and 15
Multiples of 10: 10, 20, 30, 40, 50, ...
Multiples of 15: 15, 30, 45, 60, 75, ...
The least common multiple is 30.

9. 21 and 49
Multiples of 21: 21, 42, 63, 84, 105, 126, 147, 168, ...
Multiples of 49: 49, 98, 147, 196, 245, ...
The least common multiple is 147.

11. $5 = 5$
$10 = 2 \times 5$
LCD $= 2 \times 5 = 10$

13. $7 = 7$
$4 = 2 \times 2$
LCD $= 2 \times 2 \times 7 = 28$

15. $5 = 5$
$7 = 7$
$LCD = 5 \times 7 = 35$

17. $6 = 2 \times 3$
$9 = 3 \times 3$
$LCD = 2 \times 3 \times 3 = 18$

19. $12 = 2 \times 2 \times 3$
$15 = 3 \times 5$
$LCD = 2 \times 2 \times 3 \times 5 = 60$

21. $4 = 2 \times 2$
$32 = 2 \times 2 \times 2 \times 2 \times 2$
$LCD = 2 \times 2 \times 2 \times 2 \times 2 = 32$

23. $10 = 2 \times 5$
$45 = 3 \times 3 \times 5$
$LCD = 2 \times 3 \times 3 \times 5 = 90$

25. $16 = 2 \times 2 \times 2 \times 2$
$80 = 2 \times 2 \times 2 \times 2 \times 5$
$LCD = 2 \times 2 \times 2 \times 2 \times 5 = 80$

27. $21 = 3 \times 7$
$35 = 5 \times 7$
$LCD = 3 \times 5 \times 7 = 105$

29. $24 = 2 \times 2 \times 2 \times 3$
$30 = 2 \times 3 \times 5$
$LCD = 2 \times 2 \times 2 \times 3 \times 5 = 120$

31. $3 = 3$
$2 = 2$
$6 = 2 \times 3$
$LCD = 2 \times 3 = 6$

33. $4 = 2 \times 2$
$12 = 2 \times 2 \times 3$
$6 = 2 \times 3$
$LCD = 2 \times 2 \times 3 = 12$

35. $11 = 11$
$12 = 2 \times 2 \times 3$
$6 = 2 \times 3$
$LCD = 2 \times 2 \times 3 \times 11 = 132$

37. $12 = 2 \times 2 \times 3$
$21 = 3 \times 7$
$14 = 2 \times 7$
$LCD = 2 \times 2 \times 3 \times 7 = 84$

39. $15 = 3 \times 5$
$12 = 2 \times 2 \times 3$
$8 = 2 \times 2 \times 2$
$LCD = 2 \times 2 \times 2 \times 3 \times 5 = 120$

41. $\dfrac{1}{3} = \dfrac{1}{3} \times \dfrac{3}{3} = \dfrac{3}{9}$
The numerator is 3.

43. $\dfrac{5}{7} = \dfrac{5}{7} \times \dfrac{7}{7} = \dfrac{35}{49}$
The numerator is 35.

45. $\dfrac{4}{11} = \dfrac{4}{11} \times \dfrac{5}{5} = \dfrac{20}{55}$
The numerator is 20.

47. $\dfrac{5}{12} = \dfrac{5}{12} \times \dfrac{8}{8} = \dfrac{40}{96}$
The numerator is 40.

49. $\dfrac{8}{9} = \dfrac{8}{9} \times \dfrac{12}{12} = \dfrac{96}{108}$
The numerator is 96.

51. $\dfrac{7}{20} = \dfrac{7}{20} \times \dfrac{9}{9} = \dfrac{63}{180}$
The numerator is 63.

53. $\dfrac{7}{12} = \dfrac{7 \times 3}{12 \times 3} = \dfrac{21}{36}$
$\dfrac{5}{9} = \dfrac{5 \times 4}{9 \times 4} = \dfrac{20}{36}$

55. $\dfrac{5}{16} = \dfrac{5}{16} \times \dfrac{5}{5} = \dfrac{25}{80}$
$\dfrac{17}{20} = \dfrac{17}{20} \times \dfrac{4}{4} = \dfrac{68}{80}$

57. $\dfrac{9}{10} = \dfrac{9 \times 2}{10 \times 2} = \dfrac{18}{20}$
$\dfrac{19}{20} = \dfrac{19}{20}$

59. $5 = 5$
$35 = 5 \times 7$
$LCD = 5 \times 7 = 35$
$\dfrac{2}{5} = \dfrac{2 \times 7}{5 \times 7} = \dfrac{14}{35}$
$\dfrac{14}{35}$ and $\dfrac{9}{35}$

61. $24 = 2 \times 2 \times 2 \times 3$

$8 = 2 \times 2 \times 2$

$\text{LCD} = 2 \times 2 \times 2 \times 3 = 24$

$\dfrac{5}{24}$

$\dfrac{3}{8} = \dfrac{3 \times 3}{8 \times 3} = \dfrac{9}{24}$

$\dfrac{5}{24}$ and $\dfrac{9}{24}$

63. $6 = 2 \times 3$

$15 = 3 \times 5$

$\text{LCD} = 2 \times 3 \times 5 = 30$

$\dfrac{8}{15} \times \dfrac{2}{2} = \dfrac{16}{30}$ $\dfrac{1}{6} \times \dfrac{5}{5} = \dfrac{5}{30}$

65. $15 = 3 \times 5$

$12 = 2 \times 2 \times 3$

$\text{LCD} = 3 \times 5 \times 2 \times 2 = 60$

$\dfrac{4}{15} = \dfrac{4 \times 4}{15 \times 4} = \dfrac{16}{60}$

$\dfrac{5}{12} = \dfrac{5 \times 5}{12 \times 5} = \dfrac{25}{60}$

$\dfrac{16}{60}$ and $\dfrac{25}{60}$

67. $18 = 2 \times 3 \times 3$

$36 = 2 \times 2 \times 3 \times 3$

$12 = 2 \times 2 \times 3$

$\text{LCD} = 2 \times 2 \times 3 \times 3 = 36$

$\dfrac{5}{18} \times \dfrac{2}{2} = \dfrac{10}{36}$ $\dfrac{11}{36} = \dfrac{11}{36}$ $\dfrac{7}{12} \times \dfrac{3}{3} = \dfrac{21}{36}$

$\dfrac{10}{36}, \dfrac{11}{36}, \dfrac{21}{36}$

69. $56 = 2 \times 2 \times 2 \times 7$

$8 = 2 \times 2 \times 2$

$7 = 7$

$\text{LCD} = 2 \times 2 \times 2 \times 7 = 56$

$\dfrac{3}{56}, \dfrac{7}{8} \times \dfrac{7}{7} = \dfrac{49}{56}, \dfrac{5}{7} \times \dfrac{8}{8} = \dfrac{40}{56}$

$\dfrac{3}{56}, \dfrac{49}{56}, \dfrac{40}{56}$

71. $63 = 3 \times 3 \times 7$

$21 = 3 \times 7$

$9 = 3 \times 3$

$\text{LCD} = 3 \times 3 \times 7 = 63$

$\dfrac{5}{63}, \dfrac{4}{21} \times \dfrac{3}{3} = \dfrac{12}{63}, \dfrac{8}{9} \times \dfrac{7}{7} = \dfrac{56}{63}$

$\dfrac{5}{63}, \dfrac{12}{63}, \dfrac{56}{63}$

73. a. $16 = 2 \times 2 \times 2 \times 2$

$4 = 2 \times 2$

$8 = 2 \times 2 \times 2$

$\text{LCD} = 2 \times 2 \times 2 \times 2 = 16$

b. $\dfrac{3}{16}, \dfrac{3}{4} \times \dfrac{4}{4}, \dfrac{3}{8} \times \dfrac{2}{2}$

$\dfrac{3}{16}, \dfrac{12}{16}, \dfrac{6}{16}$

Cumulative Review

75. $(5-3)^2 + 4 \times 6 - 3 = 2^2 + 4 \times 6 - 3$

$\qquad\qquad\qquad\quad = 4 + 4 \times 6 - 3$

$\qquad\qquad\qquad\quad = 4 + 24 - 3$

$\qquad\qquad\qquad\quad = 28 - 3$

$\qquad\qquad\qquad\quad = 25$

76. $4\dfrac{3}{4} \times \dfrac{2}{3} = \dfrac{19}{4} \times \dfrac{2}{3} = \dfrac{19 \times 2}{2 \times 2 \times 3} = \dfrac{19}{6}$ or $3\dfrac{1}{6}$

77. $16\dfrac{1}{2} \div \dfrac{3}{4} = \dfrac{33}{2} \div \dfrac{3}{4} = \dfrac{33}{2} \times \dfrac{4}{3} = \dfrac{3 \times 11 \times 2 \times 2}{2 \times 3} = 22$

Quick Quiz 2.6

1. $6 = 2 \times 3$

$21 = 3 \times 7$

$\text{LCD} = 2 \times 3 \times 7 = 42$

2. $28 = 2 \times 2 \times 7$

$4 = 2 \times 2$

$20 = 2 \times 2 \times 5$

$\text{LCD} = 2 \times 2 \times 5 \times 7 = 140$

3. $\dfrac{7}{26} \times \dfrac{3}{3} = \dfrac{21}{78}$

4. Answers may vary. Possible solution: Write each denominator as the product of prime factors. Form a product of all those prime factors, using each factor the greatest number of times it appears in any one denominator.

$6 = 2 \times 3$

$14 = 2 \times 7$

$15 = 3 \times 5$

$\text{LCD} = 2 \times 3 \times 5 \times 7 = 210$

2.7 Exercises

1. $\dfrac{5}{9}+\dfrac{2}{9}=\dfrac{5+2}{9}=\dfrac{7}{9}$

3. $\dfrac{7}{18}+\dfrac{15}{18}=\dfrac{22}{18}=\dfrac{11}{9}$ or $1\dfrac{2}{9}$

5. $\dfrac{19}{20}-\dfrac{11}{20}=\dfrac{8}{20}=\dfrac{2}{5}$

7. $\dfrac{53}{88}-\dfrac{19}{88}=\dfrac{34}{88}=\dfrac{17}{44}$

9. $\dfrac{1}{3}+\dfrac{1}{2}=\dfrac{1}{3}\times\dfrac{2}{2}+\dfrac{1}{2}\times\dfrac{3}{3}=\dfrac{2}{6}+\dfrac{3}{6}=\dfrac{5}{6}$

11. $\dfrac{3}{10}+\dfrac{3}{20}=\dfrac{3}{10}\times\dfrac{2}{2}+\dfrac{3}{20}=\dfrac{6}{20}+\dfrac{3}{20}=\dfrac{9}{20}$

13. $\dfrac{1}{8}+\dfrac{3}{4}=\dfrac{1}{8}+\dfrac{3}{4}\times\dfrac{2}{2}=\dfrac{1}{8}+\dfrac{6}{6}=\dfrac{7}{8}$

15. $\dfrac{4}{5}+\dfrac{7}{20}=\dfrac{4}{5}\times\dfrac{4}{4}+\dfrac{7}{20}=\dfrac{16}{20}+\dfrac{7}{20}=\dfrac{23}{20}$ or $1\dfrac{3}{20}$

17. $\dfrac{3}{10}+\dfrac{7}{100}=\dfrac{3}{10}\times\dfrac{10}{10}+\dfrac{7}{100}=\dfrac{30}{100}+\dfrac{7}{100}=\dfrac{37}{100}$

19. $\dfrac{3}{10}+\dfrac{1}{6}=\dfrac{3}{10}\times\dfrac{3}{3}+\dfrac{1}{6}\times\dfrac{5}{5}=\dfrac{9}{30}+\dfrac{5}{30}=\dfrac{14}{30}=\dfrac{7}{15}$

21. $\dfrac{7}{8}+\dfrac{5}{12}=\dfrac{7}{8}\times\dfrac{3}{3}+\dfrac{5}{12}\times\dfrac{2}{2}=\dfrac{21}{24}+\dfrac{10}{24}=\dfrac{31}{24}$ or $1\dfrac{7}{24}$

23. $\dfrac{3}{8}+\dfrac{3}{10}=\dfrac{3}{8}\times\dfrac{5}{5}+\dfrac{3}{10}\times\dfrac{4}{4}=\dfrac{15}{40}+\dfrac{12}{40}=\dfrac{27}{40}$

25. $\dfrac{29}{18}-\dfrac{5}{9}=\dfrac{29}{18}-\dfrac{5}{9}\times\dfrac{2}{2}=\dfrac{29}{18}-\dfrac{10}{18}=\dfrac{19}{18}$ or $1\dfrac{1}{18}$

27. $\dfrac{3}{7}-\dfrac{9}{21}=\dfrac{3}{7}\times\dfrac{3}{3}-\dfrac{9}{21}=\dfrac{9}{21}-\dfrac{9}{21}=0$

29. $\dfrac{5}{9}-\dfrac{5}{36}=\dfrac{5}{9}\times\dfrac{4}{4}-\dfrac{5}{36}=\dfrac{20}{36}-\dfrac{5}{36}=\dfrac{15}{36}=\dfrac{5}{12}$

31. $\dfrac{5}{12}-\dfrac{7}{30}=\dfrac{5}{12}\times\dfrac{5}{5}-\dfrac{7}{30}\times\dfrac{2}{2}=\dfrac{25}{60}-\dfrac{14}{60}=\dfrac{11}{60}$

33. $\dfrac{11}{12}-\dfrac{2}{3}=\dfrac{11}{12}-\dfrac{2}{3}\times\dfrac{4}{4}=\dfrac{11}{12}-\dfrac{8}{12}=\dfrac{3}{12}=\dfrac{1}{4}$

35. $\dfrac{17}{21}-\dfrac{1}{7}=\dfrac{17}{21}-\dfrac{1}{7}\times\dfrac{3}{3}=\dfrac{17}{21}-\dfrac{3}{21}=\dfrac{14}{21}=\dfrac{2}{3}$

37. $\dfrac{5}{12}-\dfrac{7}{18}=\dfrac{5}{12}\times\dfrac{3}{3}-\dfrac{7}{18}\times\dfrac{2}{2}=\dfrac{15}{36}-\dfrac{14}{36}=\dfrac{1}{36}$

39. $\dfrac{10}{16}-\dfrac{5}{8}=\dfrac{10}{16}-\dfrac{5}{8}\times\dfrac{2}{2}=\dfrac{10}{16}-\dfrac{10}{16}=0$

41. $\dfrac{23}{36}-\dfrac{2}{9}=\dfrac{23}{36}-\dfrac{2}{9}\times\dfrac{4}{4}=\dfrac{23}{36}-\dfrac{8}{36}=\dfrac{15}{36}=\dfrac{5}{12}$

43. $\dfrac{1}{2}+\dfrac{2}{7}+\dfrac{3}{14}=\dfrac{1}{2}\times\dfrac{7}{7}+\dfrac{2}{7}\times\dfrac{2}{2}+\dfrac{3}{14}$
$=\dfrac{7}{14}+\dfrac{4}{14}+\dfrac{3}{14}$
$=\dfrac{14}{14}$
$=1$

45. $\dfrac{5}{30}+\dfrac{3}{40}+\dfrac{1}{8}=\dfrac{5}{30}\times\dfrac{4}{4}+\dfrac{3}{40}\times\dfrac{3}{3}+\dfrac{1}{8}\times\dfrac{15}{15}$
$=\dfrac{20}{120}+\dfrac{9}{120}+\dfrac{15}{120}$
$=\dfrac{44}{120}$
$=\dfrac{11}{30}$

47. $\dfrac{7}{30}+\dfrac{2}{5}+\dfrac{5}{6}=\dfrac{7}{30}+\dfrac{2}{5}\times\dfrac{6}{6}+\dfrac{5}{6}\times\dfrac{5}{5}$
$=\dfrac{7}{30}+\dfrac{12}{30}+\dfrac{25}{30}$
$=\dfrac{44}{30}$
$=\dfrac{22}{15}$ or $1\dfrac{7}{15}$

49.
$$x + \frac{1}{7} = \frac{5}{14}$$
$$x + \frac{1}{7} \times \frac{2}{2} = \frac{5}{14}$$
$$x + \frac{2}{14} = \frac{5}{14}$$
$$\frac{3}{14} + \frac{2}{14} = \frac{5}{14}$$
$$x = \frac{3}{14}$$

51.
$$x + \frac{2}{3} = \frac{9}{11}$$
$$x + \frac{2}{3} \times \frac{11}{11} = \frac{9}{11} \times \frac{3}{3}$$
$$x + \frac{22}{33} = \frac{27}{33}$$
$$\frac{5}{33} + \frac{22}{33} = \frac{27}{33}$$
$$x = \frac{5}{33}$$

53.
$$x - \frac{3}{10} = \frac{4}{15}$$
$$x - \frac{3}{10} \times \frac{3}{3} = \frac{4}{15} \times \frac{2}{2}$$
$$x - \frac{9}{30} = \frac{8}{30}$$
$$\frac{17}{30} - \frac{9}{30} = \frac{8}{30}$$
$$x = \frac{17}{30}$$

55. $\dfrac{1}{4} + \dfrac{2}{3} = \dfrac{1}{4} \times \dfrac{3}{3} + \dfrac{2}{3} \times \dfrac{4}{4} = \dfrac{3}{12} + \dfrac{8}{12} = \dfrac{11}{12}$

She needs a total of $\dfrac{11}{12}$ cup of sugar.

57. Nuts: $\dfrac{2}{3} + \dfrac{3}{4} = \dfrac{2}{3} \times \dfrac{4}{4} + \dfrac{3}{4} \times \dfrac{3}{3}$
$$= \frac{8}{12} + \frac{9}{12}$$
$$= \frac{17}{12} \text{ or } 1\frac{5}{12} \text{ pounds}$$

Fruit: $\dfrac{1}{2} + \dfrac{3}{8} = \dfrac{1}{2} \times \dfrac{4}{4} + \dfrac{3}{8} = \dfrac{4}{8} + \dfrac{3}{8} = \dfrac{7}{8}$ pound

She mixed $\dfrac{17}{12}$ or $1\dfrac{5}{12}$ pounds of nuts and

$\dfrac{7}{8}$ pound of dried fruit.

59. $\dfrac{11}{12} - \dfrac{3}{5} = \dfrac{11}{12} \times \dfrac{5}{5} - \dfrac{3}{5} \times \dfrac{12}{12}$
$$= \frac{55}{60} - \frac{36}{60}$$
$$= \frac{19}{60}$$

Travis lost $\dfrac{19}{60}$ of the research paper.

61. Before he ate half, there were

$6 \div \dfrac{1}{2} = 6 \times 2 = 12$ chocolates. While walking, he

ate $\dfrac{1}{4}$ of the chocolates, leaving $1 - \dfrac{1}{4} = \dfrac{3}{4}$ in the

box. The box had

$12 \div \dfrac{3}{4} = \dfrac{12}{1} \times \dfrac{4}{3} = 16$ chocolates.

63. $\dfrac{7}{8} - \dfrac{7}{10} = \dfrac{7}{8} \times \dfrac{5}{5} - \dfrac{7}{10} \times \dfrac{4}{4} = \dfrac{35}{40} - \dfrac{28}{40} = \dfrac{7}{40}$

The decrease represents $\dfrac{7}{40}$ of the membership.

Cumulative Review

64. $\dfrac{15}{85} = \dfrac{15 \div 5}{85 \div 5} = \dfrac{3}{17}$

65. $\dfrac{27}{207} = \dfrac{27 \div 9}{207 \div 9} = \dfrac{3}{23}$

66.
$$14\overline{)125} \quad \begin{array}{c} 8 \\ \underline{112} \\ 13 \end{array} \qquad \frac{125}{14} = 8\frac{13}{14}$$

67. $14\dfrac{3}{7} = \dfrac{14 \times 7 + 3}{7} = \dfrac{101}{7}$

68. $4\dfrac{1}{3} \div 1\dfrac{1}{2} = \dfrac{13}{3} \div \dfrac{3}{2} = \dfrac{13}{3} \times \dfrac{2}{3} = \dfrac{26}{9}$ or $2\dfrac{8}{9}$

69. $5\dfrac{1}{2} \times 1\dfrac{3}{11} = \dfrac{11}{2} \times \dfrac{14}{11} = \dfrac{1}{1} \times \dfrac{7}{1} = 7$

Quick Quiz 2.7

1. $\dfrac{7}{16} + \dfrac{3}{4} = \dfrac{7}{16} + \dfrac{3}{4} \times \dfrac{4}{4} = \dfrac{7}{16} + \dfrac{12}{16} = \dfrac{19}{16}$ or $1\dfrac{3}{16}$

2.
$$\frac{1}{3}+\frac{5}{7}+\frac{10}{21}=\frac{1}{3}\times\frac{7}{7}+\frac{5}{7}\times\frac{3}{3}+\frac{10}{21}$$
$$=\frac{7}{21}+\frac{15}{21}+\frac{10}{21}$$
$$=\frac{32}{21}\text{ or }1\frac{11}{21}$$

3.
$$\frac{8}{9}-\frac{7}{15}=\frac{8}{9}\times\frac{5}{5}-\frac{7}{15}\times\frac{3}{3}=\frac{40}{45}-\frac{21}{45}=\frac{19}{45}$$

4. Answers may vary. Possible solution:
Find the LCD.
$9 = 3 \times 3$
$7 = 7$
$LCD = 3 \times 3 \times 7 = 63$
Rewrite the fractions with the same denominator
and then subtract.
$$\frac{8}{9}-\frac{3}{7}=\frac{56}{63}-\frac{27}{63}=\frac{29}{63}$$

2.8 Exercises

1.
$$7\frac{1}{8}$$
$$+2\frac{5}{8}$$
$$\overline{\quad 9\frac{6}{8}=9\frac{3}{4}}$$

3.
$$15\frac{3}{14}$$
$$-11\frac{1}{14}$$
$$\overline{\quad 4\frac{2}{14}=4\frac{1}{7}}$$

5.
$$12\frac{1}{3}\qquad\qquad 12\frac{2}{6}$$
$$+\ 5\frac{1}{6}\qquad\qquad +\ 5\frac{1}{6}$$
$$\overline{\qquad\qquad\quad\ 17\frac{3}{6}=17\frac{1}{2}}$$

7.
$$4\frac{3}{5}$$
$$+8\frac{2}{5}$$
$$\overline{\quad 12\frac{5}{5}=13}$$

9.
$$\begin{array}{r}1\\-\ \frac{3}{7}\\\hline\end{array}\qquad\begin{array}{r}\frac{7}{7}\\-\ \frac{3}{7}\\\hline\frac{4}{7}\end{array}$$

11.
$$\begin{array}{r}1\frac{3}{4}\\+\ \frac{5}{16}\\\hline\end{array}\qquad\begin{array}{r}1\frac{12}{16}\\+\ \frac{5}{16}\\\hline 1\frac{17}{16}=2\frac{1}{16}\end{array}$$

13.
$$\begin{array}{r}5\frac{1}{6}\\+4\frac{5}{18}\\\hline\end{array}\qquad\begin{array}{r}5\frac{3}{18}\\+4\frac{5}{18}\\\hline 9\frac{8}{18}=9\frac{4}{9}\end{array}$$

15.
$$\begin{array}{r}8\frac{1}{4}\\-8\frac{4}{16}\\\hline\end{array}\qquad\begin{array}{r}8\frac{4}{16}\\-8\frac{4}{16}\\\hline 0\end{array}$$

17.
$$\begin{array}{r}12\frac{1}{3}\\-7\frac{2}{5}\\\hline\end{array}\quad\begin{array}{r}12\frac{5}{15}\\-7\frac{6}{15}\\\hline\end{array}\quad\begin{array}{r}11\frac{20}{15}\\-7\frac{6}{15}\\\hline 4\frac{14}{15}\end{array}$$

19.
$$\begin{array}{r}30\\-15\frac{3}{7}\\\hline\end{array}\qquad\begin{array}{r}29\frac{7}{7}\\-15\frac{3}{7}\\\hline 14\frac{4}{7}\end{array}$$

21.
$$\begin{array}{r}3\\+4\frac{2}{5}\\\hline 7\frac{2}{5}\end{array}$$

23. $\begin{aligned} 14 \\ -\,3\frac{7}{10} \\ \hline \end{aligned}$ $\begin{aligned} 13\frac{10}{10} \\ -\,3\frac{7}{10} \\ \hline 10\frac{3}{10} \end{aligned}$

25. $\begin{aligned} 15\frac{4}{15} \\ +\,26\frac{8}{15} \\ \hline 41\frac{12}{15} = 41\frac{4}{5} \end{aligned}$

27. $\begin{aligned} 6\frac{1}{6} \\ +\,2\frac{1}{4} \\ \hline \end{aligned}$ $\begin{aligned} 6\frac{2}{12} \\ +\,2\frac{3}{12} \\ \hline 8\frac{5}{12} \end{aligned}$

29. $\begin{aligned} 3\frac{3}{4} \\ +\,4\frac{5}{12} \\ \hline \end{aligned}$ $\begin{aligned} 3\frac{9}{12} \\ +\,4\frac{5}{12} \\ \hline 7\frac{14}{12} = 8\frac{2}{12} = 8\frac{1}{6} \end{aligned}$

31. $\begin{aligned} 47\frac{3}{10} \\ +\,26\frac{5}{8} \\ \hline \end{aligned}$ $\begin{aligned} 47\frac{12}{40} \\ +\,26\frac{25}{40} \\ \hline 73\frac{37}{40} \end{aligned}$

33. $\begin{aligned} 19\frac{5}{6} \\ -\,14\frac{1}{3} \\ \hline \end{aligned}$ $\begin{aligned} 19\frac{5}{6} \\ -\,14\frac{2}{6} \\ \hline 5\frac{3}{6} = 5\frac{1}{2} \end{aligned}$

35. $\begin{aligned} 6\frac{1}{12} \\ -\,5\frac{10}{24} \\ \hline \end{aligned}$ $\begin{aligned} 6\frac{2}{24} \\ -\,5\frac{10}{24} \\ \hline \end{aligned}$ $\begin{aligned} 5\frac{26}{24} \\ -\,5\frac{10}{24} \\ \hline \frac{16}{24} = \frac{2}{3} \end{aligned}$

37. $\begin{aligned} 12\frac{3}{20} \\ -\,7\frac{7}{15} \\ \hline \end{aligned}$ $\begin{aligned} 12\frac{9}{60} \\ -\,7\frac{28}{60} \\ \hline \end{aligned}$ $\begin{aligned} 11\frac{69}{60} \\ -\,7\frac{28}{60} \\ \hline 4\frac{41}{60} \end{aligned}$

39. $\begin{aligned} 12 \\ -\,3\frac{7}{15} \\ \hline \end{aligned}$ $\begin{aligned} 11\frac{15}{15} \\ -\,3\frac{7}{15} \\ \hline 8\frac{8}{15} \end{aligned}$

41. $\begin{aligned} 120 \\ -\,17\frac{3}{8} \\ \hline \end{aligned}$ $\begin{aligned} 119\frac{8}{8} \\ -\,17\frac{3}{8} \\ \hline 102\frac{5}{8} \end{aligned}$

43. $\begin{aligned} 3\frac{5}{8} \\ 2\frac{2}{3} \\ +\,7\frac{3}{4} \\ \hline \end{aligned}$ $\begin{aligned} 3\frac{15}{24} \\ 2\frac{16}{24} \\ +\,7\frac{18}{24} \\ \hline 12\frac{49}{24} = 14\frac{1}{24} \end{aligned}$

45. $\begin{aligned} 20\frac{3}{4} \\ +\,22\frac{3}{8} \\ \hline \end{aligned}$ $\begin{aligned} 20\frac{6}{8} \\ +\,22\frac{3}{8} \\ \hline 42\frac{9}{8} = 43\frac{1}{8} \end{aligned}$

He biked a total of $43\frac{1}{8}$ miles.

47. $\begin{aligned} 2\frac{4}{5} \\ +\,3\frac{1}{10} \\ \hline \end{aligned}$ $\begin{aligned} 2\frac{8}{10} \\ +\,3\frac{1}{10} \\ \hline 5\frac{9}{10} \end{aligned}$ $\begin{aligned} 5\frac{9}{10} \\ +\,1 \\ \hline 6\frac{9}{10} \end{aligned}$

Lola's bike ride is $6\frac{9}{10}$ miles.

49.

$$72\frac{1}{2} \qquad 72\frac{2}{4} \qquad 71\frac{6}{4}$$
$$-69\frac{3}{4} \qquad -69\frac{3}{4} \qquad -69\frac{3}{4}$$
$$\overline{\qquad\qquad} \qquad \overline{\qquad\qquad} \qquad \overline{\quad 2\frac{3}{4}\quad}$$

Julie is $2\frac{3}{4}$ inches taller than Nina.

51. a.

$$1\frac{3}{4} \qquad\qquad 1\frac{9}{12}$$
$$+2\frac{1}{6} \qquad\qquad +2\frac{2}{12}$$
$$\overline{\qquad\qquad} \qquad\qquad \overline{\quad 3\frac{11}{12}\quad}$$

There are $3\frac{11}{12}$ pounds of haddock on the scale.

b.

$$8 \qquad\qquad 7\frac{12}{12}$$
$$-3\frac{11}{12} \qquad\qquad -3\frac{11}{12}$$
$$\overline{\qquad\qquad} \qquad\qquad \overline{\quad 4\frac{1}{12}\quad}$$

Lara needs an additional $4\frac{1}{12}$ pounds.

53. $\dfrac{379}{8} + \dfrac{89}{5} = \dfrac{1895}{40} + \dfrac{712}{40} = \dfrac{2607}{40}$ or $65\dfrac{7}{40}$

55. Estimate: $35 + 24 = 59$

Exact:
$$35\frac{1}{6} \qquad\qquad 35\frac{2}{12}$$
$$+24\frac{5}{12} \qquad\qquad +24\frac{5}{12}$$
$$\overline{\qquad\qquad} \qquad\qquad \overline{\quad 59\frac{7}{12}\quad}$$

Our estimate is very close. We are off by only $\dfrac{7}{12}$.

57.
$$\frac{6}{7} - \frac{4}{7} \times \frac{1}{3} = \frac{6}{7} - \frac{4}{21}$$
$$= \frac{6}{7} \times \frac{3}{3} - \frac{4}{21}$$
$$= \frac{18}{21} - \frac{4}{21}$$
$$= \frac{14}{21}$$
$$= \frac{2}{3}$$

59. $\dfrac{1}{2} + \dfrac{3}{8} \div \dfrac{3}{4} = \dfrac{1}{2} + \dfrac{3}{8} \times \dfrac{4}{3} = \dfrac{1}{2} + \dfrac{1}{2} = \dfrac{2}{2} = 1$

61. $\dfrac{9}{10} \div \dfrac{3}{8} \times \dfrac{5}{8} = \dfrac{9}{10} \times \dfrac{8}{3} \times \dfrac{5}{8} = \dfrac{3}{2}$ or $1\dfrac{1}{2}$

63.
$$\frac{3}{5} \times \frac{1}{2} + \frac{1}{5} \div \frac{2}{3} = \frac{3}{5} \times \frac{1}{2} + \frac{1}{5} \times \frac{3}{2}$$
$$= \frac{3}{10} + \frac{3}{10}$$
$$= \frac{6}{10}$$
$$= \frac{3}{5}$$

65. $\left(\dfrac{3}{5} - \dfrac{3}{20}\right) \times \dfrac{4}{5} = \left(\dfrac{12}{20} - \dfrac{3}{20}\right) \times \dfrac{4}{5} = \dfrac{9}{20} \times \dfrac{4}{5} = \dfrac{9}{25}$

67. $\left(\dfrac{1}{3}\right)^2 \div \dfrac{4}{9} = \dfrac{1}{9} \div \dfrac{4}{9} = \dfrac{1}{9} \times \dfrac{9}{4} = \dfrac{1}{4}$

69. $\dfrac{1}{4} \times \left(\dfrac{2}{3}\right)^2 = \dfrac{1}{4} \times \dfrac{4}{9} = \dfrac{1}{9}$

71.
$$\frac{5}{6} \div \left(\frac{2}{3} + \frac{1}{6}\right)^2 = \frac{5}{6} \div \left(\frac{4}{6} + \frac{1}{6}\right)^2$$
$$= \frac{5}{6} \div \left(\frac{5}{6}\right)^2$$
$$= \frac{5}{6} \div \frac{25}{36}$$
$$= \frac{5}{6} \times \frac{36}{25}$$
$$= \frac{6}{5} \text{ or } 1\frac{1}{5}$$

Cumulative Review

73.
$$\begin{array}{r} 1200 \\ \times\quad 400 \\ \hline 480,000 \end{array}$$

74.
$$\begin{array}{r} 4050 \\ \times\quad 2106 \\ \hline 24\ 300 \\ 405\ 00\ \ \\ 8\ 100\quad\ \ \\ \hline 8,529,300 \end{array}$$

Quick Quiz 2.8

1.
$$\begin{array}{r} 3\dfrac{4}{5} \\ +\,5\dfrac{3}{8} \\ \hline \end{array}\qquad \begin{array}{r} 3\dfrac{32}{40} \\ +\,5\dfrac{15}{40} \\ \hline 8\dfrac{47}{40}=9\dfrac{7}{40} \end{array}$$

2.
$$\begin{array}{r} 6\dfrac{5}{12} \\ -\,4\dfrac{7}{10} \\ \hline \end{array}\qquad \begin{array}{r} 6\dfrac{25}{60} \\ -\,4\dfrac{42}{60} \\ \hline \end{array}\qquad \begin{array}{r} 5\dfrac{85}{60} \\ -\,4\dfrac{42}{60} \\ \hline 1\dfrac{43}{60} \end{array}$$

3.
$$\begin{aligned} \frac{1}{5}+\frac{3}{10}\div\frac{11}{20} &= \frac{1}{5}+\frac{3}{10}\times\frac{20}{11} \\ &= \frac{1}{5}+\frac{6}{11} \\ &= \frac{11}{55}+\frac{30}{55} \\ &= \frac{41}{55} \end{aligned}$$

4. Answers may vary. Possible solution:
First, multiply. Then, rewrite fractions with
common denominators. Finally, subtract.
$$\frac{4}{5}-\frac{1}{4}\times\frac{2}{3}=\frac{4}{5}-\frac{1}{6}=\frac{24}{30}-\frac{5}{30}=\frac{19}{30}$$

2.9 Exercises

1.
$$\begin{array}{r} 8\dfrac{1}{3} \\ 5\dfrac{4}{5} \\ +\,9\dfrac{3}{10} \\ \hline \end{array}\qquad \begin{array}{r} 8\dfrac{10}{30} \\ 5\dfrac{24}{30} \\ +\,9\dfrac{9}{30} \\ \hline 22\dfrac{43}{30}=23\dfrac{13}{30} \end{array}$$

The perimeter of the triangle is $23\dfrac{13}{30}$ inches.

3. $\dfrac{5}{9}\times700=\dfrac{5}{9}\times\dfrac{700}{1}=\dfrac{3500}{9}=388\dfrac{8}{9}$

About 389 gorillas were living in this mountain
range.

5. $\dfrac{1}{16}+\dfrac{3}{4}+\dfrac{1}{16}+\dfrac{3}{16}+\dfrac{1}{2}=\dfrac{1}{16}+\dfrac{12}{16}+\dfrac{1}{16}+\dfrac{3}{16}+\dfrac{8}{16}$

$$=\dfrac{25}{16}$$
$$=1\dfrac{9}{16}$$

The minimum length of the bolt is $1\dfrac{9}{16}$ inches.

7. Sum:
$$\begin{array}{r} 6\dfrac{3}{4} \\ +\,9\dfrac{1}{2} \\ \hline \end{array}\qquad \begin{array}{r} 6\dfrac{3}{4} \\ +\,9\dfrac{2}{4} \\ \hline 15\dfrac{5}{4}=16\dfrac{1}{4} \end{array}$$

Miles to end:
$$\begin{array}{r} 26\dfrac{1}{5} \\ -\,16\dfrac{1}{4} \\ \hline \end{array}\quad \begin{array}{r} 26\dfrac{4}{20} \\ -\,16\dfrac{5}{20} \\ \hline \end{array}\quad \begin{array}{r} 25\dfrac{24}{20} \\ -\,16\dfrac{5}{20} \\ \hline 9\dfrac{19}{20} \end{array}$$

Josiah has $9\dfrac{19}{20}$ miles left to run.

9. $1\dfrac{1}{2}\times8\dfrac{1}{2}=\dfrac{3}{2}\times\dfrac{17}{2}=\dfrac{51}{4}=12\dfrac{3}{4}$

Nathaniel will need $12\dfrac{3}{4}$ cups of sugar.

$$3 \times 8\frac{1}{2} = \frac{3}{1} \times \frac{17}{2} = \frac{51}{2} = 25\frac{1}{2}$$
26 boxes will be needed.

11. $36\frac{3}{4} \times 7\frac{1}{2} = \frac{147}{4} \times \frac{15}{2} = \frac{2205}{8} = 275\frac{5}{8}$

The tank can hold $275\frac{5}{8}$ gallons.

13. $22\frac{1}{2} \times 4\frac{3}{4} = \frac{45}{2} \times \frac{19}{4}$

$\qquad = \frac{855}{8}$

$\qquad = 106\frac{7}{8}$

The *Titanic* traveled $106\frac{7}{8}$ nautical miles.

15. $\frac{1}{5} + \frac{1}{15} + \frac{1}{20} = \frac{12}{60} + \frac{4}{60} + \frac{3}{60} = \frac{12+4+3}{60} = \frac{19}{60}$

$\frac{19}{60}(660) = 209,$ so \$209 is deducted.

$660 - 209 = 451$
She has \$451 per week left.

17. a. $22 \div \frac{2}{5} = \frac{22}{1} \times \frac{5}{2} = \frac{2 \times 11 \times 5}{1 \times 2} = 55$

She can make 55 bracelets from the wire.

b. $3\frac{1}{2} \times \frac{2}{5} = \frac{7}{2} \times \frac{2}{5} = \frac{7 \times 2}{2 \times 5} = \frac{7}{5} = 1\frac{2}{5}$

A necklace would require $1\frac{2}{5}$ feet of wire.

19. a. $18\frac{1}{2} - 1\frac{1}{4} - 3\frac{1}{8} = \frac{37}{2} - \frac{5}{4} - \frac{25}{8}$

$\qquad\qquad = \frac{148}{8} - \frac{10}{8} - \frac{25}{8}$

$\qquad\qquad = \frac{148 - 10 - 25}{8}$

$\qquad\qquad = \frac{113}{8}$

$\qquad\qquad = 14\frac{1}{8}$

She actually bought $14\frac{1}{8}$ ounces of bread.

b.
$$\begin{array}{r} 14\frac{3}{4} \\ -14\frac{1}{8} \\ \hline \end{array} \qquad \begin{array}{r} 14\frac{6}{8} \\ -14\frac{1}{8} \\ \hline \frac{5}{8} \end{array}$$

The measurement was in error by $\frac{5}{8}$ of an ounce.

21. a. $160\frac{1}{8} \div 5\frac{1}{4} = \frac{1281}{8} \div \frac{21}{4}$

$\qquad\qquad = \frac{1281}{8} \times \frac{4}{21}$

$\qquad\qquad = \frac{61}{2}$

$\qquad\qquad = 30\frac{1}{2}$

The boat is traveling at $30\frac{1}{2}$ knots.

b. $213\frac{1}{2} \div \frac{61}{2} = \frac{427}{2} \div \frac{61}{2}$

$\qquad\qquad = \frac{427}{2} \times \frac{2}{61}$

$\qquad\qquad = \frac{427}{61}$

$\qquad\qquad = 7$

It will take the boat 7 hours.

23. a. $6856\frac{1}{4} \div 1\frac{1}{4} = \frac{27,425}{4} \div \frac{5}{4}$

$\qquad\qquad = \frac{27,425}{4} \times \frac{4}{5}$

$\qquad\qquad = 5485$

The bin can hold 5485 bushels.

b. $6856\frac{1}{4} \times 1\frac{3}{4} = \frac{27,425}{4} \times \frac{7}{4}$

$\qquad\qquad = \frac{191,975}{16}$

$\qquad\qquad = 11,998\frac{7}{16}$

The new bin can hold $11,998\frac{7}{16}$ cubic feet.

c. $11,998\dfrac{7}{16} \div 1\dfrac{1}{4} = \dfrac{191,975}{16} \times \dfrac{4}{5}$

$\qquad\qquad = \dfrac{38,395}{4}$

$\qquad\qquad = 9598\dfrac{3}{4}$

The new bin will hold $9598\dfrac{3}{4}$ bushels.

Cumulative Review

25. $\dfrac{17}{36} - \dfrac{2}{9} = \dfrac{17}{36} - \dfrac{8}{36} = \dfrac{17-8}{36} = \dfrac{9}{36} = \dfrac{1}{4}$

26. $\dfrac{1}{5} + \dfrac{2}{5} \times \dfrac{3}{2} - \dfrac{1}{10} = \dfrac{1}{5} + \dfrac{2\times3}{5\times2} - \dfrac{1}{10}$

$\qquad\qquad = \dfrac{1}{5} + \dfrac{3}{5} - \dfrac{1}{10}$

$\qquad\qquad = \dfrac{2}{10} + \dfrac{6}{10} - \dfrac{1}{10}$

$\qquad\qquad = \dfrac{2+6-1}{10}$

$\qquad\qquad = \dfrac{7}{10}$

27. $30 \times 4\dfrac{2}{3} = \dfrac{30}{1} \times \dfrac{14}{3} = \dfrac{3\times10\times14}{1\times3} = 140$

28. $\dfrac{15}{16} \div 1\dfrac{1}{4} = \dfrac{15}{16} \div \dfrac{5}{4} = \dfrac{15}{16} \times \dfrac{4}{5} = \dfrac{3\times5\times4}{4\times4\times5} = \dfrac{3}{4}$

Quick Quiz 2.9

1. $15\dfrac{3}{4} \times 10\dfrac{2}{3} = \dfrac{63}{4} \times \dfrac{32}{3} = \dfrac{21}{1} \times \dfrac{8}{1} = 168$

She needs 168 square feet of carpeting.

2. $41\dfrac{3}{5} \div 2\dfrac{3}{5} = \dfrac{208}{5} \div \dfrac{13}{5}$

$\qquad\qquad = \dfrac{208}{5} \times \dfrac{5}{13}$

$\qquad\qquad = \dfrac{16}{1} \times \dfrac{1}{1}$

$\qquad\qquad = 16$

She shipped out 16 packets.

3.
$$
\begin{array}{c}
1\dfrac{1}{8} \\
1\dfrac{1}{2} \\
+\;2\dfrac{3}{4} \\
\hline
\end{array}
\qquad\qquad
\begin{array}{c}
1\dfrac{1}{8} \\
1\dfrac{4}{8} \\
+\;2\dfrac{6}{8} \\
\hline
4\dfrac{11}{8} = 5\dfrac{3}{8}
\end{array}
$$

She traveled a total of $5\dfrac{3}{8}$ miles.

4. Answers may vary. Possible solution: Change each mixed number to an improper fraction, rewrite with a common denominator, subtract, and then simplify.

$3\dfrac{3}{5} - 1\dfrac{7}{8} = \dfrac{18}{5} - \dfrac{15}{8} = \dfrac{90}{40} - \dfrac{75}{40} = \dfrac{15}{40} = \dfrac{3}{8}$

He still has to go $\dfrac{3}{8}$ mile.

Use Math To Save Money

1. Tricia bought two cups of coffee each day.
$2 \times 3 \times 30 = 6 \times 30 = 180$
She spent $180 on coffee each month.

2.
$$
\begin{array}{r}
180 \\
\times\;12 \\
\hline
360 \\
180\;\; \\
\hline
2160 \\
\end{array}
$$
She would spend $2160 on coffee in 12 months.

3. $7 \times 180 = 1260$
In seven months, she would save $1260, which is more than the TV would cost.

4.
$$
\begin{array}{r}
1260 \\
-\;1000 \\
\hline
260 \\
\end{array}
$$
There would be $260 for the celebration dinner.

5. $\dfrac{3}{4} \times 1000 = \dfrac{3}{4} \times \dfrac{1000}{1} = \dfrac{3\times4\times250}{4\times1} = 750$
$$
\begin{array}{r}
1260 \\
-\;750 \\
\hline
510 \\
\end{array}
$$
There would be $510 for the celebration dinner.

6. $2 \times 30 = 60$

Tricia drinks 60 cups of coffee each month.

$60 \div 20 = 3$

She will need 3 pounds of coffee each month.

$3 \times 10 = 30$

It would cost her $30 each month to make her own coffee.

$$\begin{array}{r} 180 \\ - 30 \\ \hline 150 \end{array}$$

She would save $150 each month by making coffee.

7.

$$\begin{array}{r} 150 \\ \times 12 \\ \hline 300 \\ 150 \\ \hline 1800 \end{array}$$

She would save $1800 in a year by making coffee.

8. Answers will vary.

9. Answers will vary.

10. Answers will vary.

You Try It

1. Nine of 14 equal parts are shaded, so $\frac{9}{14}$ is shaded.

2. $\dfrac{\text{games won}}{\text{total games}} = \dfrac{85}{115} = \dfrac{5 \times 17}{5 \times 23} = \dfrac{17}{23}$

The team won $\dfrac{17}{23}$ of the games.

3. $60 = 2 \times 2 \times 3 \times 5 = 2^2 \times 3 \times 5$

4. $\dfrac{24}{80} = \dfrac{2 \times 2 \times 2 \times 3}{2 \times 2 \times 2 \times 2 \times 5} = \dfrac{3}{2 \times 5} = \dfrac{3}{10}$

5. $10\dfrac{2}{3} = \dfrac{10 \times 3 + 2}{3} = \dfrac{30 + 2}{3} = \dfrac{32}{3}$

6. $3\overline{)28}$
$$\begin{array}{r} 9 \\ 3\overline{)28} \\ \underline{27} \\ 1 \end{array}$$

$\dfrac{28}{3} = 9\dfrac{1}{3}$

7. a. $\dfrac{2}{5} \times \dfrac{2}{9} = \dfrac{2 \times 2}{5 \times 9} = \dfrac{4}{45}$

b. $\dfrac{4}{5} \times \dfrac{25}{28} = \dfrac{4 \times 5 \times 5}{5 \times 4 \times 7} = \dfrac{5}{7}$

8. $2\dfrac{1}{2} \times 4\dfrac{2}{5} = \dfrac{5}{2} \times \dfrac{22}{5} = \dfrac{5 \times 2 \times 11}{2 \times 5} = \dfrac{11}{1} = 11$

9. $\dfrac{1}{3} \div \dfrac{2}{5} = \dfrac{1}{3} \times \dfrac{5}{2} = \dfrac{1 \times 5}{3 \times 2} = \dfrac{5}{6}$

10. $7\dfrac{1}{5} \div 2\dfrac{1}{10} = \dfrac{36}{5} \div \dfrac{21}{10}$

$= \dfrac{36}{5} \times \dfrac{10}{21}$

$= \dfrac{3 \times 12 \times 5 \times 2}{5 \times 3 \times 7}$

$= \dfrac{12 \times 2}{7}$

$= \dfrac{24}{7}$ or $3\dfrac{3}{7}$

11. $6 = 2 \times 3$

$10 = 2 \times 5$

$24 = 2 \times 2 \times 2 \times 3$

$\text{LCD} = 2 \times 2 \times 2 \times 3 \times 5 = 120$

12. $\dfrac{4}{9} = \dfrac{4 \times 6}{9 \times 6} = \dfrac{24}{54}$

13. a. $\dfrac{7}{15} + \dfrac{1}{15} = \dfrac{7 + 1}{15} = \dfrac{8}{15}$

b. $\dfrac{8}{11} - \dfrac{7}{11} = \dfrac{8 - 7}{11} = \dfrac{1}{11}$

14. $\frac{1}{3}+\frac{3}{5}+\frac{9}{10}=\frac{1}{3}\times\frac{10}{10}+\frac{3}{5}\times\frac{6}{6}+\frac{9}{10}\times\frac{3}{3}$

$\qquad\qquad =\frac{10}{30}+\frac{18}{30}+\frac{27}{30}$

$\qquad\qquad =\frac{10+18+27}{30}$

$\qquad\qquad =\frac{55}{30}$

$\qquad\qquad =\frac{5\times11}{5\times6}$

$\qquad\qquad =\frac{11}{6}$ or $1\frac{5}{6}$

15.
$$8\frac{5}{6}\qquad\qquad\qquad 8\frac{5}{6}$$
$$+3\frac{1}{3}\qquad\qquad\qquad +3\frac{2}{6}$$
$$\overline{\qquad}\qquad\qquad\qquad \overline{11\frac{7}{6}}=12\frac{1}{6}$$

16.
$$10\frac{1}{4}\qquad 10\frac{5}{20}\qquad 9\frac{25}{20}$$
$$-3\frac{4}{5}\qquad -3\frac{16}{20}\qquad -3\frac{16}{20}$$
$$\overline{\qquad}\qquad \overline{\qquad}\qquad \overline{6\frac{9}{20}}$$

17. $6\times\frac{1}{2}+\left(\frac{9}{10}-\frac{2}{5}\right)=6\times\frac{1}{2}+\left(\frac{9}{10}-\frac{4}{10}\right)$

$\qquad\qquad =6\times\frac{1}{2}+\frac{5}{10}$

$\qquad\qquad =6\times\frac{1}{2}+\frac{1}{2}$

$\qquad\qquad =\frac{6}{1}\times\frac{1}{2}+\frac{1}{2}$

$\qquad\qquad =\frac{2\times3\times1}{1\times2}+\frac{1}{2}$

$\qquad\qquad =3+\frac{1}{2}$

$\qquad\qquad =\frac{6}{2}+\frac{1}{2}$

$\qquad\qquad =\frac{7}{2}$ or $3\frac{1}{2}$

Chapter 2 Review Problems

1. Three out of eight equal parts are shaded. The fraction is $\frac{3}{8}$.

2. Five out of twelve equal parts are shaded. The fraction is $\frac{5}{12}$.

3. Answers will vary.

4. Answers will vary.

5. $\dfrac{\text{number defective}}{\text{total number}}=\dfrac{9}{80}$

6. $\dfrac{\text{number who would not}}{\text{total number}}=\dfrac{87}{100}$

7. $54=2\times27=2\times3\times9=2\times3\times3\times3=2\times3^3$

8. $120=10\times12=2\times5\times2\times2\times3=2^3\times3\times5$

9. $168=8\times21=2\times2\times2\times3\times7=2^3\times3\times7$

10. 59 is prime.

11. $78=2\times39=2\times3\times13$

12. 167 is prime.

13. $\dfrac{12}{42}=\dfrac{12\div6}{42\div6}=\dfrac{2}{7}$

14. $\dfrac{13}{52}=\dfrac{13\div13}{52\div13}=\dfrac{1}{4}$

15. $\dfrac{27}{72}=\dfrac{27\div9}{72\div9}=\dfrac{3}{8}$

16. $\dfrac{168}{192}=\dfrac{168\div24}{192\div24}=\dfrac{7}{8}$

17. $4\dfrac{3}{8}=\dfrac{4\times8+3}{8}=\dfrac{35}{8}$

18. $15\dfrac{3}{4}=\dfrac{15\times4+3}{4}=\dfrac{63}{4}$

19. $6\dfrac{3}{5}=\dfrac{6\times5+3}{5}=\dfrac{33}{5}$

20. $8\overline{)45}$ with quotient 5, $\underline{40}$, remainder 5

$$\frac{45}{8} = 5\frac{5}{8}$$

21. $21\overline{)100}$ with quotient 4, $\underline{84}$, remainder 16

$$\frac{100}{21} = 4\frac{16}{21}$$

22. $7\overline{)53}$ with quotient 7, $\underline{49}$, remainder 4

$$\frac{53}{7} = 7\frac{4}{7}$$

23. $\dfrac{15}{55} = \dfrac{5 \times 3}{5 \times 11} = \dfrac{3}{11}$

$3\dfrac{15}{55} = 3\dfrac{3}{11}$

24. $\dfrac{234}{16} = \dfrac{117 \times 2}{8 \times 2} = \dfrac{117}{8}$

25. $32\overline{)132}$ with quotient 4, $\underline{128}$, remainder 4

$$\frac{132}{32} = 4\frac{4}{32} = 4\frac{1}{8}$$

26. $\dfrac{4}{7} \times \dfrac{5}{11} = \dfrac{4 \times 5}{7 \times 11} = \dfrac{20}{77}$

27. $\dfrac{7}{9} \times \dfrac{21}{35} = \dfrac{1}{3} \times \dfrac{7}{5} = \dfrac{7}{15}$

28. $12 \times \dfrac{3}{7} \times 0 = 0$

29. $\dfrac{3}{5} \times \dfrac{2}{7} \times \dfrac{10}{27} = \dfrac{1}{1} \times \dfrac{2}{7} \times \dfrac{2}{9} = \dfrac{4}{63}$

30. $5\dfrac{1}{8} \times 3\dfrac{1}{5} = \dfrac{41}{8} \times \dfrac{16}{5} = \dfrac{41}{1} \times \dfrac{2}{5} = \dfrac{82}{5}$ or $16\dfrac{2}{5}$

31. $36 \times \dfrac{4}{9} = \dfrac{36}{1} \times \dfrac{4}{9} = \dfrac{4}{1} \times \dfrac{4}{1} = 16$

32. $37\dfrac{5}{8} \times 18 = \dfrac{301}{8} \times \dfrac{18}{1} = \dfrac{301}{4} \times \dfrac{9}{1} = \dfrac{2709}{4} = 677\dfrac{1}{4}$

18 shares cost $\$677\dfrac{1}{4}$.

33. $13\dfrac{1}{2} \times 9\dfrac{2}{3} = \dfrac{27}{2} \times \dfrac{29}{3} = \dfrac{9}{2} \times \dfrac{29}{1} = \dfrac{261}{2}$ or $130\dfrac{1}{2}$

The area is $\dfrac{261}{2}$ or $130\dfrac{1}{2}$ square feet.

34. $\dfrac{3}{7} \div \dfrac{2}{5} = \dfrac{3}{7} \times \dfrac{5}{2} = \dfrac{15}{14}$ or $1\dfrac{1}{14}$

35. $900 \div \dfrac{3}{5} = \dfrac{900}{1} \times \dfrac{5}{3} = 1500$

36. $5\dfrac{3}{4} \div 11\dfrac{1}{2} = \dfrac{23}{4} \div \dfrac{23}{2} = \dfrac{23}{4} \times \dfrac{2}{23} = \dfrac{1}{2}$

37. $20 \div 2\dfrac{1}{2} = \dfrac{20}{1} \div \dfrac{5}{2} = \dfrac{20}{1} \times \dfrac{2}{5} = 8$

38. $0 \div 3\dfrac{7}{5} = 0$

39. $4\dfrac{2}{11} \div 3 = \dfrac{46}{11} \div \dfrac{3}{1} = \dfrac{46}{11} \times \dfrac{1}{3} = \dfrac{46}{33}$ or $1\dfrac{13}{33}$

40. $342 \div 28\dfrac{1}{2} = \dfrac{342}{1} \div \dfrac{57}{2} = \dfrac{342}{1} \times \dfrac{2}{57} = 6 \times 2 = 12$

12 rolls are needed.

41. $420 \div 2\dfrac{1}{4} = \dfrac{420}{1} \div \dfrac{9}{4}$

$\qquad = \dfrac{420}{1} \times \dfrac{4}{9}$

$\qquad = \dfrac{140}{1} \times \dfrac{4}{3}$

$\qquad = \dfrac{560}{3}$ or $186\dfrac{2}{3}$ calories

42. $14 = 2 \times 7$

$49 = 7 \times 7$

LCD $= 2 \times 7 \times 7 = 98$

43. $20 = 2 \times 2 \times 5$
$25 = 5 \times 5$
$\text{LCD} = 2 \times 2 \times 5 \times 5 = 100$

44. $18 = 2 \times 3 \times 3$
$6 = 2 \times 3$
$45 = 3 \times 3 \times 5$
$\text{LCD} = 2 \times 3 \times 3 \times 5 = 90$

45. $\dfrac{3}{7} = \dfrac{3}{7} \times \dfrac{8}{8} = \dfrac{24}{56}$

46. $\dfrac{11}{24} = \dfrac{11}{24} \times \dfrac{3}{3} = \dfrac{33}{72}$

47. $\dfrac{8}{15} = \dfrac{8}{15} \times \dfrac{10}{10} = \dfrac{80}{150}$

48. $\dfrac{9}{14} - \dfrac{5}{14} = \dfrac{4}{14} = \dfrac{2}{7}$

49. $\dfrac{1}{2} + \dfrac{1}{3} + \dfrac{1}{4} = \dfrac{1}{2} \times \dfrac{6}{6} + \dfrac{1}{3} \times \dfrac{4}{4} + \dfrac{1}{4} \times \dfrac{3}{3}$
$= \dfrac{6}{12} + \dfrac{4}{12} + \dfrac{3}{12}$
$= \dfrac{13}{12} \text{ or } 1\dfrac{1}{12}$

50. $\dfrac{7}{8} - \dfrac{3}{5} = \dfrac{7}{8} \times \dfrac{5}{5} - \dfrac{3}{5} \times \dfrac{8}{8} = \dfrac{35}{40} - \dfrac{24}{40} = \dfrac{11}{40}$

51. $\dfrac{7}{30} + \dfrac{2}{21} = \dfrac{7}{30} \times \dfrac{7}{7} + \dfrac{2}{21} \times \dfrac{10}{10}$
$= \dfrac{49}{210} + \dfrac{20}{210}$
$= \dfrac{69}{210}$
$= \dfrac{23}{70}$

52. $\dfrac{5}{18} + \dfrac{7}{10} = \dfrac{5}{18} \times \dfrac{5}{5} + \dfrac{7}{10} \times \dfrac{9}{9} = \dfrac{25}{90} + \dfrac{63}{90} = \dfrac{88}{90} = \dfrac{44}{45}$

53. $\dfrac{14}{15} - \dfrac{3}{25} = \dfrac{14}{15} \times \dfrac{5}{5} - \dfrac{3}{25} \times \dfrac{3}{3} = \dfrac{70}{75} - \dfrac{9}{75} = \dfrac{61}{75}$

54. $8 - 2\dfrac{3}{4} = \dfrac{32}{4} - \dfrac{11}{4} = \dfrac{21}{4} \text{ or } 5\dfrac{1}{4}$

55. $3 + 5\dfrac{2}{3} = 8\dfrac{2}{3}$

56.
$\begin{array}{r} 3\dfrac{3}{8} \\ + 2\dfrac{3}{4} \\ \hline \end{array}$
$\qquad$
$\begin{array}{r} 3\dfrac{3}{8} \\ + 2\dfrac{6}{8} \\ \hline 5\dfrac{9}{8} = 6\dfrac{1}{8} \end{array}$

57.
$\begin{array}{r} 5\dfrac{11}{16} \\ - 2\dfrac{1}{5} \\ \hline \end{array}$
$\qquad$
$\begin{array}{r} 5\dfrac{55}{80} \\ - 2\dfrac{16}{80} \\ \hline 3\dfrac{39}{80} \end{array}$

58. $\dfrac{3}{5} \times \dfrac{1}{2} + \dfrac{2}{5} \div \dfrac{2}{3} = \dfrac{3}{5} \times \dfrac{1}{2} + \dfrac{2}{5} \times \dfrac{3}{2} = \dfrac{3}{10} + \dfrac{6}{10} = \dfrac{9}{10}$

59. $\left(\dfrac{4}{5} - \dfrac{1}{2} \right)^2 \times \dfrac{10}{3} = \left(\dfrac{8}{10} - \dfrac{5}{10} \right)^2 \times \dfrac{10}{3}$
$= \left(\dfrac{3}{10} \right)^2 \times \dfrac{10}{3}$
$= \dfrac{9}{100} \times \dfrac{10}{3}$
$= \dfrac{3}{10}$

60. $1\dfrac{7}{8} + 2\dfrac{3}{4} + 4\dfrac{1}{10} = 1\dfrac{70}{80} + 2\dfrac{60}{80} + 4\dfrac{8}{80}$
$= 7\dfrac{138}{80}$
$= 8\dfrac{58}{80}$
$= 8\dfrac{29}{40}$

The total number of miles is $8\dfrac{29}{40}$ miles.

61.

$$28\frac{1}{6} \qquad\qquad 27\frac{7}{6}$$

$$-\ 1\frac{5}{6} \qquad\qquad -\ 1\frac{5}{6}$$

$$\overline{} \qquad\qquad \overline{26\frac{2}{6} = 26\frac{1}{3}}$$

Then: $26\frac{1}{3} \times 10\frac{3}{4} = \frac{79}{3} \times \frac{43}{4} = \frac{3397}{12} = 283\frac{1}{12}$

She can drive $283\frac{1}{12}$ miles.

62. $3\frac{1}{3} \times 2\frac{1}{2} = \frac{10}{3} \times \frac{1}{2} = \frac{5}{3} = 1\frac{2}{3}$ cups sugar

$4\frac{1}{4} \times \frac{1}{2} = \frac{17}{4} \times \frac{1}{2} = \frac{17}{8} = 2\frac{1}{8}$ cups flour

63. $24\frac{1}{4} \times 8\frac{1}{2} = \frac{97}{4} \times \frac{17}{2} = \frac{1649}{8} = 206\frac{1}{8}$

He can drive approximately $206\frac{1}{8}$ miles.

64. $48 \div 3\frac{1}{5} = \frac{48}{1} \div \frac{16}{5} = \frac{48}{1} \times \frac{5}{16} = \frac{3}{1} \times \frac{5}{1} = 15$

15 lengths can be cut from the pipe.

65. $15\frac{3}{4} - 6\frac{1}{8} = 15\frac{6}{8} - 6\frac{1}{8} = 9\frac{5}{8}$

It contains $9\frac{5}{8}$ liters of water.

66.

$$\begin{array}{r} 12 \\ 9 \\ +\ 14 \\ \hline 35 \end{array}$$

$35 \div 5 = 7$

$7 \times 32\frac{1}{2} = \frac{7}{1} \times \frac{65}{2} = \frac{455}{2} = 227\frac{1}{2}$

It will take $227\frac{1}{2}$ minutes or 3 hours and

$47\frac{1}{2}$ minutes.

67. $2\frac{1}{2} \times 1\frac{3}{4} = \frac{5}{2} \times \frac{7}{4} = \frac{5 \times 7}{2 \times 4} = \frac{35}{8} = 4\frac{3}{8}$

She will need $\frac{35}{8}$ or $4\frac{3}{8}$ cups of flour.

$$\begin{array}{cc} 12 & 11\frac{8}{8} \\ -\ 4\frac{3}{8} & -\ 4\frac{3}{8} \\ \hline & 7\frac{5}{8} \end{array}$$

There will be $7\frac{5}{8}$ cups of flour left in the bag.

68. $1\frac{1}{2} + \frac{1}{16} + \frac{1}{8} + \frac{1}{4} = 1\frac{8}{16} + \frac{1}{16} + \frac{2}{16} + \frac{4}{16} = 1\frac{15}{16}$

$3 - 1\frac{15}{16} = 2\frac{16}{16} - 1\frac{15}{16} = 1\frac{1}{16}$

The bolt extends $1\frac{1}{16}$ inches.

69.

$$\frac{1}{10} \times 880 = 88$$

$$\frac{1}{2} \times 880 = 440$$

$$+\ \frac{1}{8} \times 880 = +110$$

$$\overline{638}$$

Left over: 880

$$\underline{-\ 638}$$

$$242$$

She has $242 left over.

70. $460 \div 18\frac{2}{5} = \frac{460}{1} \div \frac{92}{5} = \frac{460}{1} \times \frac{5}{92} = 25$

His car gets 25 miles per gallon.

71. $\dfrac{27}{63} = \dfrac{27 \div 9}{63 \div 9} = \dfrac{3}{7}$

72. $\dfrac{7}{5} + \dfrac{11}{25} = \dfrac{35}{75} + \dfrac{33}{75} = \dfrac{68}{75}$

73.

$$\begin{array}{ccc} 4\frac{1}{3} & 4\frac{4}{12} & 3\frac{16}{12} \\ -\ 2\frac{11}{12} & -\ 2\frac{11}{12} & -\ 2\frac{11}{12} \\ \hline & & 1\frac{5}{12} \end{array}$$

74. $\dfrac{36}{49} \times \dfrac{14}{33} = \dfrac{3 \times 12 \times 2 \times 7}{3 \times 11 \times 7 \times 7} = \dfrac{24}{77}$

75. $\left(\dfrac{4}{7}\right)^3 = \dfrac{4}{7} \times \dfrac{4}{7} \times \dfrac{4}{7} = \dfrac{64}{343}$

76. $\dfrac{3}{8} \div \dfrac{1}{10} = \dfrac{3}{8} \times \dfrac{10}{1} = \dfrac{3}{4} \times \dfrac{5}{1} = \dfrac{15}{4}$ or $3\dfrac{3}{4}$

77. $5\dfrac{1}{2} \times 18 = \dfrac{11}{2} \times \dfrac{18}{1} = \dfrac{11}{1} \times \dfrac{9}{1} = 99$

78. $150 \div 3\dfrac{1}{8} = \dfrac{150}{1} \div \dfrac{25}{8} = \dfrac{150}{1} \times \dfrac{8}{25} = \dfrac{6}{1} \times \dfrac{8}{1} = 48$

How Am I Doing? Chapter 2 Test

1. $\dfrac{3}{5}$; 3 of the 5 parts are shaded.

2. $\dfrac{\text{number that went in}}{\text{total number}} = \dfrac{311}{388}$

3. $\dfrac{18}{42} = \dfrac{18 \div 6}{42 \div 6} = \dfrac{3}{7}$

4. $\dfrac{15}{70} = \dfrac{15 \div 5}{70 \div 5} = \dfrac{3}{14}$

5. $\dfrac{225}{50} = \dfrac{225 \div 25}{50 \div 25} = \dfrac{9}{2}$

6. $6\dfrac{4}{5} = \dfrac{6 \times 5 + 4}{5} = \dfrac{34}{5}$

7. $14\overline{)145}$ $\dfrac{145}{14} = 10\dfrac{5}{14}$
  10
  14
  5

8. $42 \times \dfrac{2}{7} = \dfrac{42}{1} \times \dfrac{2}{7} = \dfrac{6 \times 7 \times 2}{1 \times 7} = \dfrac{12}{1} = 12$

9. $\dfrac{7}{9} \times \dfrac{2}{5} = \dfrac{7 \times 2}{9 \times 5} = \dfrac{14}{45}$

10. $2\dfrac{2}{3} \times 5\dfrac{1}{4} = \dfrac{8}{3} \times \dfrac{21}{4} = \dfrac{2 \times 4 \times 3 \times 7}{3 \times 4} = 14$

11. $\dfrac{7}{8} \div \dfrac{5}{11} = \dfrac{7}{8} \times \dfrac{11}{5} = \dfrac{7 \times 11}{8 \times 5} = \dfrac{77}{40}$ or $1\dfrac{37}{40}$

12. $\dfrac{12}{31} \div \dfrac{8}{13} = \dfrac{12}{31} \times \dfrac{13}{8} = \dfrac{3 \times 4 \times 13}{31 \times 2 \times 4} = \dfrac{39}{62}$

13. $7\dfrac{1}{5} \div 1\dfrac{1}{25} = \dfrac{36}{5} \div \dfrac{26}{25}$
  $= \dfrac{36}{5} \times \dfrac{25}{26}$
  $= \dfrac{2 \times 18 \times 5 \times 5}{5 \times 2 \times 13}$
  $= \dfrac{18 \times 5}{13}$
  $= \dfrac{90}{13}$ or $6\dfrac{12}{13}$

14. $5\dfrac{1}{7} \div 3 = \dfrac{36}{7} \div \dfrac{3}{1} = \dfrac{36}{7} \times \dfrac{1}{3} = \dfrac{3 \times 12 \times 1}{7 \times 3} = \dfrac{12}{7}$ or $1\dfrac{5}{7}$

15. $12 = 2 \times 2 \times 3$
  $18 = 2 \times 3 \times 3$
  $\text{LCD} = 2 \times 2 \times 3 \times 3$

16. $16 = 2 \times 2 \times 2 \times 2$
  $24 = 2 \times 2 \times 2 \times 3$
  $\text{LCD} = 2 \times 2 \times 2 \times 2 \times 3 = 48$

17. $4 = 2 \times 2$
  $8 = 2 \times 2 \times 2$
  $6 = 2 \times 3$
  $\text{LCD} = 2 \times 2 \times 2 \times 3 = 24$

18. $\dfrac{5}{12} = \dfrac{5}{12} \times \dfrac{6}{6} = \dfrac{30}{72}$

19. $\dfrac{7}{9} - \dfrac{5}{12} = \dfrac{28}{36} - \dfrac{15}{36} = \dfrac{13}{36}$

20. $\dfrac{2}{15} + \dfrac{5}{12} = \dfrac{8}{60} + \dfrac{25}{60} = \dfrac{33}{60} = \dfrac{11}{20}$

21. $\dfrac{1}{4} + \dfrac{3}{7} + \dfrac{3}{14} = \dfrac{7}{28} + \dfrac{12}{28} + \dfrac{6}{28} = \dfrac{25}{28}$

22. $8\dfrac{3}{5} + 5\dfrac{4}{7} = 8\dfrac{21}{35} + 5\dfrac{20}{35} = 13\dfrac{41}{35} = 14\dfrac{6}{35}$

23. $18\dfrac{6}{7} - 13\dfrac{13}{14} = 18\dfrac{12}{14} - 13\dfrac{13}{14} = 17\dfrac{26}{14} - 13\dfrac{13}{14} = 4\dfrac{13}{14}$

24. $\dfrac{2}{9} \div \dfrac{8}{3} \times \dfrac{1}{4} = \dfrac{2}{9} \times \dfrac{3}{8} \times \dfrac{1}{4} = \dfrac{1}{48}$

25. $\left(\dfrac{1}{2}+\dfrac{1}{3}\right)\times\dfrac{7}{5}=\left(\dfrac{3}{6}+\dfrac{2}{6}\right)\times\dfrac{7}{5}=\dfrac{5}{6}\times\dfrac{7}{5}=\dfrac{7}{6}$ or $1\dfrac{1}{6}$

26. $16\dfrac{1}{2}\times9\dfrac{1}{3}=\dfrac{33}{2}\times\dfrac{28}{3}=11\times14=154$

The kitchen is 154 square feet.

27. $18\dfrac{2}{3}\div2\dfrac{1}{3}=\dfrac{56}{3}\div\dfrac{7}{3}=\dfrac{56}{3}\times\dfrac{3}{7}=\dfrac{8\times7\times3}{3\times7}=8$

He can make 8 packages.

28. $\dfrac{9}{10}-\dfrac{1}{5}=\dfrac{9}{10}-\dfrac{2}{10}=\dfrac{7}{10}$

He has $\dfrac{7}{10}$ of a mile left to walk.

29. $4\dfrac{1}{8}+3\dfrac{1}{6}+6\dfrac{3}{4}=4\dfrac{3}{24}+3\dfrac{4}{24}+6\dfrac{18}{24}$

$\qquad\qquad =13\dfrac{25}{24}$

$\qquad\qquad =14\dfrac{1}{24}$

She jogged $14\dfrac{1}{24}$ miles.

30. $\dfrac{1}{4}\times120=\dfrac{1}{4}\times\dfrac{120}{1}=30$

$\dfrac{1}{12}\times120=\dfrac{1}{12}\times\dfrac{120}{1}=10$

$\dfrac{1}{3}\times120=\dfrac{1}{3}\times\dfrac{120}{1}=40$

$120-30-10-40=40$

They shipped 40 oranges.

31. $48\dfrac{1}{8}\div\dfrac{5}{8}=\dfrac{385}{8}\times\dfrac{8}{5}=\dfrac{385}{5}=77$

They can make 77 candles.

$2\dfrac{1}{2}\times\dfrac{5}{8}=\dfrac{5}{2}\times\dfrac{5}{8}=\dfrac{5\times5}{2\times8}=\dfrac{25}{16}$ or $1\dfrac{9}{16}$

It takes $\dfrac{25}{16}$ or $1\dfrac{9}{16}$ pounds of wax to make one

pillar candle.

Chapter 3

3.1 Exercises

1. A decimal fraction is a fraction whose denominator is a power of 10. $\dfrac{23}{100}$ and $\dfrac{563}{1000}$ are decimal fractions.

3. The last decimal place on the right in 132.45678 is five places after the decimal point. This is the hundred-thousandths place.

5. 0.57 = fifty-seven hundredths

7. 3.8 = three and eight tenths

9. 7.013 = seven and thirteen thousandths

11. 28.0037 = twenty-eight and thirty-seven ten-thousandths

13. $\$124.20$ = one hundred twenty-four and $\dfrac{20}{100}$ dollars

15. $\$1236.08$ = one thousand, two hundred thirty-six and $\dfrac{8}{100}$ dollars

17. $\$18,045.19$ = eighteen thousand, forty-five and $\dfrac{19}{100}$ dollars

19. seven tenths = 0.7

21. ninety-six hundredths = 0.96

23. four hundred eighty-one thousandths = 0.481

25. six thousand one hundred fourteen millionths = 0.006114

27. $\dfrac{7}{10} = 0.7$

29. $\dfrac{76}{100} = 0.76$

31. $\dfrac{1}{100} = 0.01$

33. $\dfrac{53}{1000} = 0.053$

35. $\dfrac{2403}{10,000} = 0.2403$

37. $10\dfrac{9}{10} = 10.9$

39. $84\dfrac{13}{100} = 84.13$

41. $3\dfrac{529}{1000} = 3.529$

43. $235\dfrac{104}{10,000} = 235.0104$

45. $0.02 = \dfrac{2}{100} = \dfrac{1}{50}$

47. $3.6 = 3\dfrac{6}{10} = 3\dfrac{3}{5}$

49. $7.41 = 7\dfrac{41}{100}$

51. $12.625 = 12\dfrac{625}{1000} = 12\dfrac{5}{8}$

53. $7.0615 = 7\dfrac{615}{10,000} = 7\dfrac{123}{2000}$

55. $8.0108 = 8\dfrac{108}{10,000} = 8\dfrac{27}{2500}$

57. $235.1254 = 235\dfrac{1254}{10,000} = 235\dfrac{627}{5000}$

59. $0.0125 = \dfrac{125}{10,000} = \dfrac{1}{80}$

61. a. $\dfrac{26,300}{100,000} = \dfrac{263}{1000}$

$\dfrac{263}{1000}$ of the male population of Kentucky were smokers in 2008.

b. $\dfrac{24,200}{100,000} = \dfrac{121}{500}$

$\dfrac{121}{500}$ of the female population of Kentucky were smokers in 2008.

63. $\dfrac{4}{1,000,000} = \dfrac{1}{250,000}$

Cumulative Review

65. 56,7<u>5</u>8 rounds to 56,800 since 5 is equal to 5.

66. 8,069,<u>4</u>82 rounds to 8,069,000 since 4 is less than 5.

67. $\dfrac{36}{80} = \dfrac{4 \times 9}{4 \times 20} = \dfrac{9}{20}$

68. $\dfrac{7}{8} - \dfrac{21}{40} = \dfrac{35}{40} - \dfrac{21}{40} = \dfrac{35 - 21}{40} = \dfrac{14}{40} = \dfrac{7}{20}$

Quick Quiz 3.1

1. 5.367 = five and three hundred sixty-seven thousandths

2. $\dfrac{523}{10,000} = 0.0523$

3. $12.58 = 12\dfrac{58}{100} = 12\dfrac{29}{50}$

4. Answers may vary. Possible solution: Since the denominator has 5 zeros, the decimal will have 5 decimal places. The numerator only has 3 digits, so 2 zeros will need to be added.

$\dfrac{953}{100,000} = 0.00953$

3.2 Exercises

1. 1.3 > 1.29
The numbers to the left of the decimal points are the same. The number in tenths place of 1.3 is 3, which is greater than 2, the number in the tenths place of 1.29. Thus, 1.3 is the greater number.

3. 0.34 = 0.340
Adding a zero in the thousandths place of 0.34 on the left, we get 0.340. The two numbers are equal.

5. 18.92 < 18.93
The numbers to the left of the decimal points and in the tenths place are the same. The number in hundredths place of 18.92 is 2, which is less than 3, the number in the hundredths place of 18.93. Thus, 18.92 is the smaller number.

7. 0.00043 > 0.0004
Adding a zero in the hundred-thousandths place of 0.0004, we get 0.00040. The hundred-thousandths place is the first digit where the numbers differ. Since 3 > 0, 0.00043 is the greater number.

9. 1.002 < 1.0021
Adding a zero in the ten-thousandths place of 1.002, we get 1.0020. The ten-thousandths place is the first digit where the numbers differ. Since 0 < 1, 1.0020 is the smaller number.

11. 126.34 > 125.35
126 > 125, so 126.34 is the greater number.

13. 0.888 < 0.8888
Adding a zero in the ten-thousandths place of 0.888, we get 0.8880. The ten-thousandths place is the first digit where the numbers differ. Since 0 < 8, 0.8880 is the smaller number.

15. 0.777 > 0.7077
The hundredths place is the first digit where the numbers differ. Since 7 > 0, 0.777 is the greater number.

17. $\dfrac{72}{1000} = 0.072$

Written in decimal form, $\dfrac{72}{1000}$ is 0.072.

19. $\dfrac{8}{10} > 0.08$

Written in decimal form, $\dfrac{8}{10}$ is 0.8. The tenths place is the first place that 0.8 and 0.08 differ. Since 8 > 0, 0.8 is the greater number.

21. 12.6, 12.65, 12.8

23. 0.007, 0.0071, 0.05

25. 8.31, 8.39, 8.4, 8.41

27. 26.003, 26.033, 26.034, 26.04

29. 18.006, 18.060, 18.065, 18.066, 18.606

31. 6.92 rounds to 6.9.
Find the tenths place: 6.92. The digit to the right of the tenths place is less than 5. We round down to 6.9 and drop the digit to the right.

33. 28.98 rounds to 29.0.
Find the tenths place: 28.98. The digit to the right of the tenths place is greater than 5. We round up to 29.0 and drop the digit to the right.

35. 578.064 rounds to 578.1.
Find the tenths place: 578.064. The digit to the right of the tenths place is greater than 5. We round up to 578.1 and drop the digits to the right.

37. 2176.83 rounds to 2176.8.
Find the tenths place: 2176.83. The digit to the right of the tenths place is less than 5. We round down to 2176.8 and drop the digit to the right.

39. 26.032 rounds to 26.03.
Find the hundredths place: 26.032. The digit to the right of the hundredths place is less than 5. We round down to 26.03 and drop the digit to the right.

41. 36.997 rounds to 37.00.
Find the hundredths place: 36.997. The digit to the right of the hundredths place is greater than 5. We round up to 37.00 and drop the digit to the right.

43. 156.1749 rounds to 156.17.
Find the hundredths place: 156.1749. The digit tot he right of the hundredths place is less than 5. We round down to 156.17 and drop the digits to the right.

45. 2786.706 rounds to 2786.71.
Find the hundredths place: 2786.706. The digit to the right of the hundredths place is greater than 5. We round up to 2786.71 and drop the digit to the right.

47. 7.8155 rounds to 7.816.
Find the thousandths place: 7.8155. The digit to the right of the thousandths place is equal to 5. We round up to 7.816 and drop the digit to the right.

49. 0.05951 rounds to 0.0595.
Find the ten-thousandths place: 0.05951. The digit to the right of the ten-thousandths place is less than 5. We round down to 0.0595 and drop the digit to the right.

51. 12.0157823 rounds to 12.01578.
Find the hundred-thousandths place: 12.0157823. The digit to the right of the hundred-thousandths place is less than 5. We round down to 12.01578 and drop the digits to the right.

53. 135.564 rounds to 136.
The digit to the right of the decimal point equals 5. We round up to 136 and drop the digits to the right.

55. $788.42 rounds to $788.

57. $15,020.50 rounds to $15,021.

59. $96.3357 rounds to $96.34.

61. $5783.716 rounds to $5783.72.

63. Yankees: We round down since the digit to the right of the thousandths place, 3, is less than 5.
0.63636 rounds to 0.636.
Twins: We round up since the digit to the right of the thousandths place is 5.
0.54455 rounds to 0.545.

65. We round down since the digit to the right of the hundredths place, 1, is less than 5.
365.24122 rounds to 365.24.

67. $\dfrac{6}{100} = 0.06$ and $\dfrac{6}{10} = 0.6$

0.0059, 0.006, 0.0519, $\dfrac{6}{100}$, 0.0601, 0.0612,

0.062, $\dfrac{6}{10}$, 0.61

69. You should consider only one digit to the right of the decimal place that you wish to round to. 86.23498 is closer to 86.23 than to 86.24.

Cumulative Review

71.

$$3\frac{1}{4}$$
$$2\frac{1}{2}$$
$$+\,6\frac{3}{8}$$

$$3\frac{2}{8}$$
$$2\frac{4}{8}$$
$$+\,6\frac{3}{8}$$
$$11\frac{9}{8}=12\frac{1}{8}$$

72.
$$27\frac{1}{5}-16\frac{3}{4}=27\frac{4}{20}-16\frac{15}{20}$$
$$=26\frac{24}{20}-16\frac{15}{20}$$
$$=10\frac{9}{20}$$

73.

$$20,000$$
$$2\,000$$
$$800$$
$$+\,10,000$$
$$\overline{32,800}$$

Sales were \$32,800.

74. a. $\dfrac{7}{100}=0.07$

b. $\dfrac{145}{1000}=0.145$

Quick Quiz 3.2

1. 4.56, 4.6, 4.056, 4.559
   Add zeros to make the comparison easier.
   4.560, 4.600, 4.056, 4.559
   Rearrange with the smallest first.
   4.056, 4.559, 4.56, 4.6

2. 27.1782 rounds to 27.18.
   Find the hundredths place: 27.1782. The digit to the right of the hundredths place is greater than 5. We round up to 27.18 and drop the digits to the right.

3. 155.52525 rounds to 155.525.
   Find the thousandths place: 155.52525. The digit to the right of the thousandths place is less than 5. We round down to 155.525 and drop the digits to the right.

4. Answers may vary. Possible solution:
   Find the ten-thousandths place: 34.958365
   Look at the digit to the right of the ten-thousands place. If it is less than 5, round down. If it is 5 or greater round up. In either case, drop the digits to the right of the ten-thousands place.
   34.958365 rounds to 34.9584.

3.3 Exercises

1.

$$57.1$$
$$+\,19.7$$
$$\overline{76.8}$$

3.

$$384.25$$
$$+\,209.65$$
$$\overline{593.90}$$

5.

$$13.4$$
$$7.6$$
$$+\,275.2$$
$$\overline{296.2}$$

7.

$$4.71$$
$$+\,8.05$$
$$\overline{12.76}$$

9.

$$4.9637$$
$$28.1200$$
$$+\,3.6450$$
$$\overline{36.7287}$$

11.

$$12.00$$
$$3.62$$
$$+\,51.80$$
$$\overline{67.42}$$

13.

$$108.36$$
$$14.30$$
$$85.12$$
$$+\,28.00$$
$$\overline{235.78}$$

15.

$$753.61$$
$$28.75$$
$$162.30$$
$$100.50$$
$$+\,67.00$$
$$\overline{1112.16}$$

17. Perimeter:　　5.26
　　　　　　　　　9.28
　　　　　　　 +　6.50
　　　　　　　　　21.04
The perimeter is 21.04 feet.

19.　　1.75
　　　　2.50
　　　　1.55
　　 +　2.80
　　　　8.60
He lost a total of 8.6 pounds.

21.　　 4.99
　　　　12.50
　　　　11.85
　　　　28.50
　　　　 3.29
　 +　16.99
　　　　78.12
The bill for their day at the beach was $78.12.

23.　　46,276.0
　 +　　778.9
　　　47,054.9
The final odometer reading was 47,054.9 miles.

25.　　　18.42
　　　 706.15
　　　　21.03
　　　　45.00
　 +　 621.37
　　　1411.97
The total deposit was $1411.97.

27.　　12.8
　　 −　9.3
　　　　3.5

29.　　35.75
　　 −　9.82
　　　25.93

31.　　126.00
　　 −　76.22
　　　　49.78

33.　　586.513
　　 −　78.200
　　　508.313

35.　　220.90
　　 −　85.47
　　　135.43

37.　　24.0079
　　 −19.3614
　　　　4.6465

39.　　8.000
　　 −1.263
　　　6.737

41.　　7362.14
　　 −6173.07
　　　1189.07

43.　　1.5000
　　 −0.0365
　　　1.4635

45.　　123.621
　 +　52.960
　　　176.581

47.　　98.30
　　 −56.71
　　　41.59

49.　　0.0763
　　　2.0000
　 +　3.1600
　　　5.2363

51.　　197.600
　　 −124.375
　　　73.225

53.　　11.5830
　　 −4.0678
　　　7.5152
The lemon was 7.5152 pounds heavier than the apple.

55.　　37,026.65
　　 −　　79.49
　　　36,947.16
The professional telescope costs $36,947.16 more.

57.
$$
\begin{array}{r}
47.70 \\
+\ 7.00 \\
\hline
54.70
\end{array}
\qquad
\begin{array}{r}
100.00 \\
-\ 54.70 \\
\hline
45.30
\end{array}
$$
He received $45.30 change.

59.
$$
\begin{array}{r}
12.62 \\
-\ 0.98 \\
\hline
11.64
\end{array}
$$
The part of the wire that is not exposed is 11.64 centimeters.

61. $2.45 + 1.35 - 0.85 = 3.80 - 0.85 = 2.95$
There will be 2.95 liters left.

63.
$$
\begin{array}{r}
0.0150 \\
-\ 0.0089 \\
\hline
0.0061
\end{array}
$$
The difference is 0.0061 milligram. Yes, it is safe to drink.

65.
$$
\begin{array}{r}
43.8 \\
-\ 37.6 \\
\hline
6.2
\end{array}
$$
$6.2 billion or $6,200,000,000 more were earned in mining in 1980 than in 2000.

67.
$$
\begin{array}{r}
271.7 \\
-\ 109.6 \\
\hline
162.1
\end{array}
$$
$162.1 billion or $162,100,000,000 more were earned in communications than in agriculture, forestry, and fisheries in 2010.

69.
$$
\begin{array}{r}
2.60 \\
1.50 \\
1.30 \\
0.80 \\
+\ 2.20 \\
\hline
\$8.40
\end{array}
$$
Exact:
$$
\begin{array}{r}
2.63 \\
1.47 \\
1.26 \\
0.79 \\
+\ 2.19 \\
\hline
\$8.34
\end{array}
$$
Estimate is very close to actual amount. Difference is 6 cents.

71. $x + 7.1 = 15.5$
$$
\begin{array}{r}
15.5 \\
-\ 7.1 \\
\hline
8.4
\end{array}
$$
$x = 8.4$

73. $156.9 + x = 200.6$
$$
\begin{array}{r}
200.6 \\
-\ 156.9 \\
\hline
43.7
\end{array}
$$
$x = 43.7$

75. $4.162 = x + 2.053$
$$
\begin{array}{r}
4.162 \\
-\ 2.053 \\
\hline
2.109
\end{array}
$$
$x = 2.109$

Cumulative Review

77.
$$
\begin{array}{r}
2536 \\
\times\quad 8 \\
\hline
20,288
\end{array}
$$

78. $\dfrac{1}{4} \times 100 = \dfrac{1}{4} \times \dfrac{100}{1} = \dfrac{1}{1} \times \dfrac{25}{1} = 25$

79. $800 \times \dfrac{1}{2} = \dfrac{800}{1} \times \dfrac{1}{2} = \dfrac{400}{1} \times \dfrac{1}{1} = 400$

80. $40,000 \div 40 = 1000$

Quick Quiz 3.3

1.
$$
\begin{array}{r}
53.261 \\
1.900 \\
+\ 17.820 \\
\hline
72.981
\end{array}
$$

2.
$$
\begin{array}{r}
5.2608 \\
-\ 3.0791 \\
\hline
2.1817
\end{array}
$$

3.
$$
\begin{array}{r}
59.600 \\
-\ 3.925 \\
\hline
55.675
\end{array}
$$

4. Answers may vary. Possible solution: First add two zeros to the right of the first number, so when they are aligned vertically, they have the same number of decimal places. Then subtract all digits with the same place value, starting with

the right column and moving to the left. Borrow
when necessary.

$$
\begin{array}{r}
\overset{13\ 14\ 9}{6\ \cancel{3}\ \cancel{4}\ \cancel{10}\,10} \\
567.\cancel{4}\ \cancel{5}\ \cancel{0}\ \cancel{0} \\
-\ 345.9\ \ 8\ 7\ 2 \\
\hline
221.4\ \ 6\ 2\ 8
\end{array}
$$

3.4 Exercises

1. Each factor has two decimal places. You add the
number of decimal places to get four decimal
places. You multiply 67×8 to obtain 536. Now
you must place the decimal point four places to
the left in your answer. The result is 0.0536.

3. When you multiply a number by 100 you move
the decimal point two places to the right. The
answer would be 0.78.

5.
$$
\begin{array}{r}
0.6 \\
\times\ \ 0.2 \\
\hline
0.12
\end{array}
$$

7.
$$
\begin{array}{r}
0.12 \\
\times\ \ \ 0.5 \\
\hline
0.060 = 0.06
\end{array}
$$

9.
$$
\begin{array}{r}
0.0036 \\
\times\ \ \ \ 0.8 \\
\hline
0.00288
\end{array}
$$

11.
$$
\begin{array}{r}
452 \\
\times\ \ 0.12 \\
\hline
9\ 04 \\
45\ 2 \\
\hline
54.24
\end{array}
$$

13.
$$
\begin{array}{r}
0.043 \\
\times\ \ 0.012 \\
\hline
0086 \\
0043 \\
\hline
0.000516
\end{array}
$$

15.
$$
\begin{array}{r}
10.97 \\
\times\ \ \ 0.06 \\
\hline
0.6582
\end{array}
$$

17.
$$
\begin{array}{r}
3423 \\
\times\ \ \ 0.8 \\
\hline
2738.4
\end{array}
$$

19.
$$
\begin{array}{r}
2.163 \\
\times\ \ 0.008 \\
\hline
0.017304
\end{array}
$$

21.
$$
\begin{array}{r}
0.7613 \\
\times\ \ 1009 \\
\hline
6\ 8517 \\
761\ 300 \\
\hline
768.1517
\end{array}
$$

23.
$$
\begin{array}{r}
2350 \\
\times\ \ \ 3.6 \\
\hline
1410\ 0 \\
7050 \\
\hline
8460.0 = 8460
\end{array}
$$

25.
$$
\begin{array}{r}
4.57 \\
\times\ \ 11.8 \\
\hline
3\ 656 \\
4\ 57 \\
45\ 7 \\
\hline
53.926
\end{array}
$$

27.
$$
\begin{array}{r}
6523.7 \\
\times\ \ 0.001 \\
\hline
6.5237
\end{array}
$$

29.
$$
\begin{array}{r}
155.40 \\
\times\ \ \ \ \ \ 60 \\
\hline
9324.00
\end{array}
$$
He will have spent $9324.

31.
$$
\begin{array}{r}
10.50 \\
\times\ \ \ \ 40 \\
\hline
420.00
\end{array}
$$
He earns $420 in one week.

33.
$$
\begin{array}{r}
19.2 \\
\times\ \ 15.5 \\
\hline
9\ 60 \\
96\ 0 \\
192 \\
\hline
297.60
\end{array}
$$
The area of the bedroom is 297.6 square feet.

35.
$$\begin{array}{r} 36.90 \\ \times\ \ \ \ 18 \\ \hline 295\ 20 \\ 369\ 0 \\ \hline 664.20 \end{array}$$
He will pay a total of $664.20.

37.
$$\begin{array}{r} 26.4 \\ \times\ \ 19.5 \\ \hline 13\ 20 \\ 237\ 6 \\ 264 \\ \hline 514.80 \end{array}$$
He can travel approximately 514.8 miles.

39. Move the decimal one place to the right.
$2.86 \times 10 = 28.6$

41. Move the decimal two places to the right.
$52.125 \times 100 = 5212.5$

43. Move the decimal three places to the right.
$22.615 \times 1000 = 22,615$

45. Move the decimal four places to the right.
$5.60982 \times 10,000 = 56,098.2$

47. Move the decimal two places to the right.
$17,561.44 \times 10^2 = 1,756,144$

49. Move the decimal three places to the right.
$816.32 \times 10^3 = 816,320$

51. $5.932 \times 100 = 593.2$
There are 593.2 centimeters in 5.932 meters.

53. $2.71 \times 1000 = 2710$
He spent $2710 on the shares.

55. Graduation:
$$\begin{array}{r} 47.50 \\ \times\ \ \ 2 \\ \hline 95.00 \end{array}$$
Total spent:
$$\begin{array}{r} 95.00 \\ 95.00 \\ \times\ 117.75 \\ \hline 307.75 \end{array}$$

Birthday:
$$\begin{array}{r} 39.25 \\ \times\ \ \ \ 3 \\ \hline 117.75 \end{array}$$
Amount left:
$$\begin{array}{r} 925.75 \\ -\ 307.75 \\ \hline 618.00 \end{array}$$
Ellen has $618.00 left over.

57.
$$\begin{array}{r} 254.2 \\ \times\ \ 19.6 \\ \hline 152\ 52 \\ 2288 \\ 2542 \\ \hline 4982.32\ \text{square yards} \end{array}$$
$$\begin{array}{r} 4982.32 \\ \times\ \ \ \ \ 12.5 \\ \hline 2\ 491\ 160 \\ 9\ 964\ 64 \\ 49\ 823\ 2 \\ \hline \$62,279.000 \end{array}$$
The price of the carpet is $62,279.

59. To multiply by numbers such as 0.1, 0.01, 0.001, and 0.0001, count the number of decimal places in this first number. Then, in the other number, move the decimal point to the left from its present position the same number of decimal places as was in the first number.

Cumulative Review

61.
$$\begin{array}{r} 201\ \ \ \\ 35\overline{)7035} \\ \underline{70}\ \ \ \ \\ 035 \\ \underline{35} \\ 0 \end{array}$$
The answer is 201.

62.
$$\begin{array}{r} 451\ \ \ \ \\ 124\overline{)56,024} \\ \underline{49\ 6}\ \ \ \ \\ 6\ 42 \\ \underline{6\ 20} \\ 224 \\ \underline{124} \\ 100 \end{array}$$
The answer is 451 R 100.

63. $\dfrac{1}{3} \times 1500 = \dfrac{1}{3} \times \dfrac{1500}{1} = \dfrac{1 \times 3 \times 500}{3 \times 1} = 500$

64. $36 \div 3\dfrac{1}{3} = \dfrac{36}{1} \div \dfrac{10}{3}$
$= \dfrac{36}{1} \times \dfrac{3}{10}$
$= \dfrac{2 \times 18 \times 3}{2 \times 5}$
$= \dfrac{54}{5}$ or $10\dfrac{4}{5}$

65.
$$93.6$$
$$-\,77.5$$
$$\overline{16.1}$$
There were 16.1 million or 16,100,000 more pet cats than pet dogs.

66.
$$77.5$$
$$-\,15.0$$
$$\overline{62.5}$$
There were 62.5 million or 62,500,000 more pet dogs than pet birds.

67.
$$77.5 \qquad\qquad 182.9$$
$$15.9 \qquad\qquad -\,122.0$$
$$15.0 \qquad\qquad \overline{60.9}$$
$$+\,13.6$$
$$\overline{122.0}$$
There were 60.9 million or 60,900,000 more pet fish than pet dogs, small animals, birds, and reptiles combined.

68.
$$93.6 \qquad\qquad 186.1$$
$$77.5 \qquad\qquad -\,182.9$$
$$+\,15.0 \qquad\qquad \overline{3.2}$$
$$\overline{186.1}$$
There were 3.2 million or 3,200,000 more pet cats, dogs, and birds combined than pet fish.

Quick Quiz 3.4

1.
$$0.76$$
$$\times\ 0.04$$
$$\overline{0.0304}$$

2.
$$0.128$$
$$\times\ 25.6$$
$$\overline{768}$$
$$640$$
$$2\,56$$
$$\overline{3.2768}$$

3. Move the decimal four places to the right.
$$5.162 \times 10^4 = 51{,}620$$

4. Answers may vary. Possible solution: Count the number of decimal places in each factor and then add, $2 + 3 = 5$. Multiply and then move the decimal point 5 places to the left in the product. $3.45 \times 9.236 = 31.86420$ or 31.8642

How Am I Doing? Sections 3.1–3.4

1. $31.903 =$ thirty-one and nine hundred three thousandths

2. $\dfrac{567}{10{,}000} = 0.0567$

3. $4.09 = 4\dfrac{9}{100}$

4. $0.475 = \dfrac{475}{1000} = \dfrac{475 \div 25}{1000 \div 25} = \dfrac{19}{40}$

5. 1.6, 1.59, 1.61, 1.601
Add zeros to make the comparison easier.
1.600, 1.590, 1.610, 1.601
Rearrange with the smallest first.
1.59, 1.6, 1.601, 1.61

6. 123.49268 rounds to 123.5.
Find the tenths place: 123.49268. The digit to the right of the tenths place is greater than 5. We round up to 123.5 and drop the digits to the right.

7. 8.065447 rounds to 8.0654.
Find the ten-thousandths place: 8.065447. The digit to the right of the ten-thousandths place is less than 5. We round down to 8.0654 and drop the digits to the right.

8. 17.98523 rounds to 17.99.
Find the hundredths place; 17.98523. The digit to the right of the hundredths place is equal to 5. We round up to 17.99 and drop the digits to the right.

9.
$$5.12$$
$$4.70$$
$$8.03$$
$$+\,1.60$$
$$\overline{19.45}$$

10.
$$24.613$$
$$0.273$$
$$+\,2.305$$
$$\overline{27.191}$$

11.
$$42.16$$
$$-\,31.57$$
$$\overline{10.59}$$

12.　26.000
　　　− 18.329
　　────────
　　　　7.671

13.　11.67
　　× 0.03
　────────
　　0.3501

14. Move the decimal three places to the right.
$4.7805 \times 1000 = 4780.5$

15. Move the decimal five places to the right.
$0.0003796 \times 10^5 = 37.96$

16.　3.14
　　× 2.5
　────────
　1 570
　6 28
　────────
　7.850

17.　982
　× 0.007
　────────
　6.874

18.　0.00052
　×　0.006
　────────────
　0.00000312

3.5 Exercises

1.

```
    2.1
  6)12.6
    12
    ──
     6
     6
     ─
     0
```

3.

```
   17.83
  4)71.32
   4
   ──
   31
   28
   ──
    3 3
    3 2
    ───
     12
     12
     ──
      0
```

5.

```
    10.52
  7)73.64
    7
    ──
    3 6
    3 5
    ───
     14
     14
     ──
      0
```

7.

```
      136.5
  0.6∧)81.9∧0
       6
       ──
       21
       18
       ──
        39
        36
        ──
        3 0
        3 0
        ───
          0
```

9.

```
       5.412
  0.05)0.27∧060
       25
       ──
        2 0
        2 0
        ───
          06
           5
          ──
          10
          10
          ──
           0
```

11.

```
        53
  2.9∧)153.7∧
       145
       ───
        8 7
        8 7
        ───
          0
```

13.

```
        18
  3.8∧)68.4∧
       38
       ──
       30 4
       30 4
       ────
          0
```

　　　　　　　　71

15.
$$
\begin{array}{r}
130 \\
0.31_{\wedge}\overline{)40.30_{\wedge}} \\
\underline{31} \\
9\ 3 \\
\underline{9\ 3} \\
0
\end{array}
$$

17.
$$
\begin{array}{r}
5.25 \\
9\overline{)47.31} \\
\underline{45} \\
2\ 3 \\
\underline{1\ 8} \\
51 \\
\underline{45} \\
6
\end{array}
$$
The answer is 5.3.

19.
$$
\begin{array}{r}
1.24 \\
1.9_{\wedge}\overline{)2.3_{\wedge}60} \\
\underline{1\ 9} \\
4\ 6 \\
\underline{3\ 8} \\
80 \\
\underline{76} \\
4
\end{array}
$$
The answer is 1.2.

21.
$$
\begin{array}{r}
49.29 \\
0.85_{\wedge}\overline{)41.90_{\wedge}10} \\
\underline{34\ 0} \\
7\ 90 \\
\underline{7\ 65} \\
25\ 1 \\
\underline{17\ 0} \\
8\ 10 \\
\underline{7\ 65} \\
45
\end{array}
$$
The answer is 49.3.

23.
$$
\begin{array}{r}
94.206 \\
5\overline{)471.030} \\
\underline{45} \\
21 \\
\underline{20} \\
1\ 0 \\
\underline{1\ 0} \\
30 \\
\underline{30} \\
0
\end{array}
$$
The answer is 94.21.

25.
$$
\begin{array}{r}
13.561 \\
1.8_{\wedge}\overline{)24.4_{\wedge}100} \\
\underline{18} \\
6\ 4 \\
\underline{5\ 4} \\
1\ 0\ 1 \\
\underline{9\ 0} \\
1\ 10 \\
\underline{1\ 08} \\
20 \\
\underline{18} \\
2
\end{array}
$$
The answer is 13.56.

27.
$$
\begin{array}{r}
0.210 \\
35\overline{)7.369} \\
\underline{7\ 0} \\
36 \\
\underline{35} \\
19
\end{array}
$$
The answer is 0.21.

29.
$$
\begin{array}{r}
0.0811 \\
7\overline{)0.5681} \\
\underline{56} \\
8 \\
\underline{7} \\
11 \\
\underline{7} \\
4
\end{array}
$$
The answer is 0.081.

31.
```
             91.2643
 0.87ʌ)79.40ʌ0000
       78 3
       ─────
       1 10
         87
       ─────
       23  0
       17  4
       ─────
        5 60
        5 22
        ─────
          380
          348
          ─────
          320
          261
          ─────
           59
```
The answer is 91.264.

33.
```
          123.2
 19)2341ʌ0
    19
    ──
    44
    38
    ──
    61
    57
    ──
     4 0
     3 8
     ───
       2
```
The answer is 123.

35.
```
            213.2
 0.0046ʌ)0.9810ʌ0
         92
         ──
         61
         46
         ──
         150
         138
         ───
         120
          92
         ───
          28
```
The answer is 213.

37.
```
        82.73
 12)992.76
    96
    ──
    32
    24
    ──
     8 7
     8 4
     ───
       36
       36
       ──
        0
```
The Millers will pay $82.73 per month.

39.
```
            27.27
 13.2ʌ)360.0ʌ00
       264
       ───
        96 0
        92 4
        ────
         3 60
         2 64
         ────
          960
          924
          ───
           36
```
It will achieve approximately 27.3 miles per gallon.

41.
```
            24
 12.50ʌ)300.00ʌ
        250 0
        ─────
         50 00
         50 00
         ─────
             0
```
She must sell 24 bouquets.

43.
```
             182
 10.25ʌ)1865.50ʌ
        1025
        ────
         840 5
         820 0
         ─────
          20 50
          20 50
          ─────
              0
```
They had 182 guests.

45.

$$\begin{array}{r} 23.0 \\ 3.8_\wedge\overline{)87.4_\wedge0} \\ \underline{76} \\ 11\ 4 \\ \underline{11\ 4} \\ 0 \end{array}$$

The box contains 23 snowboards. The error was in putting in 1 less snowboard in the box than was required.

47. $0.3 \times n = 9.66$

$$\begin{array}{r} 32.2 \\ 0.3_\wedge\overline{)9.6_\wedge6} \\ \underline{9} \\ 06 \\ \underline{6} \\ 0\ 6 \\ \underline{6} \\ 0 \end{array}$$

$n = 32.2$

49. $1.3 \times n = 1267.5$

$$\begin{array}{r} 975 \\ 1.3_\wedge\overline{)1267.5_\wedge} \\ \underline{117} \\ 97 \\ \underline{91} \\ 65 \\ \underline{65} \\ 0 \end{array}$$

$n = 975$

51. $n \times 0.098 = 4.312$

$$\begin{array}{r} 44 \\ 0.098_\wedge\overline{)4.312_\wedge} \\ \underline{3\ 92} \\ 392 \\ \underline{392} \\ 0 \end{array}$$

$n = 44$

53. Yes, multiplying and dividing the numerator and denominator by 10,000 is the same as multiplying by $\dfrac{10,000}{10,000}$, which is 1.

$$\frac{2.9356}{0.0716} \times \frac{10,000}{10,000} = \frac{29,356}{716}$$

$$\begin{array}{r} 41 \\ 716\overline{)29356} \\ \underline{2864} \\ 716 \\ \underline{716} \\ 0 \end{array}$$

Cumulative Review

54. $\dfrac{3}{8} + 2\dfrac{4}{5} = \dfrac{15}{40} + 2\dfrac{32}{40} = 2\dfrac{47}{40} = 3\dfrac{7}{40}$

55. $2\dfrac{13}{16} - 1\dfrac{7}{8} = 2\dfrac{13}{16} - 1\dfrac{14}{16} = 1\dfrac{29}{16} - 1\dfrac{14}{16} = \dfrac{15}{16}$

56. $3\dfrac{1}{2} \times 2\dfrac{1}{6} = \dfrac{7}{2} \times \dfrac{13}{6} = \dfrac{91}{12}$ or $7\dfrac{7}{12}$

57. $7\dfrac{1}{2} \div \dfrac{1}{2} = \dfrac{15}{2} \div \dfrac{1}{2} = \dfrac{15}{2} \times \dfrac{2}{1} = \dfrac{15}{1} \times \dfrac{1}{1} = 15$

58.

$$\begin{array}{r} 45.0 \\ -\ 12.5 \\ \hline 32.5 \end{array}$$

There was $32.5 billion or $32,500,000,000 more property damage from Hurricane Andrew than Hurricane Hugo.

59.

$$\begin{array}{r} 12.5 \\ -\ 11.1 \\ \hline 1.4 \end{array}$$

There was $1.4 billion or $1,400,000,000 more property damage from Hurricane Hugo than Hurricane Agnes.

60.

$$\begin{array}{r} 3.60 \\ 45\overline{)162.20} \\ \underline{135} \\ 27\ 2 \\ \underline{27\ 0} \\ 20 \end{array}$$

There was about 3.6 times more property damage from Hurricane Katrina than Hurricane Andrew.

61.
$$
\begin{array}{r}
1\ 4.61 \\
11.1_\wedge\overline{)162.2_\wedge00} \\
\underline{111} \\
51\ 2 \\
\underline{44\ 4} \\
6\ 8\ 0 \\
\underline{6\ 6\ 6} \\
1\ 40 \\
\underline{1\ 11} \\
29
\end{array}
$$

There was about 14.6 times more property damage from Hurricane Katrina than Hurricane Agnes.

Quick Quiz 3.5

1.
$$
\begin{array}{r}
0.658 \\
0.07_\wedge\overline{)0.04_\wedge606} \\
\underline{4\ 2} \\
40 \\
\underline{35} \\
56 \\
\underline{56} \\
0
\end{array}
$$

2.
$$
\begin{array}{r}
3.258 \\
0.52_\wedge\overline{)1.69_\wedge416} \\
\underline{1\ 56} \\
13\ 4 \\
\underline{10\ 4} \\
3\ 01 \\
\underline{2\ 60} \\
416 \\
\underline{416} \\
0
\end{array}
$$

3.
$$
\begin{array}{r}
6.580 \\
8\overline{)52.643} \\
\underline{48} \\
4\ 6 \\
\underline{4\ 0} \\
64 \\
\underline{64} \\
03 \\
\end{array}
$$

The answer is 6.58.

4. Answers may vary. Possible solution: Move the decimal point three places to the right in the divisor and dividend. Then divide as whole

numbers, adding zeros to the right of the dividend as needed.
$$
\begin{array}{r}
.2993 \\
0.578_\wedge\overline{)0.173_\wedge0000} \\
\underline{115\ 6} \\
57\ 40 \\
\underline{52\ 02} \\
5\ 380 \\
\underline{5\ 202} \\
1780 \\
\underline{1734} \\
46
\end{array}
$$

So, $0.173 \div 0.578 \approx 0.299$.

3.6 Exercises

1. 0.75 and $\dfrac{3}{4}$ are different ways to express the same quantity.

3. The digits 8942 repeat.

5.
$$
\begin{array}{r}
0.25 \\
4\overline{)1.00} \\
\underline{8} \\
20 \\
\underline{20} \\
0
\end{array}
$$

$\dfrac{1}{4} = 0.25$

7.
$$
\begin{array}{r}
0.8 \\
5\overline{)4.0} \\
\underline{4\ 0} \\
0
\end{array}
$$

$\dfrac{4}{5} = 0.8$

9.
$$
\begin{array}{r}
0.125 \\
8\overline{)1.000} \\
\underline{8} \\
20 \\
\underline{16} \\
40 \\
\underline{40} \\
0
\end{array}
$$

$\dfrac{1}{8} = 0.125$

11. $20\overline{)7.00}$ with quotient 0.35
$$\underline{6\ 0}$$
$$1\ 00$$
$$\underline{1\ 00}$$
$$0$$

$$\frac{7}{20} = 0.35$$

13. $50\overline{)31.00}$ with quotient 0.62
$$\underline{30\ 0}$$
$$1\ 00$$
$$\underline{1\ 00}$$
$$0$$

$$\frac{31}{50} = 0.62$$

15. $4\overline{)9.00}$ with quotient 2.25
$$\underline{8}$$
$$10$$
$$\underline{8}$$
$$20$$
$$\underline{20}$$
$$0$$

$$\frac{9}{4} = 2.25$$

17. $2\frac{7}{8}$ means $2 + \frac{7}{8}$.

$8\overline{)7.000}$ with quotient 0.875
$$\underline{6\ 4}$$
$$60$$
$$\underline{56}$$
$$40$$
$$\underline{40}$$
$$0$$

$$\frac{7}{8} = 0.875 \text{ and } 2\frac{7}{8} = 2.875$$

19. $5\frac{3}{16}$ means $5 + \frac{3}{16}$

$16\overline{)3.0000}$ with quotient 0.1875
$$\underline{1\ 6}$$
$$1\ 40$$
$$\underline{1\ 28}$$
$$120$$
$$\underline{112}$$
$$80$$
$$\underline{80}$$
$$0$$

$$\frac{3}{16} = 0.1875 \text{ and } 5\frac{3}{16} = 5.1875$$

21. $3\overline{)2.000}$ with quotient 0.666
$$\underline{1\ 8}$$
$$20$$
$$\underline{18}$$
$$20$$
$$\underline{18}$$
$$2$$

$$\frac{2}{3} = 0.\overline{6}$$

23. $11\overline{)5.000}$ with quotient 0.454
$$\underline{4\ 4}$$
$$60$$
$$\underline{55}$$
$$50$$
$$\underline{44}$$
$$6$$

$$\frac{5}{11} = 0.\overline{45}$$

25. $3\frac{7}{12}$ means $3+\frac{7}{12}$.

$$
\begin{array}{r}
0.5833 \\
12\overline{)7.0000} \\
6\,0 \\
\hline
1\,00 \\
96 \\
\hline
40 \\
36 \\
\hline
40 \\
36 \\
\hline
4
\end{array}
$$

$\frac{7}{12}=0.58\overline{3}$ and $3\frac{7}{12}=3.58\overline{3}$

27. $4\frac{2}{9}$ means $4+\frac{2}{9}$.

$$
\begin{array}{r}
0.222 \\
9\overline{)2.000} \\
1\,8 \\
\hline
20 \\
18 \\
\hline
20 \\
18 \\
\hline
2
\end{array}
$$

$\frac{2}{9}=0.\overline{2}$ and $4\frac{2}{9}=4.\overline{2}$

29.

$$
\begin{array}{r}
0.3076 \\
13\overline{)4.0000} \\
3\,9 \\
\hline
100 \\
91 \\
\hline
90 \\
78 \\
\hline
12
\end{array}
$$

$\frac{4}{13}$ rounds to 0.308.

31.

$$
\begin{array}{r}
0.9047 \\
21\overline{)19.0000} \\
18\,9 \\
\hline
10 \\
0 \\
\hline
100 \\
84 \\
\hline
160 \\
147 \\
\hline
13
\end{array}
$$

$\frac{19}{21}$ rounds to 0.905.

33.

$$
\begin{array}{r}
0.1458 \\
48\overline{)7.0000} \\
4\,8 \\
\hline
2\,20 \\
1\,92 \\
\hline
280 \\
240 \\
\hline
400 \\
384 \\
\hline
16
\end{array}
$$

$\frac{7}{48}$ rounds to 0.146.

35.

$$
\begin{array}{r}
2.0357 \\
28\overline{)57.0000} \\
56 \\
\hline
1\,00 \\
84 \\
\hline
160 \\
140 \\
\hline
200 \\
196 \\
\hline
4
\end{array}
$$

$\frac{57}{28}$ rounds to 2.036.

37.
$$\begin{array}{r} 0.4038 \\ 52\overline{)21.0000} \\ \underline{20\ 8} \\ 20 \\ \underline{00} \\ 200 \\ \underline{156} \\ 440 \\ \underline{416} \\ 24 \end{array}$$

$\dfrac{21}{52}$ rounds to 0.404.

39.
$$\begin{array}{r} 0.944 \\ 18\overline{)17.0} \\ \underline{16\ 2} \\ 80 \\ \underline{72} \\ 80 \\ \underline{72} \\ 8 \end{array}$$

$\dfrac{17}{18}$ rounds to 0.944.

41.
$$\begin{array}{r} 3.1428 \\ 7\overline{)22.0000} \\ \underline{21} \\ 1\ 0 \\ \underline{7} \\ 30 \\ \underline{28} \\ 20 \\ \underline{14} \\ 60 \\ \underline{56} \\ 4 \end{array}$$

$\dfrac{22}{7}$ rounds to 3.143.

43. $3\dfrac{9}{19}$ means $3 + \dfrac{9}{19}$.

$$\begin{array}{r} 0.4736 \\ 19\overline{)9.0000} \\ \underline{7\ 6} \\ 1\ 40 \\ \underline{1\ 33} \\ 70 \\ \underline{57} \\ 130 \\ \underline{114} \\ 16 \end{array}$$

$\dfrac{9}{19}$ rounds to 0.474, so $3\dfrac{9}{19}$ rounds to 3.474.

45.
$$\begin{array}{r} 0.875 \\ 8\overline{)7.000} \\ \underline{6\ 4} \\ 60 \\ \underline{56} \\ 40 \\ \underline{40} \\ 0 \end{array}$$

$\dfrac{7}{8} = 0.875 < 0.88$

47.
$$\begin{array}{r} 0.0625 \\ 16\overline{)1.0000} \\ \underline{96} \\ 40 \\ \underline{32} \\ 80 \\ \underline{80} \\ 0 \end{array}$$

$0.07 > 0.0625 = \dfrac{1}{16}$

49.

$$16\overline{)5.0000}$$

with quotient 0.3125

$$
\begin{array}{r}
0.3125 \\
16\overline{)5.0000} \\
4\,8 \\
\hline
20 \\
16 \\
\hline
40 \\
32 \\
\hline
80 \\
80 \\
\hline
0
\end{array}
$$

$$\frac{5}{16} = 0.3125$$

$\dfrac{5}{16}$ inch is 0.3125 inch.

51. $9\dfrac{1}{2}$ means $9 + \dfrac{1}{2}$.

$$
\begin{array}{r}
0.5 \\
2\overline{)1.0} \\
1\,0 \\
\hline
0
\end{array}
$$

$$9\frac{1}{2} = 9.5$$

$$
\begin{array}{r}
9\frac{1}{2} \\
-9.31
\end{array}
\qquad
\begin{array}{r}
9.50 \\
-9.31 \\
\hline
0.19
\end{array}
$$

The difference is 0.19 inch.

53.

$$
\begin{array}{r}
0.375 \\
8\overline{)3.000} \\
2\,4 \\
\hline
60 \\
56 \\
\hline
40 \\
40 \\
\hline
0
\end{array}
\qquad
\begin{array}{r}
2.400 \\
-2.375 \\
\hline
0.025
\end{array}
$$

$$2\frac{3}{8} = 2.375 \qquad \text{Yes; it is 0.025 inch}$$

too wide.

55. $2.4 + (0.5)^2 - 0.35 = 2.4 + 0.25 - 0.35$
$$= 2.65 - 0.35$$
$$= 2.30 \text{ or } 2.3$$

57. $2.3 \times 3.2 - 5 \times 0.8 = 7.36 - 4.00 = 3.36$

59. $12 \div 0.03 - 50 \times (0.5 + 1.5)^3 = 12 \div 0.03 - 50 \times (2)^3$
$$= 12 \div 0.03 - 50 \times (8)$$
$$= 400 - 400$$
$$= 0$$

61. $(1.1)^3 + 2.6 \div 0.13 + 0.083 = 1.331 + 20 + 0.083$
$$= 21.414$$

63. $(14.73 - 14.61)^2 \div (1.18 + 0.82) = (0.12)^2 \div 2$
$$= 0.0144 \div 2$$
$$= 0.0072$$

65. $(0.5)^3 + (3 - 2.6) \times 0.5 = (0.5)^3 + 0.4 \times 0.5$
$$= 0.125 + 0.20$$
$$= 0.325$$

67. $(0.76 + 4.24) \div 0.25 + 8.6 = 5.00 \div 0.25 + 8.6$
$$= 20.0 + 8.6$$
$$= 28.6$$

69. $(1.6)^3 + (2.4)^2 + 18.666 \div 3.05 + 4.86$
$$= 4.096 + 5.76 + 6.12 + 4.86$$
$$= 20.836$$

71. From the calculator, $\dfrac{5236}{8921} = 0.586930$.

73. a.
$$
\begin{array}{r}
0.16\overline{16} \\
-0.00\overline{16} \\
\hline
0.16
\end{array}
$$

b.
$$
\begin{array}{r}
0.1616\overline{16} \\
-0.016666\overline{} \\
\hline
0.144949\overline{}
\end{array}
$$

c. (b) is a repeating and (a) is a nonrepeating decimal.

Cumulative Review

75. $\dfrac{25}{2} = 12\dfrac{1}{2}$

$$
\begin{array}{ccc}
12\dfrac{1}{2} & 12\dfrac{2}{4} & 11\dfrac{6}{4} \\[2mm]
-\ 6\dfrac{3}{4} & -\ 6\dfrac{3}{4} & -\ 6\dfrac{3}{4} \\[2mm]
\hline
& & 5\dfrac{3}{4}
\end{array}
$$

The water is $5\dfrac{3}{4}$ feet deep.

76.
$$
\begin{array}{cc}
6\dfrac{1}{2} & 6\dfrac{5}{10} \\[2mm]
+\ 25\dfrac{4}{5} & +\ 25\dfrac{8}{10} \\[2mm]
\hline
& 31\dfrac{13}{10} = 32\dfrac{3}{10}
\end{array}
$$

The depth of the water is $32\dfrac{3}{10}$ feet.

Quick Quiz 3.6

1. $3\dfrac{9}{16}$ means $3 + \dfrac{9}{16}$.

$$
\begin{array}{r}
0.5625 \\
16\overline{)9.0000} \\
\underline{8\ 0} \\
1\ 00 \\
\underline{96} \\
40 \\
\underline{32} \\
80 \\
\underline{80} \\
0
\end{array}
$$

$\dfrac{9}{16} = 0.5625$ and $3\dfrac{9}{16} = 3.5625$

2.
$$
\begin{array}{r}
0.294 \\
17\overline{)5.000} \\
\underline{3\ 4} \\
1\ 60 \\
\underline{1\ 53} \\
70 \\
\underline{68} \\
2
\end{array}
$$

$\dfrac{5}{17}$ rounds to 0.29.

3. $(0.7)^2 + 1.92 \div 0.3 - 0.79$
   $= 0.49 + 1.92 \div 0.3 - 0.79$
   $= 0.49 + 6.4 - 0.79$
   $= 6.89 - 0.79$
   $= 6.1$

4. Answers may vary. Possible solution: Parentheses, exponent, multiplication, subtraction.
$45.78 - (3.42 - 2.09)^2 \times 0.4$
$= 45.78 - 1.33^2 \times 0.4$
$= 45.78 - 1.7689 \times 0.4$
$= 45.78 - 0.70756$
$= 45.07244$

3.7 Exercises

1. $238{,}598{,}980 + 487{,}903{,}870$
   $\approx 200{,}000{,}000 + 500{,}000{,}000$
   $= 700{,}000{,}000$

3. $56{,}789.345 - 33{,}875.125 \approx 60{,}000 - 30{,}000$
   $= 30{,}000$

5. $12{,}638 \times 0.7892 \approx 10{,}000 \times 0.8 = 8000$

7. $879.654 \div 56.82 \approx 900 \div 60 = 15$

9. $11{,}760{,}770 \div 483 \approx 10{,}000{,}000 \div 500 = 20{,}000$
   The average price was about $20,000.

11. Estimate:

```
     500
   ×   6
   ─────
    3000
```

```
     525
   × 5.68
   ──────
    42 00
   315 0
  2625
  ───────
  2982.00
```

She will receive 2982 kroner.

13. Estimate:

```
      50
    × 60
   ─────
    3000
```

```
    48.3
  × 56.9
  ──────
   43 47
  289 8
 2415
 ───────
 2748.27
```

15. Estimate:

```
        10 0
  0.1∧)10∧0
        1
      ─────
        00 0
```

```
       96
 0.12∧)11.52∧
      108
      ────
       72
       72
      ────
        0
```

Hans can make 96 molds.

17. Estimate:

$$10 + 10 + 10 = 30$$
$$30 \div 3 = 10$$

```
   11.68
   10.42
 + 12.67
 ───────
   34.77
```

```
    11.59
  3)34.77
    3
    ─
    4
    3
    ─
    1 7
    1 5
    ───
     27
     27
     ──
      0
```

The average is 11.59 meters of rainfall per year.

19. Estimate:

```
      20
    1)20
      2
      ─
      00
```

```
        24.0
  0.75∧)18.00∧
        15 0
        ────
         3 00
         3 00
         ────
            0
```

A jumbo bag has 24 servings.

21. Estimate:

$$40 + 10 + 60 = 110$$
$$110 \times 10 = 1100$$

```
   43.9
   11.3
 + 63.4
 ──────
  118.6
```

```
   118.6
 × 10.65
 ───────
   5 930
  71 16
 1186 0
 ────────
 1263.090
```

The total cost is $1263.09.

23. Estimate:

$$6 \times 8 \times 9 = 432$$
$$2 \times 9 \times 8 = 144$$
$$400 + 100 = 500$$

```
        6×8×8.50 = 408
 + 1.5×8.50×8 = 102
 ──────────────────
              $510
```

She earned $510 for that week.

25. Estimate: $3 - 0.1 = 2.9$
The year 2015 is 10 years after 2005.
$10 \times 0.011 = 0.11$
The amount of deforestation during this time
period was 0.11 million square kilometer.

$$
\begin{array}{r}
3.413 \\
- 0.110 \\
\hline
3.303
\end{array}
$$

Thus, 3.303 million square kilometers, or
3,303,000 square kilometers, of rainforest will
remain.

27. Estimate:
$300 \times 60 = 18,000$
$18,000 - 10,000 = 8000$

$$
\begin{array}{r}
288.65 \\
\times \quad 60 \\
\hline
17,319
\end{array}
$$

$$
\begin{array}{r}
17,319 \\
- 11,500 \\
\hline
5\,819
\end{array}
$$

He will pay \$17,319 over 5 years.
He will pay \$5819 more than the loan.

29. Estimate:
$10 \div 10 = 1$
$1.3 - 1 = 0.3$

$$
\begin{array}{r}
1.151 \\
7\overline{)8.060} \\
\underline{7} \\
1\ 0 \\
\underline{7} \\
36 \\
\underline{35} \\
10 \\
\underline{7} \\
3
\end{array}
$$

$$
\begin{array}{r}
1.300 \\
- 1.151 \\
\hline
0.149
\end{array}
$$

Yes, by 0.149 milligram per liter.

31. Estimate: $20,000 \div 100 = 200$

$$
\begin{array}{r}
137 \\
126.4_\wedge \overline{)17316.8_\wedge} \\
\underline{1264} \\
4676 \\
\underline{3792} \\
884\ 8 \\
\underline{884\ 8} \\
0
\end{array}
$$

It will take 137 minutes.

33. Estimate: $90 - 70 = 20$

$$
\begin{array}{r}
89.3 \\
- 66.4 \\
\hline
22.9
\end{array}
$$

The consumption was more by
22.9 quadrillion Btu's.

35. Estimate:
$40 + 70 + 80 = 190$
$200 \div 3 = 66.\overline{6}$
$(43.8 + 66.4 + 76.0) \div 3 = 186.2 \div 3 \approx 62.1$
The average consumption for 1960, 1970, and
1980 was approximately 62.1 quadrillion Btu;
62,100,000,000,000,000 Btu.

Cumulative Review

37. $7\dfrac{5}{6} = \dfrac{7 \times 6 + 5}{6} = \dfrac{42 + 5}{6} = \dfrac{47}{6}$

38.
$$
\begin{array}{r}
7 \\
5\overline{)37} \\
\underline{35} \\
2
\end{array}
$$

$\dfrac{37}{5} = 7\dfrac{2}{5}$

39. $\dfrac{7}{25} \times \dfrac{15}{42} = \dfrac{7 \times 3 \times 5}{5 \times 5 \times 2 \times 3 \times 7} = \dfrac{1}{10}$

40. $\dfrac{4}{7} + \dfrac{1}{2} \times \dfrac{2}{3} = \dfrac{4}{7} + \dfrac{1}{3} = \dfrac{12}{21} + \dfrac{7}{21} = \dfrac{19}{21}$

Quick Quiz 3.7

1. Total Rainfall: Difference:
   1.23   8.50
   2.58   − 7.48
   <u>3.67</u>   ─────────
   7.48 inches   1.02 inches
   The rainfall was 1.02 inches less.

2. Distance traveled:
   87,929.2
   − 87,569.2
   ─────────────
      360.0 miles

$$
\begin{array}{r}
23.22 \\
15.5_\wedge\overline{)360.0_\wedge00} \\
\underline{310} \\
50\ 0 \\
\underline{46\ 5} \\
3\ 5\ 0 \\
\underline{3\ 1\ 0} \\
4\ 00 \\
\underline{3\ 10} \\
90
\end{array}
$$

   They achieved 23.2 miles per gallon.

3.   275.50
   ×   36
   ───────────
      1653 00
      8265 0
   ───────────
      9918.00
   He will make $9918 in car payments.

      9918
   − 8000
   ──────────
      1918
   He will pay back $1918 more than the original amount.

4. Answers may vary. Possible solution:
   Divide 24.7 by 1.3.
   $24.7 \div 1.3 = 19$
   They will need 19 boxes.

Use Math To Save Money

1. SHELL: $4.55
   ARCO: $4.43 + 0.45 = $4.88

2. SHELL: 3($4.55) = $13.65
   ARCO: 3($4.43) + $0.45 = $13.29 + $0.45
   $ = \$13.74$

3. SHELL: 4($4.55) = $18.20
   ARCO: 4($4.43) + $0.45 = $17.72 + $0.45
   $ = \$18.17$

4. SHELL: 10($4.55) = $45.50
   ARCO: 10($4.43) + $0.45 = $44.30 + $0.45
   $ = \$44.75$

5. $4.55x = 4.43x + 0.45$
   $0.12x = 0.45$
   $x = 3.75$
   The price is the same for 3.75 gallons of gas.

6. For less than four gallons, the SHELL station is less expensive.

7. For more than four gallons, the ARCO station is less expensive.

8. Answers will vary.

9. Answers will vary.

10. Answers will vary.

You Try It

1. 332.194 = three hundred thirty-two and one hundred ninety-four thousandths

2. $\dfrac{54}{1000} = 0.054$

3. $0.844 = \dfrac{844}{1000} = \dfrac{4 \times 211}{4 \times 250} = \dfrac{211}{250}$

4. 5.73 = 5.730; 5.7 = 5.700
   5.7, 5.713, 5.73, 5.735

5. a. 1.354 rounds to 1.35.
      Find the hundredths place: 1.3<u>5</u>4. The digit to the right of the hundredths place is less than 5. We round down to 1.35 and drop the digits to the right.

   b. 9.077641 rounds to 9.0776.
      Find the ten-thousandths place: 9.077<u>6</u>41. The digit to the right of the ten-thousandths place is less than 5. We round down to 9.0776 and drop the digits to the right.

6. a.   24.350
          0.017
       + 9.440
       ──────────
         33.807

b.   172.200
          $-$ 42.186
          130.014

7. **a.**   0.95
                 0.3
                 0.285

   b.   1.225
                 2.8
                 9800
                 2 450
                 3.4300 = 3.43

8. **a.** Move the decimal point two places to the right.
$7.93 \times 100 = 793$

   b. Move the decimal point three places to the right.
$0.00015 \times 1000 = 0.15$

   c. Move the decimal point three places to the right.
$0.3125 \times 10^3 = 312.5$

   d. Move the decimal point four places to the right.
$0.4119 \times 10^4 = 4119$

   e. Move the decimal point five places to the right.
$2.375 \times 10^5 = 237,500$

9. **a.**
$$0.04_\wedge \overline{)0.15_\wedge 2} \quad \begin{array}{r} 3\,.8 \\ \hline \end{array}$$
          12
          3 2
          3 2
          0

   b.
$$0.007_\wedge \overline{)13.300_\wedge} \quad 1\,900.$$
          7
          6 3
          6 3
          0

10. **a.**
$$25\overline{)11.00} \quad 0.44$$
          10 0
          1 00
          1 00
          0

$$\frac{11}{25} = 0.44$$

   b.
$$14\overline{)9.0000} \quad 0.6428$$
          8 4
          60
          56
          40
          28
          120
          112
          8

$\frac{9}{14}$ rounds to 0.643.

11. $3.5 - (0.3)^2 \div 0.45 + (9.5 - 8.1)$
$= 3.5 - (0.3)^2 \div 0.45 + 1.4$
$= 3.5 - 0.09 \div 0.45 + 1.4$
$= 3.5 - 0.2 + 1.4$
$= 3.3 + 1.4$
$= 4.7$

Chapter 3 Review Problems

1. 13.672 = thirteen and six hundred seventy-two thousandths

2. 0.00084 = eighty-four hundred-thousandths

3. $\dfrac{7}{10} = 0.7$

4. $\dfrac{81}{100} = 0.81$

5. $1\dfrac{523}{1000} = 1.523$

6. $\dfrac{79}{10,000} = 0.0079$

7. $0.17 = \dfrac{17}{100}$

8. $0.036 = \dfrac{36}{1000} = \dfrac{9}{250}$

9. $34.24 = 34\dfrac{24}{100} = 34\dfrac{6}{25}$

10. $1.00025 = 1\dfrac{25}{100,000} = 1\dfrac{1}{4000}$

11. Since $\dfrac{9}{100} = 0.09,\ 2\dfrac{9}{100} = 2 + 0.09 = 2.09.$

$2\dfrac{9}{100} = 2.09$

12. $0.716 > 0.706$
The hundredths place is the first digit where the numbers differ. Since $1 > 0$, 0.716 is the greater number.

13. $\dfrac{65}{100} < 0.655$

This is true because $\dfrac{65}{100} = 0.65$ and

$0.65 < 0.655.$

14. 0.981, 0.918, 0.98, 0.901
Add zeros to make the comparison easier.
0.981, 0.918, 0.980, 0.901
Rearrange with the smallest first.
0.901, 0.918, 0.98, 0.981

15. 5.62, 5.2, 5.6, 5.26, 5.59
Add zeros to make the comparison easier.
5.62, 5.20, 5.60, 5.26, 5.59
Rearrange with the smallest first.
5.2, 5.26, 5.59, 5.6, 5.62

16. 2.36, 2.3, 2.362, 2.302
Add zeros to make the comparison easier.
2.360, 2.300, 2.362, 2.302
Rearrange with the smallest first.
2.3, 2.302, 2.36, 2.362

17. 0.613 rounds to 0.6
Find the tenths place: 0.613. The digit to the right of the tenths place is less than 5. We round down to 0.6 and drop the digits to the right.

18. 19.2076 rounds to 19.21.
Find the hundredths place: 19.2076. The digit to the right of the hundredths place is greater than 5. We round up to 19.21 and drop the digits to the right.

19. 9.85215 rounds to 9.8522.
Find the ten-thousandths place: 9.85215. The digit to the right of the ten-thousandths place is equal to 5. We round up to 9.8522 and drop the digit to the right.

20. $156.48 rounds to $156.

21.
```
   9.6
  11.5
  21.8
+ 34.7
------
  77.6
```

22.
```
   2.50
  32.70
 116.94
+  0.67
-------
 152.81
```

23.
```
  17.030
-  2.448
-------
  14.582
```

24.
```
 182.422
- 68.550
-------
 113.872
```

25.
```
    0.098
 ×  0.032
 --------
    0196
    0294
 --------
 0.003136
```

26.
```
 126.83
 ×    7
 ------
 887.81
```

27.
```
  7053
 × 0.34
 ------
 282 12
 2115 9
 ------
 2398.02
```

28. Move the decimal point three places to the right.
$0.000613 \times 10^3 = 0.613$

29. Move the decimal point five places to the right.
$1.2354 \times 10^5 = 123,540$

30. $\begin{array}{r} 3.49 \\ \times\ 2.5 \\ \hline 1\ 745 \\ 6\ 98 \\ \hline 8.725 \end{array}$

The cost is \$8.73.

31. $5.2_\wedge\overline{)191.3_\wedge 6}$
$\begin{array}{r} 36.8 \\ \hline 156 \\ \hline 35\ 3 \\ 31\ 2 \\ \hline 4\ 1\ 6 \\ 4\ 1\ 6 \\ \hline 0 \end{array}$

32. $8\overline{)1863.2}$
$\begin{array}{r} 232.9 \\ \hline 16 \\ \hline 26 \\ 24 \\ \hline 23 \\ 16 \\ \hline 7\ 2 \\ 7\ 2 \\ \hline 0 \end{array}$

33. $1.3_\wedge\overline{)746.7_\wedge 50}$
$\begin{array}{r} 574.42 \\ \hline 65 \\ \hline 96 \\ 91 \\ \hline 57 \\ 52 \\ \hline 5\ 5 \\ 5\ 2 \\ \hline 30 \\ 26 \\ \hline 4 \end{array}$

574.42 rounds to 574.4.

34. $0.06_\wedge\overline{)0.00_\wedge 3539}$
$\begin{array}{r} 0.0589 \\ \hline 30 \\ \hline 53 \\ 48 \\ \hline 59 \\ 54 \\ \hline 5 \end{array}$

0.0589 rounds to 0.059.

35. $12\overline{)11.0000}$
$\begin{array}{r} 0.9166 \\ \hline 10\ 8 \\ \hline 20 \\ 12 \\ \hline 80 \\ 72 \\ \hline 80 \\ 72 \\ \hline 8 \end{array}$

$\dfrac{11}{12} = 0.91\overline{6}$

36. $20\overline{)17.00}$
$\begin{array}{r} 0.85 \\ \hline 160 \\ \hline 1\ 00 \\ 1\ 00 \\ \hline 0 \end{array}$

$\dfrac{17}{20} = 0.85$

37. $1\dfrac{5}{6}$ means $1 + \dfrac{5}{6}$.

$6\overline{)5.000}$
$\begin{array}{r} 0.833 \\ \hline 4\ 8 \\ \hline 20 \\ 18 \\ \hline 20 \\ 18 \\ \hline 2 \end{array}$

$\dfrac{5}{6} = 0.8\overline{3}$ and $1\dfrac{5}{6} = 1.8\overline{3}$

38.

$$
\begin{array}{r}
0.7857 \\
14\overline{)11.0000} \\
9\ 8 \\
\hline
1\ 20 \\
1\ 12 \\
\hline
80 \\
70 \\
\hline
100 \\
98 \\
\hline
2
\end{array}
$$

$\dfrac{11}{14}$ rounds to 0.786.

39.

$$
\begin{array}{r}
0.3448 \\
29\overline{)10.0000} \\
8\ 7 \\
\hline
1\ 30 \\
1\ 16 \\
\hline
140 \\
116 \\
\hline
240 \\
232 \\
\hline
8
\end{array}
$$

$\dfrac{10}{29}$ rounds to 0.345.

40.

$$
\begin{array}{r}
0.3913 \\
23\overline{)9.0000} \\
6\ 9 \\
\hline
2\ 10 \\
2\ 07 \\
\hline
30 \\
23 \\
\hline
70 \\
69 \\
\hline
1
\end{array}
$$

$\dfrac{9}{23}$ rounds to 0.391, and $3\dfrac{9}{23}$ rounds to 3.391.

41. $2.3 \times 1.82 + 3 \times 5.12 = 4.186 + 15.36 = 19.546$

42. $3.57 - (0.4)^3 \times 2.5 \div 5 = 3.57 - 0.064 \times 2.5 \div 5$
$$= 3.57 - 0.16 \div 5$$
$$= 3.57 - 0.032$$
$$= 3.538$$

43. $2.4 \div (2 - 1.6)^2 + 8.13 = 2.4 \div (0.4)^2 + 8.13$
$$= 2.4 \div 0.16 + 8.13$$
$$= 15 + 8.13$$
$$= 23.13$$

44.

$$
\begin{array}{r}
2398.26 \\
- 1959.07 \\
\hline
439.19
\end{array}
$$

45. $32.15 \times 0.02 \times 10^2 = 32.15 \times 0.02 \times 100$
$$= 0.643 \times 100$$
$$= 64.3$$

46. $1.809 - 0.62 + 3.27 = 1.189 + 3.27 = 4.459$

47.

$$
\begin{array}{r}
0.904 \\
2.3_{\wedge}\overline{)2.0_{\wedge}792} \\
2\ 0\ 7 \\
\hline
92 \\
92 \\
\hline
0
\end{array}
$$

48. $8 \div 0.4 + 0.1 \times (0.2)^2 = 20 + 0.1 \times 0.04$
$$= 20 + 0.004$$
$$= 20.004$$

49. $(3.8 - 2.8)^3 \div (0.5 + 0.3) = 1^3 \div 0.8 = 1 \div 0.8 = 1.25$

50. Tickets $= 228 + 2.5 \times 388 + 3 \times 430$
$$= 228 + 970 + 1290$$
$$= 2488$$
Not tickets $= 2600 - 2488 = 112$
112 people still have not received their tickets.

51.

$$
\begin{array}{r}
26325.8 \\
- 26005.8 \\
\hline
320.0
\end{array}
$$

$$
\begin{array}{r}
24.80 \\
12.9_{\wedge}\overline{)320.0_{\wedge}00} \\
258 \\
\hline
62\ 0 \\
51\ 6 \\
\hline
10\ 4\ 0 \\
10\ 3\ 2 \\
\hline
80 \\
0 \\
\hline
80
\end{array}
$$

His car got 24.8 miles per gallon.

52.

$$
\begin{array}{r}
0.0025 \\
12\overline{)0.0300} \\
\underline{24} \\
60 \\
\underline{60} \\
0
\end{array}
$$

$$
\begin{array}{r}
0.0025 \\
-\ 0.0020 \\
\hline
0.0005
\end{array}
$$

No; it is not safe by 0.0005 milligram per liter.

53.

$$
\begin{array}{r}
15.748 \\
2.54\overline{)40.00000} \\
\underline{25\ 4} \\
14\ 60 \\
\underline{12\ 70} \\
1\ 900 \\
\underline{1\ 778} \\
1220 \\
\underline{1016} \\
2040 \\
\underline{2032} \\
8
\end{array}
$$

This measurement was 15.75 inches.

54. a. Fence $= 2 \times 18.3 + 2 \times 9.6$
$\qquad = 36.6 + 19.2$
$\qquad = 55.8$
He needs 55.8 feet.

b.

$$
\begin{array}{r}
18.3 \\
\times\ 9.6 \\
\hline
10\ 98 \\
164\ 7 \\
\hline
175.68
\end{array}
$$

The area is 175.68 square feet.

55.

$$
\begin{array}{r}
75.5 \\
\times\ 18.5 \\
\hline
37\ 75 \\
604\ 0 \\
755 \\
\hline
1396.75
\end{array}
$$

The area is 1396.75 square feet.

56. Galeton to Wellsboro
$5.7 + 18.4 = 24.1$ miles
Coudersport to Gaines
$16.3 + 8.2 + 5.7 = 30.2$ miles
Difference
$30.2 - 24.1 = 6.1$
The distance is 6.1 miles longer.

57.

$$
\begin{array}{r}
118.9 \\
25.6 \\
18.9 \\
43.9 \\
22.6 \\
13.8 \\
+\ 16.2 \\
\hline
259.9
\end{array}
$$

It is 259.9 feet around the field.

58.

$$
\begin{array}{r}
212.50 \\
\times\ \ \ \ 60 \\
\hline
\$12,750.00
\end{array}
$$

$$
\begin{array}{r}
199.50 \\
\times\ \ \ \ 60 \\
\hline
11,970.00 \\
+\ \ \ \ 285.00 \\
\hline
\$12,255.00
\end{array}
$$

They should change to the new loan.

59.

$$
\begin{array}{r}
950 \\
-\ 720 \\
\hline
230
\end{array}
$$

The average monthly benefit increased \$230 from 1995 to 2005.

60.

$$
\begin{array}{r}
31.666 \\
30\overline{)950.000} \\
\underline{90} \\
50 \\
\underline{30} \\
20\ 0 \\
\underline{18\ 0} \\
2\ 00 \\
\underline{1\ 80} \\
200 \\
\underline{180} \\
20
\end{array}
$$

$\dfrac{950}{30}$ rounds to 31.67.

The average daily benefit was \$31.67 in 2005.

61.
$$
\begin{array}{r}
1164 \\
-\ 720 \\
\hline
444
\end{array}
$$

The average monthly benefit increased $444 from 1995 to 2010.

$1164 + 444 = 1608$

$$
\begin{array}{r}
53.6 \\
30\overline{)1608.0} \\
\underline{150} \\
108 \\
\underline{90} \\
18\ 0 \\
\underline{18\ 0} \\
0
\end{array}
$$

The average daily benefit will be $53.60 in 2025.

62.
$$
\begin{array}{r}
1164 \\
-\ 810 \\
\hline
354
\end{array}
$$

The average monthly benefit increased $354 from 2000 to 2010.

$1164 + 354 = 1518$

$$
\begin{array}{r}
50.6 \\
30\overline{)1518.0} \\
\underline{150} \\
18\ 0 \\
\underline{18\ 0} \\
0
\end{array}
$$

The average daily benefit will be $50.60 in 2020.

How Am I Doing? Chapter 3 Test

1. $12.043 =$ twelve and forty-three thousandths

2. $\dfrac{3977}{10,000} = 0.3977$

3. $7.\dot{1}5 = 7\dfrac{15}{100} = 7\dfrac{3}{20}$

4. $0.261 = \dfrac{261}{1000}$

5. 2.19, 2.91, 2.9, 2.907
Add zeros to make the comparison easier.
2.190, 2.910, 2.900, 2.907
Rearrange with the smallest first.
2.19, 2.9, 2.907, 2.91

6. 78.6562 rounds to 78.66.
Find the hundredths place: 78.6$\underline{5}$62. The digit to the right of the hundredths place is greater than 5. We round up to 78.66 and drop the digits to the right.

7. 0.0341752 rounds to 0.0342.
Find the ten-thousandths place: 0.034$\underline{1}$752. The digit to the right of the ten-thousandths place is greater than 5. We round up to 0.0342 and drop the digits to the right.

8.
$$
\begin{array}{r}
96.200 \\
1.348 \\
+\ 2.150 \\
\hline
99.698
\end{array}
$$

9.
$$
\begin{array}{r}
17.00 \\
2.10 \\
16.80 \\
0.04 \\
+\ 1.59 \\
\hline
37.53
\end{array}
$$

10.
$$
\begin{array}{r}
1.0075 \\
-\ 0.9096 \\
\hline
0.0979
\end{array}
$$

11.
$$
\begin{array}{r}
72.300 \\
-\ 1.145 \\
\hline
71.155
\end{array}
$$

12.
$$
\begin{array}{r}
8.31 \\
\times\ 0.07 \\
\hline
0.5817
\end{array}
$$

13. Move the decimal three places to the right.
$2.189 \times 10^3 = 2189$

14.
$$
\begin{array}{r}
0.1285 \\
0.08_\wedge\overline{)0.01_\wedge0280} \\
\underline{8} \\
22 \\
\underline{16} \\
68 \\
\underline{64} \\
40 \\
\underline{40} \\
0
\end{array}
$$

15.

$$\begin{array}{r} 47 \\ 0.69_\wedge \overline{\smash{)}32.43_\wedge} \\ \underline{27\ 6} \\ 4\ 83 \\ \underline{4\ 83} \\ 0 \end{array}$$

16.

$$\begin{array}{r} 1.2 \\ 9\overline{\smash{)}11.0} \\ \underline{9} \\ 2\ 0 \\ \underline{1\ 8} \\ 2 \end{array}$$

$$\frac{11}{9} = 1.\overline{2}$$

17.

$$\begin{array}{r} 0.875 \\ 8\overline{\smash{)}7.000} \\ \underline{6\ 4} \\ 60 \\ \underline{56} \\ 40 \\ \underline{40} \\ 0 \end{array}$$

$$\frac{7}{8} = 0.875$$

18. $(0.3)^3 + 1.02 \div 0.5 - 0.58$
$= 0.027 + 1.02 \div 0.5 - 0.58$
$= 0.027 + 2.04 - 0.58$
$= 2.067 - 0.58$
$= 1.487$

19. $19.36 \div (0.24 + 0.26) \times (0.4)^2 = 19.36 \div 0.5 \times 0.16$
$\qquad\qquad\qquad\qquad\qquad = 19.36 \div 0.5 \times 0.16$
$\qquad\qquad\qquad\qquad\qquad = 38.72 \times 0.16$
$\qquad\qquad\qquad\qquad\qquad = 6.1952$

20.

$$\begin{array}{r} 3.17 \\ \times\ \ 8.5 \\ \hline 1\ 585 \\ 25\ 36 \\ \hline 26.945 \end{array}$$

Peter spent $26.95.

21.

$$\begin{array}{r} 42780.5 \\ -\ 42620.5 \\ \hline 160.0 \end{array}$$

$$\begin{array}{r} 18.82 \\ 8.5\overline{\smash{)}160.000} \\ \underline{85} \\ 75\ 0 \\ \underline{68\ 0} \\ 7\ 00 \\ \underline{6\ 80} \\ 200 \\ \underline{170} \\ 30 \end{array}$$

His car achieved 18.8 miles per gallon.

22.

$$\begin{array}{r} 8.01 \\ 5.03 \\ +\ \ 8.53 \\ \hline 21.57 \end{array}$$

$$\begin{array}{r} 25.00 \\ -\ 21.57 \\ \hline 3.43 \end{array}$$

It is 3.43 centimeters less.

23. Time $= 40 + 1.5 \times 9 = 40 + 13.5 = 53.5$ hours
Salary $= \$7.30 \times 53.5 = \$390.55$
She earned $390.55.

Cumulative Test for Chapters 1–3

1. 38,056,954 = thirty-eight million, fifty-six thousand, nine hundred fifty-four

2.

$$\begin{array}{r} 156,028 \\ 301,579 \\ +\ \ 21,980 \\ \hline 479,587 \end{array}$$

3.

$$\begin{array}{r} 1,091,000 \\ -\ 1,036,520 \\ \hline 54,480 \end{array}$$

4.

$$\begin{array}{r} 589 \\ \times\ \ \ 67 \\ \hline 4\ 123 \\ 35\ 34 \\ \hline 39,463 \end{array}$$

5.
$$\begin{array}{r} 1200 \\ \times\quad 40 \\ \hline 48{,}000 \end{array}$$

6.
$$\begin{array}{r} 316 \\ 15\overline{)4740} \\ \underline{45} \\ 24 \\ \underline{15} \\ 90 \\ \underline{90} \\ 0 \end{array}$$

7. $20 \div 4 + 2^5 - 7 \times 3 = 20 \div 4 + 32 - 7 \times 3$
$$= 5 + 32 - 21$$
$$= 37 - 21$$
$$= 16$$

8. Locate the digit to the right of the thousands place: 236,813. Since this digit is 5 or greater, round up to 237,000.

9. $58,216 \times 438,207 \approx 60,000 \times 400,000$
$$= 24,000,000,000$$

10. $\dfrac{18}{45} = \dfrac{18 \div 9}{45 \div 9} = \dfrac{2}{5}$

11. $4\dfrac{3}{8} = \dfrac{4 \times 8 + 3}{8} = \dfrac{32 + 3}{8} = \dfrac{35}{8}$

12. $2\dfrac{1}{5} \times 3\dfrac{1}{3} = \dfrac{11}{5} \times \dfrac{10}{3} = \dfrac{11 \times 2 \times 5}{5 \times 3} = \dfrac{22}{3}$ or $7\dfrac{1}{3}$

13.
$$\begin{array}{r} 5\dfrac{3}{8} \\ + 2\dfrac{11}{12} \\ \hline \end{array} \qquad \begin{array}{r} 5\dfrac{9}{24} \\ + 2\dfrac{22}{24} \\ \hline 9\dfrac{31}{24} = 8\dfrac{7}{24} \end{array}$$

14. $\dfrac{23}{35} - \dfrac{2}{5} = \dfrac{23}{35} - \dfrac{2}{5} \times \dfrac{7}{7} = \dfrac{23}{35} - \dfrac{14}{35} = \dfrac{9}{35}$

15. $\dfrac{5}{16} \times \dfrac{4}{5} + \dfrac{9}{10} \times \dfrac{2}{3} = \dfrac{1}{4} + \dfrac{3}{5} = \dfrac{5}{20} + \dfrac{12}{20} = \dfrac{17}{20}$

16. $52 \div 3\dfrac{1}{4} = 52 \div \dfrac{13}{4} = \dfrac{52}{1} \times \dfrac{4}{13} = 16$

17. $1\dfrac{3}{8} \div \dfrac{5}{12} = \dfrac{11}{8} \div \dfrac{5}{12}$
$$= \dfrac{11}{8} \times \dfrac{12}{5}$$
$$= \dfrac{11 \times 4 \times 3}{4 \times 2 \times 5}$$
$$= \dfrac{33}{10}$$
$$= 3\dfrac{3}{10}$$

18. $\dfrac{39}{1000} = 0.039$

19. 2.1, 20.1, 2.01, 2.12, 2.11
Add zeros to make the comparison easier.
2.10, 20.10, 2.01, 2.12, 2.11
Rearrange with the smallest first.
2.01, 2.1, 2.11, 2.12, 20.1

20. 26.07984 rounds to 26.080.
Find the thousandths place: 26.07984. The digit to the right of the thousandths place is greater than 5. We round up to 26.080 and drop the digits to the right.

21.
$$\begin{array}{r} 3.126 \\ 8.400 \\ 10.330 \\ + \quad 0.090 \\ \hline 21.946 \end{array}$$

22.
$$\begin{array}{r} 28.100 \\ - 14.982 \\ \hline 13.118 \end{array}$$

23.
$$\begin{array}{r} 28.7 \\ \times 0.05 \\ \hline 1.435 \end{array}$$

24. Move the decimal point three places to the right.
$0.1823 \times 1000 = 182.3$

25.

$$
\begin{array}{r}
1.058 \\
0.06_\wedge \overline{)0.06_\wedge 348} \\
\underline{6} \\
3 \\
\underline{0} \\
34 \\
\underline{30} \\
48 \\
\underline{48} \\
0
\end{array}
$$

26.

$$
\begin{array}{r}
0.8125 \\
16\overline{)13.0000} \\
\underline{12\ 8} \\
20 \\
\underline{16} \\
40 \\
\underline{32} \\
80 \\
\underline{80} \\
0
\end{array}
$$

$$\frac{13}{16} = 0.8125$$

27. a.

$$
\begin{array}{r}
10.5 \\
\times\ 10.5 \\
\hline
5\ 25 \\
105\ 0 \\
\hline
110.25
\end{array}
$$

The area is 110.25 square feet.

b.

$$
\begin{array}{r}
10.5 \\
\times\ \ 4 \\
\hline
42.0
\end{array}
$$

The perimeter is 42 feet.

28.

$$
\begin{array}{r}
60 \\
320.50_\wedge \overline{)19,230.00_\wedge} \\
\underline{19,230.0} \\
0
\end{array}
$$

It will take 60 months.

Chapter 4

1. A <u>ratio</u> is a comparison of two quantities that have the same units.

3. The ratio 5 : 8 is read <u>5 to 8</u>.

5. $6:18 = \dfrac{6}{18} = \dfrac{6 \div 6}{18 \div 6} = \dfrac{1}{3}$

7. $21:18 = \dfrac{21}{18} = \dfrac{21 \div 3}{18 \div 3} = \dfrac{7}{6}$

9. $150:225 = \dfrac{150}{225} = \dfrac{150 \div 75}{225 \div 75} = \dfrac{2}{3}$

11. $165 \text{ to } 90 = \dfrac{165}{90} = \dfrac{165 \div 15}{90 \div 15} = \dfrac{11}{6}$

13. $60 \text{ to } 72 = \dfrac{60}{72} = \dfrac{60 \div 12}{72 \div 12} = \dfrac{5}{6}$

15. $28 \text{ to } 42 = \dfrac{28}{42} = \dfrac{28 \div 14}{42 \div 14} = \dfrac{2}{3}$

17. $32 \text{ to } 20 = \dfrac{32}{20} = \dfrac{32 \div 4}{20 \div 4} = \dfrac{8}{5}$

19. $8 \text{ ounces to } 12 \text{ ounces} = \dfrac{8}{12} = \dfrac{8 \div 4}{12 \div 4} = \dfrac{2}{3}$

21. $39 \text{ kilograms to } 26 \text{ kilograms} = \dfrac{39}{26}$
$$= \dfrac{39 \div 13}{26 \div 13}$$
$$= \dfrac{3}{2}$$

23. $\$75 \text{ to } \$95 = \dfrac{75}{95} = \dfrac{75 \div 5}{95 \div 5} = \dfrac{15}{19}$

25. $312 \text{ yards to } 24 \text{ yards} = \dfrac{312}{24} = \dfrac{312 \div 24}{24 \div 24} = \dfrac{13}{1}$

27. $2\dfrac{1}{2} \text{ pounds to } 4\dfrac{1}{4} \text{ pounds} = \dfrac{2\frac{1}{2}}{4\frac{1}{4}}$
$$= 2\dfrac{1}{2} \div 4\dfrac{1}{4}$$
$$= \dfrac{5}{2} \div \dfrac{17}{4}$$
$$= \dfrac{5}{2} \times \dfrac{4}{17}$$
$$= \dfrac{10}{17}$$

29. $\dfrac{165}{285} = \dfrac{165 \div 15}{285 \div 15} = \dfrac{11}{19}$

31. $\dfrac{35}{165} = \dfrac{35 \div 5}{165 \div 5} = \dfrac{7}{33}$

33. $\dfrac{205}{1225} = \dfrac{205 \div 5}{1225 \div 5} = \dfrac{41}{245}$

35. $\dfrac{450}{205} = \dfrac{450 \div 5}{205 \div 5} = \dfrac{90}{41}$

37. $\dfrac{44}{704} = \dfrac{44 \div 44}{704 \div 44} = \dfrac{1}{16}$

39. $\dfrac{\$42}{12 \text{ pairs of socks}} = \dfrac{\$42 \div 6}{12 \text{ pairs of socks} \div 6}$
$$= \dfrac{\$7}{2 \text{ pairs of socks}}$$

41. $\dfrac{\$170}{12 \text{ bushes}} = \dfrac{\$170 \div 2}{12 \text{ bushes} \div 2} = \dfrac{\$85}{6 \text{ bushes}}$

43. $\dfrac{\$114}{12 \text{ CDs}} = \dfrac{\$114 \div 3}{12 \text{ CDs} \div 3} = \dfrac{\$38}{4 \text{ CDs}} = \dfrac{\$19}{2 \text{ CDs}}$

45. $\dfrac{6150 \text{ rev}}{15 \text{ miles}} = \dfrac{6150 \text{ rev} \div 15}{15 \text{ miles} \div 15}$
$$= \dfrac{410 \text{ rev}}{1 \text{ mile}}$$
$$= 410 \text{ rev/mile}$$

47. $\dfrac{\$330,000}{12 \text{ employees}} = \dfrac{\$330,000 \div 12}{12 \text{ employees} \div 12}$

$\quad = \dfrac{\$27,500}{1 \text{ employee}}$

$\quad = \$27,500/\text{employee}$

49. $\dfrac{\$600}{40 \text{ hours}} = \dfrac{\$600 \div 40}{40 \text{ hours} \div 40} = \$15/\text{hr}$

51. $\dfrac{308 \text{ miles}}{11 \text{ gallons}} = \dfrac{308 \text{ miles} \div 11}{11 \text{ gallons} \div 11} = 28 \text{ mi/gal}$

53. $\dfrac{1120 \text{ people}}{16 \text{ square miles}} = \dfrac{1120 \text{ people} \div 16}{16 \text{ square miles} \div 16}$

$\quad = 70 \text{ people/sq mi}$

55. $\dfrac{840 \text{ books}}{12 \text{ libraries}} = \dfrac{840 \text{ books} \div 12}{12 \text{ libraries} \div 12}$

$\quad = 70 \text{ books/library}$

57. $\dfrac{297 \text{ miles}}{4.5 \text{ hours}} = \dfrac{297 \text{ miles} \div 4.5}{4.5 \text{ hours} \div 4.5} = 66 \text{ mi/hour}$

59. $\dfrac{475 \text{ patients}}{25 \text{ doctors}} = \dfrac{475 \text{ patients} \div 25}{25 \text{ doctors} \div 25}$

$\quad = 19 \text{ patients/doctor}$

61. $\dfrac{84 \text{ geraniums}}{28 \text{ pots}} = \dfrac{84 \text{ geraniums} \div 28}{28 \text{ pots} \div 28}$

$\quad = 3 \text{ geraniums/pot}$

63. $\dfrac{\$2970}{135 \text{ shares}} = \$22/\text{share}$

The cost was $22/share.

65. Profit = $1485 − $1080 = $405

$\dfrac{\$405 \text{ profit}}{90 \text{ puppets}} = \$4.50 \text{ profit per puppet}$

She made a profit of $4.50 per puppet.

67. a. 16-ounce box: $\dfrac{\$1.28}{16 \text{ ounces}} = \$0.08 \text{ oz}$

24-ounce box: $\dfrac{\$1.68}{24 \text{ ounces}} = \$0.07/\text{oz}$

b. $\begin{array}{r} 0.08 \\ -\ 0.07 \\ \hline 0.01 \end{array}$

You save $0.01/ounce.

c. $48(0.01) = 0.48$
The consumer saves $0.48.

69. a. $\dfrac{3978 \text{ moose}}{306 \text{ acres}} = 13 \text{ moose/acre}$
There were 13 moose/acre on the North Slope.

b. $\dfrac{5520 \text{ moose}}{460 \text{ acres}} = 12 \text{ moose/acre}$
There were 12 moose/acre on the South Slope.

c. The moose are more closely crowded together on the North Slope.

71. a. $\dfrac{\$14,332.50}{350 \text{ shares}} = \$40.95/\text{share}$
She paid $40.95/share.

b. $\dfrac{\$11,088}{210 \text{ shares}} = \$52.80/\text{share}$
He paid $52.80/share.

c. $\begin{array}{r} \$52.80 \\ -\ \$40.95 \\ \hline \$11.85/\text{share} \end{array}$
He paid $11.85/share more.

73. Design: $\dfrac{750 \text{ meters per second}}{330 \text{ meters per second}} = \text{Mach } 2.3$

Modify: $\dfrac{810 \text{ meters per second}}{330 \text{ meters per second}} = \text{Mach } 2.5$

$\begin{array}{r} 2.5 \\ -\ 2.3 \\ \hline 0.2 \end{array}$

It was increased by Mach 0.2.

Cumulative Review

75. $\begin{array}{r} 2\frac{1}{4} \\ +\ \frac{3}{8} \\ \hline \end{array}$ $\qquad$ $\begin{array}{r} 2\frac{2}{8} \\ +\ \frac{3}{8} \\ \hline 2\frac{5}{8} \end{array}$

76. $\dfrac{5}{7} \div \dfrac{3}{21} = \dfrac{5}{7} \times \dfrac{21}{3} = \dfrac{5 \times 21}{7 \times 3} = 5$

77. $\dfrac{3}{5} \times \dfrac{5}{8} - \dfrac{2}{3} \times \dfrac{1}{4} = \dfrac{3}{8} - \dfrac{2}{12} = \dfrac{9}{24} - \dfrac{4}{24} = \dfrac{5}{24}$

78. $3\dfrac{1}{16} - 2\dfrac{1}{24} = 3\dfrac{3}{48} - 2\dfrac{2}{48} = 1\dfrac{1}{48}$

79. $12 \times 5.2 = 62.4$ sq yd

$\dfrac{\$764.40}{62.4 \text{ sq yd}} = \$12.25/\text{sq yd}$

The cost of the carpet is \$12.25/square yard.

80. $1050 \times \$23 = \$24,150$ total
$1050 \times (39 - 23) = 1050 \times 16 = \$16,800$ profit
The store paid \$24,150 and the profit was \$16,800.

Quick Quiz 4.1

1. 51 to $85 = \dfrac{51}{85} = \dfrac{51 \div 17}{85 \div 17} = \dfrac{3}{5}$

2. $\dfrac{1700 \text{ square feet}}{55 \text{ pounds}} = \dfrac{1700 \text{ square feet} \div 5}{55 \text{ pounds} \div 5}$
$= \dfrac{340 \text{ square feet}}{11 \text{ pounds}}$

3. $\dfrac{462 \text{ trees}}{17 \text{ acres}} = 27.18$ trees/acre

4. Answers may vary. Possible solution: Divide number of cans by number of people. Then simplify.
$\dfrac{663 \text{ cans}}{231 \text{ people}} = \dfrac{221 \text{ cans}}{77 \text{ people}}$

4.2 Exercises

1. A proportion states that two ratios or rates are <u>equal</u>.

3. 6 is to 8 as 3 is to 4.
$\dfrac{6}{8} = \dfrac{3}{4}$

5. 20 is to 36 as 5 is to 9.
$\dfrac{20}{36} = \dfrac{5}{9}$

7. 220 is to 11 as 400 is to 20.
$\dfrac{220}{11} = \dfrac{400}{20}$

9. $4\dfrac{1}{3}$ is to 13 as $5\dfrac{2}{3}$ is to 17.

$\dfrac{4\frac{1}{3}}{13} = \dfrac{5\frac{2}{3}}{17}$

11. 6.5 is to 14 as 13 is to 28.
$\dfrac{6.5}{14} = \dfrac{13}{28}$

13. $\dfrac{3 \text{ inches}}{40 \text{ miles}} = \dfrac{27 \text{ inches}}{360 \text{ miles}}$

15. $\dfrac{\$40}{12 \text{ cars}} = \dfrac{\$60}{18 \text{ cars}}$

17. $\dfrac{3 \text{ hours}}{\$525} = \dfrac{7 \text{ hours}}{\$1225}$

19. $\dfrac{3 \text{ teaching assistants}}{40 \text{ children}} = \dfrac{21 \text{ teaching assistants}}{280 \text{ children}}$

21. $\dfrac{4800 \text{ people}}{3 \text{ restaurants}} = \dfrac{11,200 \text{ people}}{7 \text{ restaurants}}$

23. $\dfrac{10}{25} \overset{?}{=} \dfrac{6}{15}$
$10 \times 15 \overset{?}{=} 25 \times 6$
$150 = 150$
It is a proportion.

25. $\dfrac{8}{10} \overset{?}{=} \dfrac{13}{15}$
$8 \times 15 \overset{?}{=} 10 \times 13$
$120 \neq 130$
It is not a proportion.

27. $\dfrac{17}{75} \overset{?}{=} \dfrac{22}{100}$
$17 \times 100 \overset{?}{=} 75 \times 22$
$1700 \neq 1650$
It is not a proportion.

29. $\dfrac{102}{120} \overset{?}{=} \dfrac{85}{100}$
$102 \times 100 \overset{?}{=} 120 \times 85$
$10,200 = 10,200$
It is a proportion.

31.
$$\frac{2.5}{4} \overset{?}{=} \frac{7.5}{12}$$
$$2.5 \times 12 \overset{?}{=} 4 \times 7.5$$
$$30 = 30$$
It is a proportion.

33.
$$\frac{3}{17} \overset{?}{=} \frac{4.5}{24.5}$$
$$3 \times 24.5 \overset{?}{=} 17 \times 4.5$$
$$73.5 \neq 76.5$$
It is not a proportion.

35.
$$\frac{6}{2\frac{1}{2}} \overset{?}{=} \frac{12}{5}$$
$$6 \times 5 \overset{?}{=} 2\frac{1}{2} \times 12$$
$$30 \overset{?}{=} \frac{5}{2} \times 12$$
$$30 = 30$$
It is a proportion.

37.
$$\frac{7\frac{1}{3}}{3} \overset{?}{=} \frac{23}{9}$$
$$7\frac{1}{3} \times 9 \overset{?}{=} 3 \times 23$$
$$66 \neq 69$$
It is not a proportion.

39.
$$\frac{\frac{1}{4}}{2} \overset{?}{=} \frac{\frac{7}{20}}{2.8}$$
$$\frac{1}{4} \times 2.8 \overset{?}{=} 2 \times \frac{7}{20}$$
$$0.7 = 0.7$$
It is a proportion.

41.
$$\frac{135 \text{ miles}}{3 \text{ hours}} \overset{?}{=} \frac{225 \text{ miles}}{5 \text{ hours}}$$
$$135 \times 5 \overset{?}{=} 3 \times 225$$
$$675 = 675$$
It is a proportion.

43.
$$\frac{166 \text{ gallons}}{14 \text{ acres}} \overset{?}{=} \frac{249 \text{ gallons}}{21 \text{ acres}}$$
$$166 \times 21 \overset{?}{=} 14 \times 249$$
$$3486 = 3486$$
It is a proportion.

45.
$$\frac{21 \text{ homeruns}}{96 \text{ games}} \overset{?}{=} \frac{18 \text{ homeruns}}{81 \text{ games}}$$
$$21 \times 81 \overset{?}{=} 96 \times 18$$
$$1701 \neq 1728$$
It is not a proportion.

47.
$$\frac{22}{132} \overset{?}{=} \frac{32}{160}$$
$$22 \times 160 \overset{?}{=} 132 \times 32$$
$$3520 \neq 4224 \quad \text{no}$$
The ratio is not the same for both schools.

49. a.
$$\frac{675}{18} \overset{?}{=} \frac{820}{20}$$
$$675 \times 20 \overset{?}{=} 18 \times 820$$
$$13,500 \neq 14,760 \quad \text{no}$$
They do not travel at the same rate.

b. The van traveled at a faster rate.

51.
$$\frac{75 \text{ feet}}{20 \text{ feet}} \overset{?}{=} \frac{105 \text{ feet}}{28 \text{ feet}}$$
$$75 \times 28 \overset{?}{=} 20 \times 105$$
$$2100 = 2100 \quad \text{Yes}$$
They have the same ratio.

53. a.
$$\frac{169}{221} = \frac{169 \div 13}{221 \div 13} = \frac{13}{17}$$
$$\frac{247}{323} = \frac{247 \div 19}{323 \div 19} = \frac{13}{17}$$
True
Yes, the fractions are equal.

b.
$$\frac{169}{221} \overset{?}{=} \frac{247}{323}$$
$$169 \times 323 \overset{?}{=} 221 \times 247$$
$$54,587 = 54,587 \quad \text{True}$$

c. The equality test for fractions; for most students it is faster to multiply than to reduce fractions.

Cumulative Review

54. $9.6 + 7.8 + 2.56 + 3.004 + 0.1765 = 23.1405$

55.
$$\begin{array}{r} 3.04 \\ \times\ 5.92 \\ \hline 608 \\ 2\ 736 \\ 15\ 20 \\ \hline 17.9968 \end{array}$$

56.
$$\begin{array}{r} 29,366,215 \\ -\ 28,963,807 \\ \hline 402,408 \end{array}$$

57.
$$\begin{array}{r} 25.8 \\ 7.03_\wedge \overline{\smash{\big)}\,181.37_\wedge 4} \\ \underline{140\ 6} \\ 40\ 77 \\ \underline{35\ 15} \\ 5\ 62\ 4 \\ \underline{5\ 62\ 4} \\ 0 \end{array}$$

58.

$$\begin{array}{r} 3\frac{1}{4} \\ +\ 4\frac{3}{8} \\ \hline \end{array} \qquad \begin{array}{r} 3\frac{2}{8} \\ +\ 4\frac{3}{8} \\ \hline 7\frac{5}{8} \end{array}$$

$$\begin{array}{r} 20 \\ -\ 7\frac{5}{8} \\ \hline \end{array} \qquad \begin{array}{r} 19\frac{8}{8} \\ -\ 7\frac{5}{8} \\ \hline 12\frac{3}{8} \end{array}$$

She needs to walk $12\frac{3}{8}$ miles.

Quick Quiz 4.2

1. 8 is to 18 as 28 is to 63.
$$\frac{8}{18} = \frac{28}{63}$$

2. 13 is to 32 as $3\frac{1}{4}$ is to 8.

$$\frac{13}{32} = \frac{3\frac{1}{4}}{8}$$

3. $\dfrac{15 \text{ shots}}{4 \text{ goals}} \overset{?}{=} \dfrac{75 \text{ shots}}{22 \text{ goals}}$

$15 \times 22 \overset{?}{=} 4 \times 75$

$330 \neq 300$

It is not a proportion.

4. Answers may vary. Possible solution: Cross multiply. If the products are equal, it is a proportion.

$33 \times 225 \overset{?}{=} 45 \times 165$

$7425 = 7425$

Yes, it is a proportion.

How Am I Doing? Sections 4.1–4.2

1. 13 to $18 = \dfrac{13}{18}$

2. 44 to $220 = \dfrac{44}{220} = \dfrac{1}{5}$

3. \$72 to $\$16 = \dfrac{72}{16} = \dfrac{9}{2}$

4. 135 meters to 165 meters $= \dfrac{135}{165} = \dfrac{9}{11}$

5. **a.** $\dfrac{\$60}{\$330} = \dfrac{2}{11}$

     b. $\dfrac{\$32}{\$330} = \dfrac{16}{165}$

6. $\dfrac{9}{300} = \dfrac{3 \text{ flight attendants}}{100 \text{ passengers}}$

7. $\dfrac{620}{840} = \dfrac{31 \text{ gallons}}{42 \text{ square feet}}$

8. $\dfrac{65}{4} = 16.25$ miles/hour

9. $\dfrac{\$630}{18} = \$35$ per MP3 player

10. $\dfrac{2400}{15} = 160$ cookies/pound of cookie dough

11. 13 is to 40 as 39 is to 120.
$$\frac{13}{40} = \frac{39}{120}$$

12. 116 is to 148 as 29 is to 37.
$$\frac{116}{148} = \frac{29}{37}$$

13. $\dfrac{33 \text{ nautical miles}}{2 \text{ hours}} = \dfrac{49.5 \text{ nautical miles}}{3 \text{ hours}}$

14. $\dfrac{3000 \text{ shoes}}{\$370} = \dfrac{7500 \text{ shoes}}{\$925}$

15. $\dfrac{14}{31} \overset{?}{=} \dfrac{42}{93}$
$14(93) \overset{?}{=} 31(42)$
$1302 = 1302$
It is a proportion.

16. $\dfrac{17}{33} \overset{?}{=} \dfrac{19}{45}$
$45(17) \overset{?}{=} 33(19)$
$765 \neq 627$
It is not a proportion.

17. $\dfrac{6.5}{4.8} \overset{?}{=} \dfrac{120}{96}$
$6.5 \times 96 \overset{?}{=} 4.8 \times 120$
$624 \neq 576$
It is not a proportion.

18. $\dfrac{15}{24} \overset{?}{=} \dfrac{1\frac{5}{8}}{2\frac{3}{5}}$
$15 \times 2\frac{3}{5} \overset{?}{=} 24 \times 1\frac{5}{8}$
$39 = 39$
It is a proportion.

19. $\dfrac{670}{1541} \overset{?}{=} \dfrac{820}{1886}$
$670(1886) \overset{?}{=} 1541(820)$
$1,263,620 = 1,263,620$
It is a proportion.

20. $\dfrac{60}{8} \overset{?}{=} \dfrac{3000}{400}$
$60(400) \overset{?}{=} 8(3000)$
$24,000 = 24,000$
It is a proportion.

4.3 Exercises

1. Divide each side of the equation by the number
a. Calculate $\dfrac{b}{a}$. The value of n is $\dfrac{b}{a}$.

3. $8 \times n = 72$
$\dfrac{8 \times n}{8} = \dfrac{72}{8}$
$n = 9$

5. $3 \times n = 16.8$
$\dfrac{3 \times n}{3} = \dfrac{16.8}{3}$
$n = 5.6$

7. $n \times 6.7 = 134$
$\dfrac{n \times 6.7}{6.7} = \dfrac{134}{6.7}$
$n = 20$

9. $50.4 = 6.3 \times n$
$\dfrac{50.4}{6.3} = \dfrac{6.3 \times n}{6.3}$
$8 = n$

11. $\dfrac{4}{9} \times n = 22$
$\dfrac{\frac{4}{9} \times n}{\frac{4}{9}} = \dfrac{22}{\frac{4}{9}}$
$n = 22 \div \dfrac{4}{9}$
$n = 22 \times \dfrac{9}{4}$
$n = \dfrac{198}{4}$
$n = 49\frac{1}{2}$

13. $\dfrac{n}{20} = \dfrac{3}{4}$
$n \times 4 = 20 \times 3$
$\dfrac{n \times 4}{4} = \dfrac{60}{4}$
$n = 15$
Check: $\dfrac{15}{20} \overset{?}{=} \dfrac{3}{4}$
$4 \times 15 \overset{?}{=} 20 \times 3$
$60 = 60$

15.

$$\frac{6}{n} = \frac{3}{8}$$

$$6 \times 8 = n \times 3$$

$$\frac{48}{3} = \frac{n \times 3}{3}$$

$$16 = n$$

Check: $\frac{6}{16} \stackrel{?}{=} \frac{3}{8}$

$$6 \times 8 \stackrel{?}{=} 16 \times 3$$

$$48 = 48$$

17.

$$\frac{12}{40} = \frac{n}{25}$$

$$12 \times 25 = 40 \times n$$

$$\frac{300}{40} = \frac{40 \times n}{40}$$

$$7.5 = n$$

Check: $\frac{12}{40} \stackrel{?}{=} \frac{7.5}{25}$

$$12 \times 25 \stackrel{?}{=} 40 \times 7.5$$

$$300 = 300$$

19.

$$\frac{50}{100} = \frac{2.5}{n}$$

$$50 \times n = 100 \times 2.5$$

$$\frac{50 \times n}{50} = \frac{250}{50}$$

$$n = 5$$

Check: $\frac{50}{100} \stackrel{?}{=} \frac{2.5}{5}$

$$50 \times 5 \stackrel{?}{=} 100 \times 2.5$$

$$250 = 250$$

21.

$$\frac{n}{6} = \frac{150}{12}$$

$$n \times 12 = 6 \times 150$$

$$\frac{n \times 12}{12} = \frac{900}{12}$$

$$n = 75$$

Check: $\frac{75}{6} \stackrel{?}{=} \frac{150}{12}$

$$75 \times 12 \stackrel{?}{=} 6 \times 150$$

$$900 = 900$$

23.

$$\frac{15}{4} = \frac{n}{6}$$

$$15 \times 6 = 4 \times n$$

$$\frac{90}{4} = \frac{4 \times n}{4}$$

$$22.5 = n$$

Check: $\frac{15}{4} \stackrel{?}{=} \frac{22.5}{6}$

$$15 \times 6 \stackrel{?}{=} 4 \times 22.5$$

$$90 = 90$$

25.

$$\frac{240}{n} = \frac{5}{4}$$

$$240 \times 4 = n \times 5$$

$$960 = n \times 5$$

$$\frac{960}{5} = \frac{n \times 5}{5}$$

$$n = 192$$

Check: $\frac{240}{192} \stackrel{?}{=} \frac{5}{4}$

$$240 \times 4 \stackrel{?}{=} 192 \times 5$$

$$960 = 960$$

27.

$$\frac{21}{n} = \frac{2}{3}$$

$$21 \times 3 = n \times 2$$

$$\frac{63}{2} = \frac{n \times 2}{2}$$

$$31.5 = n$$

29.

$$\frac{9}{26} = \frac{n}{52}$$

$$9 \times 52 = 26 \times n$$

$$\frac{468}{26} = \frac{26 \times n}{26}$$

$$18 = n$$

31.

$$\frac{15}{12} = \frac{10}{n}$$

$$15 \times n = 12 \times 10$$

$$\frac{15 \times n}{15} = \frac{120}{15}$$

$$n = 8$$

33.

$$\frac{n}{36} = \frac{4.5}{1}$$

$$n \times 1 = 36 \times 4.5$$

$$n = 162$$

35.
$$\frac{1.8}{n} = \frac{0.7}{12}$$
$$1.8 \times 12 = 0.7n$$
$$\frac{21.6}{0.7} = \frac{0.7n}{0.7}$$
$$30.9 \approx n$$

37.
$$\frac{11}{12} = \frac{n}{32.8}$$
$$11 \times 32.8 = 12 \times n$$
$$\frac{360.8}{12} = \frac{12 \times n}{12}$$
$$n \approx 30.1$$

39.
$$\frac{13.8}{15} = \frac{n}{6}$$
$$13.8 \times 6 = 15 \times n$$
$$82.8 = 15 \times n$$
$$\frac{82.8}{15} = \frac{15 \times n}{15}$$
$$5.5 \approx n$$

41.
$$\frac{3}{n} = \frac{6\frac{1}{4}}{100}$$
$$3 \times 100 = 6\frac{1}{4} \times n$$
$$\frac{300}{\frac{25}{4}} = \frac{\frac{25}{4} \times n}{\frac{25}{4}}$$
$$48 = n$$

43.
$$\frac{n \text{ pounds}}{20 \text{ ounces}} = \frac{2 \text{ pounds}}{32 \text{ ounces}}$$
$$32 \times n = 20 \times 2$$
$$\frac{32 \times n}{32} = \frac{40}{32}$$
$$n = 1.25$$

45.
$$\frac{145 \text{ kilometers}}{2 \text{ hours}} = \frac{220 \text{ kilometers}}{n \text{ hours}}$$
$$145 \times n = 2 \times 220$$
$$\frac{145 \times n}{145} = \frac{440}{145}$$
$$n \approx 3.03$$

47.
$$\frac{32 \text{ meters}}{5 \text{ yards}} = \frac{24 \text{ meters}}{n \text{ yards}}$$
$$32 \times n = 5 \times 24$$
$$\frac{32 \times n}{32} = \frac{120}{32}$$
$$n = 3.75$$

49.
$$\frac{36.4 \text{ feet}}{5 \text{ meters}} = \frac{n \text{ feet}}{12 \text{ meters}}$$
$$36.4 \times 12 = 5 \times n$$
$$\frac{436.8}{5} = \frac{5 \times n}{5}$$
$$n = 87.36$$

51.
$$\frac{35 \text{ dimes}}{3.5 \text{ dollars}} = \frac{n \text{ dimes}}{8 \text{ dollars}}$$
$$35 \times 8 = 3.5 \times n$$
$$\frac{280}{3.5} = \frac{3.5 \times n}{3.5}$$
$$80 = n$$

53.
$$\frac{3\frac{1}{4} \text{ feet}}{8 \text{ pounds}} = \frac{n \text{ feet}}{12 \text{ pounds}}$$
$$3\frac{1}{4} \times 12 = 8 \times n$$
$$\frac{39}{8} = \frac{8 \times n}{8}$$
$$4\frac{7}{8} = n$$

55.
$$\frac{n \text{ inches}}{5 \text{ inches}} = \frac{6.6 \text{ inches}}{3 \text{ inches}}$$
$$3 \times n = 5 \times 6.6$$
$$\frac{3 \times n}{3} = \frac{33}{3}$$
$$n = 11$$
The enlargement will be 11 inches wide.

57.
$$\frac{n}{2\frac{1}{3}} = \frac{4\frac{5}{6}}{3\frac{1}{9}}$$
$$3\frac{1}{9} \times n = 2\frac{1}{3} \times 4\frac{5}{6}$$
$$\frac{28}{9} \times n = \frac{7}{3} \times \frac{29}{6}$$
$$\frac{\frac{28}{9} \times n}{\frac{28}{9}} = \frac{\frac{203}{18}}{\frac{28}{9}}$$
$$n = \frac{203}{18} \times \frac{9}{28}$$
$$n = \frac{29}{8} \text{ or } 3\frac{5}{8}$$

59.

$$\frac{8\frac{1}{6}}{n} = \frac{5\frac{1}{2}}{7\frac{1}{3}}$$

$$7\frac{1}{3} \times 8\frac{1}{6} = 5\frac{1}{2} \times n$$

$$\frac{22}{3} \times \frac{49}{6} = \frac{11}{2} \times n$$

$$\frac{\frac{539}{9}}{\frac{11}{2}} = \frac{\frac{11}{2} \times n}{\frac{11}{2}}$$

$$\frac{539}{9} \times \frac{2}{11} = n$$

$$\frac{98}{9} = 10\frac{8}{9} = n$$

Cumulative Review

60. $4^3 + 20 \div 5 + 6 \times 3 - 5 \times 2$

$= 64 + 20 \div 5 + 6 \times 3 - 5 \times 2$

$= 64 + 4 + 18 - 10$

$= 76$

61. $(3+1)^3 - 30 \div 6 - 144 \div 12$

$= 4^3 - 30 \div 6 - 144 \div 12$

$= 64 - 30 \div 6 - 144 \div 12$

$= 64 - 5 - 12$

$= 47$

62. $0.563 =$ five hundred sixty-three thousandths

63. thirty-four ten-thousandths $= 0.0034$

64. Purchase:

$$\begin{array}{r} 156 \\ \times\ 32 \\ \hline 312 \\ 468 \\ \hline 4992 \end{array}$$

Sales:

$$\begin{array}{r} 78 \\ \times\ 45 \\ \hline 390 \\ 312 \\ \hline 3510 \end{array} \qquad \begin{array}{r} 78 \\ \times\ 39 \\ \hline 702 \\ 234 \\ \hline 3042 \end{array}$$

Profit $= 3510 + 3042 - 4992 = 1560$
He will make a profit of $1560.

65. Team 1 will play each of the other 7 teams 2 times = 14. Team 2 will play each of the remaining 6 teams 2 times = 12. Team 3 will play each of the remaining 5 teams = 10.
$14 + 12 + 10 + 8 + 6 + 4 + 2 = 56$
A total of 56 games will have been played.

Quick Quiz 4.3

1. $\dfrac{n}{26} = \dfrac{9}{130}$

$n \times 130 = 26 \times 9$

$\dfrac{n \times 130}{130} = \dfrac{234}{130}$

$n = 1.8$

2. $\dfrac{8}{6} = \dfrac{2\frac{2}{3}}{n}$

$8 \times n = 6 \times 2\frac{2}{3}$

$\dfrac{8 \times n}{8} = \dfrac{16}{8}$

$n = 2$

3. $\dfrac{17 \text{ hits}}{93 \text{ pitches}} = \dfrac{n \text{ hits}}{62 \text{ pitches}}$

$17 \times 62 = 93 \times n$

$\dfrac{1054}{93} = \dfrac{93 \times n}{93}$

$n \approx 11.3$

4. Answers may vary. Possible solution: Cross multiply. Divide both sides by $2\frac{1}{2}$. Then simplify.

$$\frac{2\frac{1}{2}}{3\frac{3}{4}} = \frac{16\frac{1}{2}}{n}$$

$$2\frac{1}{2} \times n = 3\frac{3}{4} \times 16\frac{1}{2}$$

$$2\frac{1}{2} \times n = \frac{15}{4} \times \frac{33}{2}$$

$$\frac{2\frac{1}{2} \times n}{2\frac{1}{2}} = \frac{\frac{495}{8}}{2\frac{1}{2}}$$

$$n = \frac{495}{8} \div \frac{5}{2}$$

$$n = \frac{495}{8} \times \frac{2}{5}$$

$$n = \frac{99}{4} \text{ or } 24\frac{3}{4}$$

4.4 Exercises

1. He should continue with people on the top of the fraction. That would be 60 people that he observed on Saturday night. He does not know the number of dogs, so this would be *n*. The proportion would be

$$\frac{12 \text{ people}}{5 \text{ dogs}} = \frac{60 \text{ people}}{n \text{ dogs}}.$$

3. $\dfrac{140 \text{ sold}}{23 \text{ returned}} = \dfrac{980 \text{ sold}}{n \text{ returned}}$

$140 \times n = 23 \times 980$

$\dfrac{140 \times n}{n} = \dfrac{22{,}540}{140}$

$n = 161$

Approximately 161 cars will be brought back for repairs.

5. $\dfrac{\frac{3}{4} \text{ cup}}{1 \text{ gallon}} = \dfrac{n \text{ cups}}{4 \text{ gallons}}$

$\dfrac{3}{4} \times 4 = 1 \times n$

$3 = n$

He will need 3 cups of bleach.

7. $\dfrac{n \text{ kilometers}}{5 \text{ miles}} = \dfrac{1\frac{1}{2} \text{ kilometers}}{1 \text{ mile}}$

$n \times 1 = 5 \times 1\dfrac{1}{2}$

$n = 7\dfrac{1}{2}$

Approximately $7\dfrac{1}{2}$ kilometers are in 5 miles.

9. $\dfrac{230 \text{ Indian rupees}}{5 \text{ U.S. dollars}} = \dfrac{n \text{ Indian rupees}}{120 \text{ U.S. dollars}}$

$230 \times 120 = 5 \times n$

$\dfrac{27{,}600}{5} = \dfrac{5 \times n}{5}$

$5520 = n$

Anna received 5520 Indian rupees.

11. $\dfrac{6.5 \text{ feet}}{5 \text{ feet}} = \dfrac{n \text{ feet}}{152 \text{ feet}}$

$6.5 \times 152 = 5 \times n$

$\dfrac{988}{5} = \dfrac{5 \times n}{5}$

$197.6 = n$

The stadium is 197.6 feet tall.

13. $\dfrac{3 \text{ inches}}{125 \text{ miles}} = \dfrac{5.2 \text{ inches}}{n \text{ miles}}$

$3 \times n = 125 \times 5.2$

$\dfrac{3 \times n}{3} = \dfrac{650}{3}$

$n = 216.\overline{6}$

The approximate distance between the two beaches is 217 miles.

15. $\dfrac{3 \text{ cups}}{8 \text{ people}} = \dfrac{n \text{ cups}}{34 \text{ people}}$

$3 \times 34 = 8 \times n$

$\dfrac{102}{8} = \dfrac{8 \times n}{8}$

$n = 12\dfrac{3}{4}$

He will need to make $12\dfrac{3}{4}$ cups of curry sauce.

17. $\dfrac{17 \text{ made}}{25 \text{ free throws}} = \dfrac{n \text{ made}}{150 \text{ free throws}}$

$17 \times 150 = 25 \times n$

$\dfrac{2250}{25} = \dfrac{25 \times n}{25}$

$102 = n$

They will make 102 free throws.

19. $\dfrac{n \text{ gallons}}{600 \text{ miles}} = \dfrac{6 \text{ gallons}}{192 \text{ miles}}$

$192 \times n = 600 \times 6$

$\dfrac{192 \times n}{192} = \dfrac{3600}{192}$

$n = 18.75$

She will use 18.75 gallons.

21. $\dfrac{24 \text{ tagged}}{n \text{ total}} = \dfrac{12 \text{ tagged}}{20 \text{ total}}$

$24 \times 20 = 12 \times n$

$\dfrac{480}{12} = \dfrac{12 \times n}{12}$

$40 = n$

The estimate is 40 hawks.

23. $\dfrac{425 \text{ pounds}}{3 \text{ acres}} = \dfrac{n \text{ pounds}}{14 \text{ acres}}$ $1983\dfrac{1}{3}$ pounds

$425 \times 14 = 3 \times n$

$\dfrac{5950}{3} = \dfrac{3 \times n}{3}$ $1983\dfrac{1}{3} \times 1.8 = 3570$

$1983.\overline{3} = n$

The Grants will get \$3570 from the crop.

25. $\dfrac{5 \text{ defective}}{100 \text{ made}} = \dfrac{n \text{ defective}}{5400 \text{ made}}$

$5 \times 5400 = 100 \times n$

$\dfrac{27{,}000}{100} = \dfrac{100 \times n}{n}$

$270 = n$

The company should expect 270 defective chips.

27. Water: $\dfrac{n \text{ cups}}{3 \text{ servings}} = \dfrac{2 \text{ cups}}{6 \text{ servings}}$

$n \times 6 = 3 \times 2$

$\dfrac{n \times 6}{6} = \dfrac{6}{6}$

$n = 1$

1 cup of water

Milk: $\dfrac{n \text{ cups}}{3 \text{ servings}} = \dfrac{\frac{3}{4} \text{ cup}}{6 \text{ servings}}$

$n \times 6 = 3 \times \dfrac{3}{4}$

$n \times 6 = \dfrac{9}{4}$

$\dfrac{n \times 6}{6} = \dfrac{\frac{9}{4}}{6}$

$n = \dfrac{9}{4} \div 6 = \dfrac{9}{4} \times \dfrac{1}{6} = \dfrac{3}{8}$

$\dfrac{3}{8}$ cup of milk

You need 1 cup of water and $\dfrac{3}{8}$ cup of milk.

29. Water: $\dfrac{n \text{ cups}}{8 \text{ servings}} = \dfrac{2 \text{ cups}}{6 \text{ servings}}$

$n \times 6 = 8 \times 2$

$\dfrac{n \times 6}{6} = \dfrac{16}{6}$

$n = \dfrac{16}{6} = \dfrac{8}{3}$

High altitude: $\dfrac{8}{3} \times 1\dfrac{1}{4} = \dfrac{8}{3} \times \dfrac{5}{4} = \dfrac{10}{3} = 3\dfrac{1}{3}$

$3\dfrac{1}{3}$ cups of water

Milk: $\dfrac{n \text{ cups}}{8 \text{ servings}} = \dfrac{\frac{3}{4} \text{ cups}}{6 \text{ servings}}$

$n \times 6 = 8 \times \dfrac{3}{4}$

$n \times 6 = 6$

$\dfrac{n \times 6}{6} = \dfrac{6}{6}$

$n = 1$

1 cup of milk

You need 2 cups of water and 1 cup of milk.

31. Pujols: $\dfrac{\$14{,}427{,}326}{47 \text{ home runs}} \approx \$306{,}964/\text{home run}$

Howard: $\dfrac{\$15{,}000{,}000}{45 \text{ home runs}} \approx \$333{,}333/\text{home run}$

Albert Pujols, approximately \$306,964 for each home run; Ryan Howard, approximately \$333,333 for each home run.

33. Bryant:

$\dfrac{\$23{,}034{,}375}{617 \text{ two-point shots}} \approx \$37{,}333/\text{two-point shot}$

Duncan:

$\dfrac{\$22{,}183{,}218}{559 \text{ two-point shots}} \approx \$39{,}684/\text{two-point shot}$

Kobe Bryant, approximately \$37,333 for each two-point shot; Tim Duncan, approximately \$39,684 for each two-point shot.

Cumulative Review

35. $\begin{array}{r} 5.75 \\ \times\ \ \ 3 \\ \hline 17.25 \end{array}$

He spent \$17.25.

36. $\begin{array}{r} 12.5 \\ \times\ \ 2.8 \\ \hline 10\ 00 \\ 25\ 0\ \ \\ \hline 35.00 \end{array}$

She spent \$35.

37. 56.148 rounds to 56.1

38. 2.7489<u>5</u> rounds to 2.7490

39. a. $1\dfrac{3}{16}\times\dfrac{4}{5}=\dfrac{19}{16}\times\dfrac{4}{5}=\dfrac{76}{80}=\dfrac{19}{20}$

One frame requires $\dfrac{19}{20}$ of a square foot.

b. $\dfrac{19}{20}\times1500=19\times75=1452$

A total of 1452 square feet are needed.

Quick Quiz 4.4

1. $\dfrac{36\text{ feet}}{160\text{ pounds}}=\dfrac{54\text{ feet}}{n\text{ pounds}}$

$36\times n=160\times54$

$\dfrac{36\times n}{36}=\dfrac{8640}{36}$

$n=240$

It will weigh 240 pounds.

2. $\dfrac{11\text{ inches}}{64\text{ miles}}=\dfrac{5\text{ inches}}{n\text{ miles}}$

$11\times n=64\times5$

$\dfrac{11\times n}{11}=\dfrac{320}{11}$

$n\approx29.09$

5 inches represents 29.09 miles.

3. $\dfrac{7}{16}=\dfrac{n}{100}$

$7\times100=16\times n$

$\dfrac{700}{16}=\dfrac{16\times n}{16}$

$n=43.75$ free throws

He will expect to make 44 free throws.

4. Answers may vary. Possible solution: Write a proportion and solve for *n*.

$\dfrac{70\text{ euros}}{104\text{ dollars}}=\dfrac{n\text{ euros}}{400\text{ dollars}}$

$70\times400=104\times n$

$28,000=104\times n$

$\dfrac{28,000}{104}=\dfrac{104\times n}{104}$

$269.23\approx n$

It is worth about 269 euros.

Use Math to Save Money

1. Gold Plan: No extra meals, $2205.
Silver Plan: Four extra meals in each of the 15 weeks,
$1764 + 15 × 4 × $10 = $1764 + $600 = $2364.
Bronze Plan: Eight extra meals in each of the

15 weeks, $1386+15\times8\times\$10 = \$1386+\$1200$
$= \$2586.$
The Gold Plan is best for 20 dining-hall meals each week.

2. Gold Plan: No extra meals, $2205.
Silver Plan: Two extra meals in each of the 15 weeks,
$1764 + 15 × 2 × $10 = $1764 + $300 = $2064.
Bronze Plan: Six extra meals in each of the 15 weeks,
$1386 + 15 × 6 × $10 = $1386 + $900 = $2286.
The Silver Plan is best for 18 dining-hall meals each week.

3. Gold Plan: no extra meals, $2205.
Silver Plan: no extra meals, $1764.
Bronze Plan: Four extra meals in each of the 15 weeks,
$1386 + 15 × 4 × $10 = $1386 + $600 = $1986.
The Silver Plan is best for 18 dining-hall meals each week.

4. Gold Plan: no extra meals, $2205
Silver Plan: no extra meals, $1764.
Bronze Plan: Two extra meals in each of the 15 weeks,
$1386 + 15 × 2 × $10 = $1386 + $300 = $1686.
The Bronze Plan is best for 14 dining-hall meals each week.

5. If he spends $20 each week, he will spend
15 × $20 = $300 over the course of the semester.

The amount saved is $\dfrac{1}{10}$ of $300:

$\dfrac{1}{10}\times\$300=\$30.$

He saves $30.

6. If he spends $40 each week, he will spend
15 × $40 = $600 over the course of the semester.

The amount saved is $\dfrac{1}{10}$ of $600:

$\dfrac{1}{10}\times\$600=\$60.$

He saves $60.

7. If he spends $60 each week, he will spend
15 × $60 = $900 over the course of the semester.

The amount saved is $\dfrac{1}{10}$ of $900:

$\dfrac{1}{10}\times\$900=\$90.$

He saves $90.

8. Gold Plan: no extra meals, $2205
Silver Plan: One extra meal in each of the
15 weeks,
$1764 + 15 \times 1 \times $10 = $1764 + $150 = $1914.
Bronze Plan: Five extra meals in each of the
15 weeks,
$1386 + 15 \times 5 \times $10 = $1386 + $750 = $2136.
Since he plans to spend $30 each week at other
establishments regardless of which dining plan
he chooses, he should choose the Silver Plan.

9. He will spend $1914 for the dining plan, and $30
each week at other establishments less the $\frac{1}{10}$
discount.

$$\$1914 + 15 \times \$30 - \frac{1}{10} \times 15 \times \$30$$
$$= \$1914 + \$450 - \$45$$
$$= \$2319$$

He will spend $2319 for food during the
semester.

You Try It

1. a. 15 cars to 45 cars $= \frac{15}{45} = \frac{1 \times 15}{3 \times 15} = \frac{1}{3}$

   b. $64 : 80 = \frac{64}{80} = \frac{4 \times 16}{5 \times 16} = \frac{4}{5}$

   c. $\$70$ to $\$320 = \frac{70}{320} = \frac{7 \times 10}{32 \times 10} = \frac{7}{32}$

2. $\frac{27 \text{ teachers}}{720 \text{ students}} = \frac{27 \text{ teachers} \div 9}{720 \text{ students} \div 9} = \frac{3 \text{ teachers}}{80 \text{ students}}$

3. a. $\frac{756 \text{ miles}}{12 \text{ hours}} = \frac{756 \text{ miles} \div 12}{12 \text{ hours} \div 12} = 63 \text{ mi/hour}$

   b. $\frac{1550 \text{ pounds}}{620 \text{ square feet}} = \frac{1550 \text{ pounds} \div 620}{620 \text{ square feet} \div 620}$
   $$= 2.5 \text{ pounds/square foot}$$

4. 15 is to 60 as 13 is to 52.
$$\frac{15}{60} = \frac{13}{52}$$

5. a. $\frac{6}{70} \stackrel{?}{=} \frac{2}{24}$
$$6 \times 24 \stackrel{?}{=} 70 \times 2$$
$$144 \neq 140$$
It is not a proportion.

b. $\frac{45 \text{ pounds}}{30 \text{ square feet}} \stackrel{?}{=} \frac{54 \text{ pounds}}{36 \text{ square feet}}$
$$45 \times 36 \stackrel{?}{=} 30 \times 54$$
$$1620 = 1620$$
It is a proportion.

6. $\frac{n}{4} = \frac{49}{28}$
$$n \times 28 = 4 \times 49$$
$$\frac{n \times 28}{28} = \frac{196}{28}$$
$$n = 7$$

7. $\frac{10 \text{ pencils}}{\$24} = \frac{15 \text{ pencils}}{\$n}$
$$10 \times n = 24 \times 15$$
$$\frac{10 \times n}{10} = \frac{360}{10}$$
$$n = 36$$
15 pencils would cost $36.

Chapter 4 Review Problems

1. $88 : 40 = \frac{88}{40} = \frac{88 \div 8}{40 \div 8} = \frac{11}{5}$

2. $28 : 35 = \frac{28}{35} = \frac{28 \div 7}{35 \div 7} = \frac{4}{5}$

3. $250 : 475 = \frac{250}{475} = \frac{250 \div 25}{475 \div 25} = \frac{10}{19}$

4. $2\frac{1}{3}$ to $4\frac{1}{4} = \frac{2\frac{1}{3}}{4\frac{1}{4}}$
$$= 2\frac{1}{3} \div 4\frac{1}{4}$$
$$= \frac{7}{3} \div \frac{17}{4}$$
$$= \frac{7}{3} \times \frac{4}{17}$$
$$= \frac{28}{51}$$

5. $\frac{168}{300} = \frac{168 \div 12}{300 \div 12} = \frac{14}{25}$

6. $\frac{26 \text{ tons}}{65 \text{ tons}} = \frac{26 \div 13}{65 \div 13} = \frac{2}{5}$

7. $\dfrac{\$60}{\$480} = \dfrac{60 \div 60}{480 \div 60} = \dfrac{1}{8}$

8. $\begin{array}{r} 60 \\ + \ 45 \\ \hline 105 \end{array}$ $\dfrac{\$105}{\$480} = \dfrac{105 \div 15}{480 \div 15} = \dfrac{7}{32}$

9. $\dfrac{\$75}{6 \text{ people}} = \dfrac{\$25}{2 \text{ people}}$

10. $\dfrac{44 \text{ revolutions}}{121 \text{ minutes}} = \dfrac{4 \text{ revolusions}}{11 \text{ minutes}}$

11. $\dfrac{75 \text{ heartbeats}}{60 \text{ seconds}} = \dfrac{5 \text{ heartbeats}}{4 \text{ seconds}}$

12. $\dfrac{\$2125}{125 \text{ shares}} = \dfrac{\$17}{1 \text{ share}} = \$17/\text{share}$

13. $\dfrac{\$1344}{12 \text{ credit-hours}} = \$112/\text{credit-hour}$

14. $\dfrac{\$742.50}{55 \text{ sq yd}} = \dfrac{\$13.50}{1 \text{ yd}} = \$13.50/\text{square yard}$

15. a. $\dfrac{\$2.96}{4 \text{ oz}} = \dfrac{\$2.96 \div 4}{4 \text{ oz} \div 4} = \$0.74/\text{oz}$
    The cost of the small jar is \$0.74/oz.

    b. $\dfrac{\$5.22}{9 \text{ oz}} = \dfrac{\$5.22 \div 9}{9 \text{ oz} \div 9} = \$0.58/\text{oz}$
    The cost of the large jar is \$0.58/oz.

    c. $\begin{array}{r} \$0.74 \\ - \ 0.58 \\ \hline \$0.16/\text{oz} \end{array}$
    The savings is \$0.16/oz.

16. $\dfrac{12}{48} = \dfrac{7}{28}$

17. $\dfrac{1\frac{1}{2}}{5} = \dfrac{4}{13\frac{1}{3}}$

18. $\dfrac{3 \text{ buses}}{138 \text{ passengers}} = \dfrac{5 \text{ buses}}{230 \text{ passengers}}$

19. $\dfrac{15 \text{ pounds}}{\$4.50} = \dfrac{27 \text{ pounds}}{\$8.10}$

20. $\dfrac{16}{48} \overset{?}{=} \dfrac{2}{12}$
    $16 \times 12 \overset{?}{=} 48 \times 2$
    $192 \neq 96$
    It is not a proportion.

21. $\dfrac{36}{30} \overset{?}{=} \dfrac{60}{50}$
    $36 \times 50 \overset{?}{=} 30 \times 60$
    $1800 = 1800$
    It is a proportion.

22. $\dfrac{37}{33} \overset{?}{=} \dfrac{22}{19}$
    $37 \times 19 \overset{?}{=} 33 \times 22$
    $703 \neq 726$
    It is not a proportion.

23. $\dfrac{84 \text{ miles}}{7 \text{ gallons}} \overset{?}{=} \dfrac{108 \text{ miles}}{9 \text{ gallons}}$
    $84 \times 9 \overset{?}{=} 7 \times 108$
    $756 = 756$
    It is a proportion.

24. $\dfrac{156 \text{ rev}}{6 \text{ min}} \overset{?}{=} \dfrac{181 \text{ rev}}{7 \text{ min}}$
    $156 \times 7 \overset{?}{=} 6 \times 181$
    $1092 \neq 1086$
    It is not a proportion.

25. $9 \times n = 162$
    $\dfrac{9 \times n}{9} = \dfrac{162}{9}$
    $n = 18$

26. $5 \times n = 38$
    $\dfrac{5 \times n}{5} = \dfrac{38}{5}$
    $n = 7\dfrac{3}{5} \text{ or } 7.6$

27. $442 = 20 \times n$
    $\dfrac{442}{20} = \dfrac{20 \times n}{20}$
    $22.1 = n \text{ or } n = 22\dfrac{1}{10}$

28.
$$\frac{3}{11} = \frac{9}{n}$$
$$3 \times n = 11 \times 9$$
$$\frac{3 \times n}{3} = \frac{99}{3}$$
$$n = 33$$

29.
$$\frac{n}{28} = \frac{6}{24}$$
$$24 \times n = 28 \times 6$$
$$\frac{24 \times n}{24} = \frac{168}{24}$$
$$n = 7$$

30.
$$\frac{n}{32} = \frac{15}{20}$$
$$n \times 20 = 32 \times 15$$
$$\frac{n \times 20}{20} = \frac{480}{20}$$
$$n = 24$$

31.
$$\frac{3\frac{1}{3}}{2\frac{2}{3}} = \frac{7}{n}$$
$$3\frac{1}{3} \times n = 2\frac{2}{3} \times 7$$
$$\frac{10}{3} \times n = \frac{8}{3} \times 7$$
$$\frac{\frac{10}{3} \times n}{\frac{10}{3}} = \frac{\frac{56}{3}}{\frac{10}{3}}$$
$$n = \frac{56}{3} \div \frac{10}{3} = \frac{56}{3} \times \frac{3}{10} = \frac{56}{10} = \frac{28}{5} = 5\frac{3}{5} \text{ or } 5.6$$

32.
$$\frac{42}{50} = \frac{n}{6}$$
$$6 \times 42 = 50 \times n$$
$$\frac{252}{50} = \frac{50 \times n}{50}$$
$$5.0 \approx n$$

33.
$$\frac{2.25}{9} = \frac{4.75}{n}$$
$$2.25 \times n = 9 \times 4.75$$
$$\frac{2.25 \times n}{2.25} = \frac{42.75}{2.25}$$
$$n = 19$$

34.
$$\frac{36}{n} = \frac{109}{18}$$
$$36 \times 18 = n \times 109$$
$$\frac{648}{109} = \frac{n \times 109}{109}$$
$$n \approx 5.9$$

35.
$$\frac{35 \text{ miles}}{28 \text{ gallons}} = \frac{15 \text{ miles}}{n \text{ gallons}}$$
$$35 \times n = 28 \times 15$$
$$\frac{35 \times n}{35} = \frac{420}{35}$$
$$n = 12$$

36.
$$\frac{8 \text{ defective}}{100 \text{ perfect}} = \frac{44 \text{ defective}}{n \text{ perfect}}$$
$$8 \times n = 100 \times 44$$
$$\frac{8 \times n}{8} = \frac{4400}{8}$$
$$n = 550$$

37.
$$\frac{3 \text{ gallons}}{2 \text{ rooms}} = \frac{n \text{ gallons}}{10 \text{ rooms}}$$
$$3 \times 10 = 2 \times n$$
$$\frac{30}{2} = \frac{2 \times n}{2}$$
$$15 = n$$
They need 15 gallons.

38.
$$\frac{49 \text{ coffee}}{100 \text{ adults}} = \frac{n \text{ coffee}}{3450 \text{ adults}}$$
$$49 \times 3450 = 100 \times n$$
$$\frac{169,050}{100} = \frac{100 \times n}{100}$$
$$1691 \approx n$$
1691 employees drink coffee.

39.
$$\frac{24 \text{ francs}}{5 \text{ dollars}} = \frac{n \text{ francs}}{420 \text{ dollars}}$$
$$24 \times 420 \approx 5 \times n$$
$$\frac{10,080}{5} = \frac{5 \times n}{5}$$
$$2016 = n$$
She received 2016 francs.

40. $\dfrac{6 \text{ kronor}}{0.83 \text{ dollar}} = \dfrac{n \text{ kronor}}{125 \text{ dollars}}$

$6 \times 125 = 0.83 \times n$

$\dfrac{750}{0.83} = \dfrac{0.83 \times n}{0.83}$

$903.61 \approx n$

They received 903.61 Swedish kronor.

41. $\dfrac{225 \text{ miles}}{3 \text{ inches}} = \dfrac{n \text{ miles}}{8 \text{ inches}}$

$225 \times 8 = 3 \times n$

$\dfrac{1800}{3} = \dfrac{3 \times n}{3}$

$600 = n$

The cities are 600 miles apart.

42. $\dfrac{6 \text{ feet}}{16 \text{ feet}} = \dfrac{n \text{ feet}}{320 \text{ feet}}$

$6 \times 320 = 16 \times n$

$\dfrac{1920}{16} = \dfrac{16 \times n}{16}$

$120 = n$

The height of the building is 120 feet.

43. a. $\dfrac{680 \text{ miles}}{26 \text{ gallons}} = \dfrac{200 \text{ miles}}{n \text{ gallons}}$

$680 \times n = 26 \times 200$

$\dfrac{680 \times n}{680} = \dfrac{5200}{680}$

$n \approx 7.65$

They will need an additional 7.65 gallons.

b.
$$\begin{array}{r} 7.65 \\ \times \ \ 3.20 \\ \hline 1\,5300 \\ 22\,95 \ \ \ \\ \hline 24.4800 \end{array}$$

The additional fuel will cost $24.48.

44. $\dfrac{3.5 \text{ cm}}{2.5 \text{ cm}} = \dfrac{8 \text{ cm}}{n \text{ cm}}$

$3.5 \times n = 2.5 \times 8$

$\dfrac{3.5 \times n}{3.5} = \dfrac{20}{3.5}$

$n \approx 5.71$

The print will be 5.71 centimeters tall.

45. $\dfrac{3 \text{ grams}}{50 \text{ pounds}} = \dfrac{n \text{ grams}}{125 \text{ pounds}}$

$3 \times 125 = 50 \times n$

$\dfrac{375}{50} = \dfrac{50 \times n}{50}$

$7.5 = n$

She should take 7.5 grams.

46. $\dfrac{21 \text{ eat at cafeteria}}{35 \text{ students}} = \dfrac{n \text{ eat at cafeteria}}{2800 \text{ students}}$

$21 \times 2800 = 35 \times n$

$\dfrac{58,800}{35} = \dfrac{35 \times n}{35}$

$n = 1680$

1680 students eat at least once a week in the cafeteria.

47. $\dfrac{3 \text{ gallons}}{500 \text{ sq ft}} = \dfrac{n \text{ gallons}}{1400 \text{ sq ft}}$

$3 \times 1400 = 500 \times n$

$\dfrac{4200}{500} = \dfrac{500 \times n}{500}$

$8.4 = n$

He needs 8.4 gallons which equals 9 gallons in real life.

48. $\dfrac{n \text{ liters}}{1250 \text{ runners}} = \dfrac{2 \text{ liters}}{3 \text{ runners}}$

$3 \times n = 1250 \times 2$

$\dfrac{3 \times n}{3} = \dfrac{2500}{3}$

$n \approx 833.33$

Technically, they need 833.33 liters, but in real life, 834 liters will be needed.

49. $\dfrac{14 \text{ cm}}{145 \text{ feet}} = \dfrac{11 \text{ cm}}{n \text{ feet}}$

$14 \times n = 145 \times 11$

$\dfrac{14 \times n}{14} = \dfrac{1595}{14}$

$n \approx 113.93$

Width is approximately 113.93 feet.

50. $\dfrac{68 \text{ goals}}{27 \text{ games}} = \dfrac{n \text{ goals}}{34 \text{ games}}$

$68 \times 34 = 27 \times n$

$\dfrac{2312}{27} = \dfrac{27 \times n}{27}$

$86 \approx n$

She is expected to score 86 goals.

51. $\dfrac{n \text{ people}}{45,600 \text{ residents}} = \dfrac{3 \text{ people}}{10 \text{ residents}}$

$10 \times n = 45,600 \times 3$

$\dfrac{10 \times n}{10} = \dfrac{136,800}{10}$

$n \approx 13,680$

You would expect 13,680 people to read the *Boston Globe*.

52. $\dfrac{n \text{ trips}}{240 \text{ trips}} = \dfrac{13 \text{ trips}}{16 \text{ trips}}$

$16 \times n = 240 \times 13$

$\dfrac{16 \times n}{16} = \dfrac{3120}{16}$

$n = 195$

195 trips will have the passengers spotting at least one whale.

How Am I Doing? Chapter 4 Test

1. $\dfrac{18}{52} = \dfrac{18 \div 2}{52 \div 2} = \dfrac{9}{26}$

2. $\dfrac{70}{185} = \dfrac{70 \div 5}{185 \div 5} = \dfrac{14}{37}$

3. $\dfrac{784 \text{ miles}}{24 \text{ gallons}} = \dfrac{784 \text{ miles} \div 8}{24 \text{ gallons} \div 8} = \dfrac{98 \text{ miles}}{3 \text{ gallons}}$

4. $\dfrac{2100 \text{ square feet}}{45 \text{ pounds}} = \dfrac{2100 \text{ square feet} \div 15}{45 \text{ pounds} \div 15}$

$= \dfrac{140 \text{ square feet}}{3 \text{ pounds}}$

5. $\dfrac{19 \text{ tons}}{5 \text{ days}} = \dfrac{19 \text{ tons} \div 5}{5 \text{ days} \div 5} = 3.8 \text{ tons/day}$

6. $\dfrac{\$57.96}{7 \text{ hours}} = \dfrac{\$57.96 \div 7}{7 \text{ hours} \div 7} = \$8.28/\text{hour}$

7. $\dfrac{5400 \text{ feet}}{22 \text{ poles}} = \dfrac{5400 \text{ feet} \div 22}{22 \text{ poles} \div 22} = 245.45 \text{ feet/pole}$

8. $\dfrac{\$9373}{110 \text{ shares}} = \dfrac{\$9373 \div 110}{110 \text{ shares} \div 110} = \$85.21/\text{share}$

9. $\dfrac{17}{29} = \dfrac{51}{87}$

10. $\dfrac{2\frac{1}{2}}{10} = \dfrac{6}{24}$

11. $\dfrac{490 \text{ miles}}{21 \text{ gallons}} = \dfrac{280 \text{ miles}}{12 \text{ gallons}}$

12. $\dfrac{3 \text{ hours}}{180 \text{ miles}} = \dfrac{5 \text{ hours}}{300 \text{ miles}}$

13. $\dfrac{50}{24} \overset{?}{=} \dfrac{34}{16}$

$50 \times 16 \overset{?}{=} 24 \times 34$

$800 \neq 816$

It is not a proportion.

14. $\dfrac{3\frac{1}{2}}{14} = \dfrac{5}{20}$

$3\dfrac{1}{2} \times 20 \overset{?}{=} 14 \times 5$

$\dfrac{7}{2} \times \dfrac{20}{1} \overset{?}{=} 70$

$70 = 70$

It is a proportion.

15. $\dfrac{32 \text{ smokers}}{46 \text{ nonsmokers}} \overset{?}{=} \dfrac{160 \text{ smokers}}{230 \text{ nonsmokers}}$

$32 \times 230 \overset{?}{=} 46 \times 160$

$7360 = 7360$

It is a proportion.

16. $\dfrac{\$0.74}{16 \text{ ounces}} \overset{?}{=} \dfrac{\$1.84}{40 \text{ ounces}}$

$0.74 \times 40 \overset{?}{=} 16 \times 1.84$

$29.6 \neq 29.44$

It is not a proportion.

17. $\dfrac{n}{20} = \dfrac{4}{5}$

$n \times 5 = 20 \times 4$

$\dfrac{n \times 5}{5} = \dfrac{80}{5}$

$n = 16$

18. $\dfrac{8}{3} = \dfrac{60}{n}$

$8 \times n = 3 \times 60$

$\dfrac{8 \times n}{8} = \dfrac{180}{8}$

$n = 22.5$

19.
$$\frac{2\frac{2}{3}}{8} = \frac{6\frac{1}{3}}{n}$$

$$2\frac{2}{3} \times n = 8 \times 6\frac{1}{3}$$

$$\frac{8}{3} \times n = 8 \times \frac{19}{3}$$

$$\frac{\frac{8}{3} \times n}{\frac{8}{3}} = \frac{\frac{152}{3}}{\frac{8}{3}}$$

$$n = \frac{152}{3} \div \frac{8}{3} = \frac{152}{3} \times \frac{3}{8} = 19$$

20.
$$\frac{4.2}{11} = \frac{n}{77}$$

$$4.2 \times 77 = 11 \times n$$

$$\frac{323.4}{11} = \frac{11 \times n}{11}$$

$$29.4 = n$$

21.
$$\frac{45 \text{ women}}{15 \text{ men}} = \frac{n \text{ women}}{40 \text{ men}}$$

$$45 \times 40 = 15 \times n$$

$$\frac{1800}{15} = \frac{15 \times n}{15}$$

$$120 = n$$

22.
$$\frac{5 \text{ kilograms}}{11 \text{ pounds}} = \frac{32 \text{ kilograms}}{n \text{ pounds}}$$

$$5 \times n = 11 \times 32$$

$$\frac{5 \times n}{5} = \frac{352}{5}$$

$$n = 70.4$$

23.
$$\frac{n \text{ inches of snow}}{14 \text{ inches of rain}} = \frac{12 \text{ inches of snow}}{1.4 \text{ inches of rain}}$$

$$n \times 1.4 = 14 \times 12$$

$$\frac{n \times 1.4}{1.4} = \frac{168}{1.4}$$

$$n = 120$$

24.
$$\frac{5 \text{ pounds of coffee}}{\$n} = \frac{\frac{1}{2} \text{ pound of coffee}}{\$5.20}$$

$$5 \times 5.20 = \frac{1}{2} \times n$$

$$\frac{26}{\frac{1}{2}} = \frac{\frac{1}{2} \times n}{\frac{1}{2}}$$

$$52 = n$$

25.
$$\frac{3 \text{ eggs}}{11 \text{ people}} = \frac{n \text{ eggs}}{22 \text{ people}}$$

$$3 \times 22 = 11 \times n$$

$$\frac{66}{11} = \frac{11 \times n}{11}$$

$$6 = n$$

He will need 6 eggs.

26.
$$\frac{42 \text{ ft}}{170 \text{ lb}} = \frac{20 \text{ ft}}{n \text{ lb}}$$

$$42 \times n = 170 \times 20$$

$$\frac{42 \times n}{42} = \frac{3400}{42}$$

$$n \approx 80.95$$

This cable will weigh 80.95 pounds.

27.
$$\frac{9 \text{ inches}}{57 \text{ miles}} = \frac{3 \text{ inches}}{n \text{ miles}}$$

$$9 \times n = 57 \times 3$$

$$\frac{9 \times n}{9} = \frac{171}{9}$$

$$n = 19$$

3 inches represents 19 miles.

28.
$$\frac{\$240}{4000 \text{ sq ft}} = \frac{\$n}{6000 \text{ sq ft}}$$

$$240 \times 6000 = 4000 \times n$$

$$\frac{1,440,000}{4000} = \frac{4000 \times n}{4000}$$

$$360 = n$$

It will cost $360.

29.
$$\frac{n \text{ miles}}{220 \text{ km}} = \frac{1 \text{ mile}}{1.61 \text{ km}}$$

$$1.61 \times n = 220 \times 1$$

$$\frac{1.61 \times n}{1.61} = \frac{220}{1.61}$$

$$n \approx 136.646$$

The distance is about 136.6 miles.

30.
$$\frac{570 \text{ km}}{9 \text{ hr}} = \frac{n \text{ km}}{11 \text{ hr}}$$

$$570 \times 11 = 9 \times n$$

$$6270 = 9 \times n$$

$$\frac{6270}{9} = \frac{9 \times n}{9}$$

$$696.67 \approx n$$

He could go 696.67 kilometers.

31. $\dfrac{n \text{ free throws made}}{120 \text{ free throws}} = \dfrac{11 \text{ free throws made}}{15 \text{ free throws}}$

$$15 \times n = 120 \times 11$$

$$\frac{15 \times n}{15} = \frac{1320}{15}$$

$$n = 88$$

You would expect 88 more free throws made.

32. $\dfrac{7 \text{ hits}}{34 \text{ bats}} = \dfrac{n \text{ hits}}{155 \text{ bats}}$

$$7 \times 155 = 34 \times n$$

$$\frac{1085}{34} = \frac{34 \times n}{34}$$

$$32 \approx n$$

She would have 32 hits.

Chapter 5

5.1 Exercises

1. In this section we introduced percent, which means "per centum" or "per <u>hundred</u>."

3. Move the decimal point <u>two</u> places to the <u>left</u>, <u>drop</u> the % symbol.

5. $\dfrac{59}{100} = 59\%$

7. $\dfrac{4}{100} = 4\%$

9. $\dfrac{80}{100} = 80\%$

11. $\dfrac{245}{100} = 245\%$

13. $\dfrac{12.5}{100} = 12.5\%$

15. $\dfrac{4\frac{1}{3}}{100} = 4\frac{1}{3}\%$

17. $\dfrac{13}{100} = 13\%$

19. $\dfrac{9}{100} = 9\%$

21. $51\% = 0.51$

23. $7\% = 0.07$

25. $20\% = 0.20 = 0.2$

27. $43.6\% = 0.436$

29. $0.03\% = 0.0003$

31. $0.72\% = 0.0072$

33. $1.25\% = 0.0125$

35. $275\% = 2.75$

37. $0.74 = 74\%$

39. $0.50 = 50\%$

41. $0.08 = 8\%$

43. $0.563 = 56.3\%$

45. $0.002 = 0.2\%$

47. $0.0057 = 0.57\%$

49. $1.35 = 135\%$

51. $5.16 = 516\%$

53. $0.27 = 27\%$

55. $0.2 = 20\%$

57. $0.94 = 94\%$

59. $2.31 = 231\%$

61. $\dfrac{10}{100} = 10\%$

63. $0.089 = 8.9\%$

65. $62\% = \dfrac{62}{100} = 0.62$

67. $138\% = \dfrac{138}{100} = 1.38$

69. $\dfrac{0.3}{100} = 0.003$

71. $\dfrac{75}{100} = 0.75$

73. $\dfrac{57}{100} = 57\%$
57% voted for Barack Obama.

75. $115\% = 1.15$
The value increased by 1.15.

77. $0.6\% = 0.006$
0.006 are valued at more than $1 million.

79. $19\% = 0.19; \ 0.32\% = 0.0032$
First-year students were expected to spend 0.19 less and overall spending was expected to drop 0.0032.

81. 36% = 36 percent

= 36 "per one hundred"

$= 36 \times \dfrac{1}{100}$

$= \dfrac{36}{100}$

$= 0.36$

The rule is using the fact that 36% means 36 per one hundred.

83. a. $1562\% = 15.62$

b. $1562\% = \dfrac{1562}{100}$

c. $1562\% = \dfrac{1562}{100} = \dfrac{1562 \div 2}{100 \div 2} = \dfrac{781}{50}$

Cumulative Review

85. $0.56 = \dfrac{56}{100} = \dfrac{56 \div 4}{100 \div 4} = \dfrac{14}{25}$

86. $0.78 = \dfrac{78}{100} = \dfrac{78 \div 2}{100 \div 2} = \dfrac{39}{50}$

87.
```
      0.6875
 16)11.000
    9 6
    ───
    1 40
    1 28
    ────
      120
      112
      ───
       80
       80
       ──
        0
```

$\dfrac{11}{16} = 0.6875$

88.
```
     0.875
 8)7.000
   6 4
   ───
    60
    56
    ──
    40
    40
    ──
     0
```

$\dfrac{7}{8} = 0.875$

89. $3 \times 246 + 7 \times 380 + 5 \times 168 + 9 \times 122$

$= 738 + 2660 + 840 + 1098$

$= 5336$

There are a total of 5336 vases.

Quick Quiz 5.1

1. $0.007 = 0.7\%$

2. $\dfrac{4.5}{100} = 4.5\%$

3. $1.25\% = 0.0125$

4. Answers may vary. Possible solution: Move the decimal place two places to the left and drop the % symbol.

$0.00072\% = 0.0000072$

5.2 Exercises

1. Write the number in front of the percent symbol as the numerator of the fraction. Write the number 100 as the denominator of the fraction. Reduce the fraction if possible.

3. $6\% = \dfrac{6}{100} = \dfrac{3}{50}$

5. $33\% = \dfrac{33}{100}$

7. $55\% = \dfrac{55}{100} = \dfrac{11}{20}$

9. $75\% = \dfrac{75}{100} = \dfrac{3}{4}$

11. $20\% = \dfrac{20}{100} = \dfrac{1}{5}$

13. $9.5\% = 0.095 = \dfrac{95}{1000} = \dfrac{19}{200}$

15. $22.5\% = 0.225 = \dfrac{225}{1000} = \dfrac{9}{40}$

17. $64.8\% = .648 = \dfrac{648}{1000} = \dfrac{81}{125}$

19. $71.25\% = 0.7125 = \dfrac{7125}{10,000} = \dfrac{57}{80}$

21. $168\% = \dfrac{168}{100} = \dfrac{42}{25} = 1\dfrac{17}{25}$

23. $340\% = \dfrac{340}{100} = \dfrac{17}{5} = 3\dfrac{2}{5}$

25. $1200\% = \dfrac{1200}{100} = 12$

27. $3\dfrac{5}{8}\% = \dfrac{\frac{29}{8}}{100} = \dfrac{29}{800}$

29. $12\dfrac{1}{2}\% = \dfrac{\frac{25}{2}}{100} = \dfrac{25}{200} = \dfrac{1}{8}$

31. $8\dfrac{4}{5}\% = \dfrac{\frac{44}{5}}{100} = \dfrac{44}{500} = \dfrac{11}{125}$

33. $10.1\% = \dfrac{10.1}{100} = \dfrac{10.1 \times 10}{100 \times 10} = \dfrac{101}{1000}$

The number of crimes increased by $\dfrac{101}{1000}$.

35. $4\dfrac{4}{5}\% = \dfrac{\frac{24}{5}}{100} = \dfrac{24}{500} = \dfrac{6}{125}$

This was an increase of $\dfrac{6}{125}$.

37. $\dfrac{3}{4} = 0.75 = 75\%$

39. $\dfrac{7}{10} = 0.7 = 70\%$

41. $\dfrac{7}{20} = 0.35 = 35\%$

43. $\dfrac{18}{25} = 0.72 = 72\%$

45. $\dfrac{11}{40} = 0.275 = 27.5\%$

47. $\dfrac{18}{5} = 3.6 = 360\%$

49. $2\dfrac{1}{2} = 2.5 = 250\%$

51. $4\dfrac{1}{8} = \dfrac{33}{8} = 4.125 = 412.5\%$

53. $\dfrac{1}{3} = 0.3333 = 33.33\%$

55. $\dfrac{5}{12} \approx 0.4167 = 41.67\%$

57. $\dfrac{17}{4} = 4.25 = 425\%$

59. $\dfrac{26}{50} = 0.52 = 52\%$

61.
$$
\begin{array}{r}
0.025 \\
40\overline{)1.000} \\
\underline{8\,0} \\
200 \\
\underline{200} \\
0
\end{array}
$$

$\dfrac{1}{40} = 0.025 = 2.5\%$

The brain represents approximately 2.5% of an average person's weight.

63. $\dfrac{119}{2000} = 0.0595 = 5.95\%$

It comprises 5.95% of the earth's total surface area.

65.
$$
\begin{array}{r}
0.37 \\
8\overline{)3.00} \\
\underline{2\,4} \\
60 \\
\underline{56} \\
4
\end{array}
$$

$\dfrac{3}{8} = 0.37\dfrac{4}{8} = 0.37\dfrac{1}{2} = 37\dfrac{1}{2}\%$

67.
$$
\begin{array}{r}
0.07 \\
40\overline{)3.00} \\
\underline{2\,80} \\
20
\end{array}
$$

$\dfrac{3}{40} = 0.07\dfrac{20}{40} = 0.07\dfrac{1}{2} = 7\dfrac{1}{2}\%$

69. $15\overline{)4.00}$ quotient 0.26

$\phantom{15\overline{)}}\underline{3\,0}$

$\phantom{15\overline{)}}1\,00$

$\phantom{15\overline{)1}}\underline{90}$

$\phantom{15\overline{)1}}10$

$\dfrac{4}{15} = 0.26\dfrac{10}{15} = 0.26\dfrac{2}{3} = 26\dfrac{2}{3}\%$

71. $9\overline{)2.00}$ quotient 0.22

$\phantom{9\overline{)}}\underline{1\,8}$

$\phantom{9\overline{)}}2$

$\dfrac{2}{9} = 0.22\dfrac{2}{9} = 22\dfrac{2}{9}\%$

73. $12\overline{)11.00000}$ quotient 0.9166

$\phantom{12\overline{)}}\underline{10\,8}$

$\phantom{12\overline{)}}20$

$\phantom{12\overline{)}}\underline{12}$

$\phantom{12\overline{)}}80$

$\phantom{12\overline{)}}\underline{72}$

$\phantom{12\overline{)}}80$

$\phantom{12\overline{)}}\underline{72}$

$\phantom{12\overline{)}}80$

$\phantom{12\overline{)}}\underline{72}$

$\phantom{12\overline{)}}8$

$\dfrac{11}{12} \approx 0.9167 = 91.67\%$

$\dfrac{11}{12};\ 0.9167;\ 91.67\%$

75. $0.56 = 56\%$

$0.56 = \dfrac{56}{100} = \dfrac{14}{25}$

$\dfrac{14}{25};\ 0.56;\ 56\%$

77. $0.005 = \dfrac{5}{1000} = \dfrac{1}{200}$

$0.005 = 0.5\%$

$\dfrac{1}{200};\ 0.005;\ 0.5\%$

79. $9\overline{)5.0}$ quotient 0.555

$\phantom{9\overline{)}}\underline{4\,5}$

$\phantom{9\overline{)}}50$

$\phantom{9\overline{)}}\underline{45}$

$\phantom{9\overline{)}}50$

$\phantom{9\overline{)}}\underline{45}$

$\phantom{9\overline{)}}5$

$\dfrac{5}{9} = 0.\overline{5}$

$\dfrac{5}{9};\ 0.5556;\ 55.56\%$

81. $\dfrac{1}{8} = 0.125$

$3\dfrac{1}{8}\% = 0.3125 \approx 0.0313$

$0.03125 = \dfrac{3125}{100,000} = \dfrac{1}{32}$

$\dfrac{1}{32};\ 0.0313;\ 3\dfrac{1}{8}\%$

83. $28\dfrac{15}{16} = \dfrac{28\frac{15}{16}}{100} = 28\dfrac{15}{16} \div 100 = \dfrac{463}{16} \times \dfrac{1}{100} = \dfrac{463}{1600}$

85. $\dfrac{123}{800} = \dfrac{n}{100}$

$800 \times n = 123 \times 100$

$\dfrac{800 \times n}{800} = \dfrac{12,300}{800}$

$n = 15.375$

$\dfrac{123}{800} = 15.375\%$

Cumulative Review

87. $\dfrac{15}{n} = \dfrac{8}{3}$

$15 \times 3 = n \times 8$

$\dfrac{45}{8} = \dfrac{n \times 8}{8}$

$n = 5.625$

88. $\dfrac{32}{24} = \dfrac{n}{3}$

$32 \times 3 = 24 \times n$

$96 = 24 \times n$

$\dfrac{96}{24} = \dfrac{24 \times n}{24}$

$4 = n$

89. $10,041$
986
$4,283$
$+ \ 533,855$
$\overline{549,165}$

There were a total of 549,165 documents.

90. $2\dfrac{1}{2} \times 1800 = \dfrac{5}{2} \times \dfrac{1800}{1} = 4500$

The new steak house has an area of 4500 square feet.

Quick Quiz 5.2

1. $45\% = \dfrac{45}{100} = \dfrac{9}{20}$

2. $7\dfrac{3}{5}\% = \dfrac{\frac{38}{5}}{100} = \dfrac{38}{500} = \dfrac{19}{250}$

3. $\dfrac{23}{25} = \dfrac{92}{100} = 92\%$

4. Answers may vary. Possible solution: First change $\dfrac{3}{8}$ to a decimal by dividing 3 by 8. Then add 8. Finally, move the decimal point two places to the left and drop the % symbol.

$\dfrac{3}{8} = 0.375$

$8\dfrac{3}{8}\% = 8.375\% = 0.08375$

5.3A Exercises

1. What is 20% of $300?

3. 20 baskets out of 25 shots is what percent?

5. This type is called "a percent problem when we do not know the base." We can translate this into an equation.

$108 = 18\% \times n$

$108 = 0.18n$

$\dfrac{108}{0.18} = \dfrac{0.18n}{0.18}$

$600 = n$

7. What is 5% of 90?

$n = 5\% \times 90$

9. 30% of what is 5?

$30\% \times n = 5$

11. 17 is what percent of 85?

$17 = n \times 85$

13. What is 20% of 140?

$n = 20\% \times 140$

$n = 0.20 \times 140$

$n = 28$

15. Find 40% of 140.

$n = 40\% \times 140$

$n = 0.4 \times 140$

$n = 56$

17. What is 6% of $850?

$n = 6\% \times 850$

$n = 0.06 \times 850$

$n = 51$

The tax was $51.

19. 2% of what is 26?

$2\% \times n = 26$

$0.02 \times n = 26$

$\dfrac{0.02 \times n}{0.02} = \dfrac{26}{0.02}$

$n = 1300$

21. 52 is 4% of what?

$52 = 4\% \times n$

$52 = 0.04 \times n$

$\dfrac{52}{0.04} = \dfrac{0.04n}{0.04}$

$1300 = n$

23. 22% of what is $33?

$22\% \times n = 33$

$0.22 \times n = 33$

$\dfrac{0.22 \times n}{0.22} = \dfrac{33}{0.22}$

$n = 150$

The before-tax price is $150.

25. What percent of 200 is 168?

$$n \times 200 = 168$$

$$\frac{n \times 200}{200} = \frac{168}{200}$$

$$n = 0.84$$

$$n = 84\%$$

27. 33 is what percent of 300?

$$33 = n \times 300$$

$$\frac{33}{300} = \frac{n \times 300}{300}$$

$$0.11 = n$$

$$11\% = n$$

29. 78 is what percent of 120?

$$78 = n \times 120$$

$$\frac{78}{120} = \frac{n \times 120}{120}$$

$$0.65 = n$$

$$65\% = n$$

65% of the points were scored by the winning team.

31. 20% of 155 is what?

$$20\% \times 155 = n$$

$$0.20 \times 155 = n$$

$$31 = n$$

33. 170% of what is 144.5?

$$170\% \times n = 144.5$$

$$1.70 \times n = 144.5$$

$$\frac{1.70 \times n}{1.70} = \frac{144.5}{1.70}$$

$$n = 85$$

35. 84 is what percent of 700?

$$84 = n \times 700$$

$$\frac{84}{700} = \frac{n \times 700}{700}$$

$$0.12 = n$$

$$12\% = n$$

37. Find 0.4% of 820.

$$n = 0.4\% \times 820$$

$$n = 0.004 \times 820$$

$$n = 3.28$$

39. What percent of 35 is 22.4?

$$n \times 35 = 22.4$$

$$\frac{n \times 35}{35} = \frac{22.4}{35}$$

$$n = 0.64$$

$$n = 64\%$$

41. 15 is 20% of what?

$$15 = 20\% \times n$$

$$15 = 0.2 \times n$$

$$\frac{15}{0.2} = \frac{0.2 \times n}{0.2}$$

$$75 = n$$

43. 8 is what percent of 1000?

$$8 = n \times 1000$$

$$\frac{8}{1000} = \frac{n \times 1000}{1000}$$

$$0.008 = n$$

$$0.8\% = n$$

45. What is 10.5% of 180?

$$n = 10.5\% \times 180$$

$$n = 0.105 \times 180$$

$$n = 18.9$$

47. 44 is what percent of 55?

$$44 = n \times 55$$

$$\frac{44}{55} = \frac{n \times 55}{55}$$

$$0.8 = n$$

$$n = 80\%$$

49. 493 million is what percent of 800 million?

$$493 = n \times 800$$

$$\frac{493}{800} = \frac{n \times 800}{800}$$

$$0.6163 \approx n$$

$$61.63\% = n$$

About 61.63% of oil spilled is caused by natural seepage.

51. 62% of 1070 is what?

$$62\% \times 1070 = n$$

$$0.62 \times 1070 = n$$

$$663.4 = n$$

663 students are taking a composition course.

53. 60% of what is 24?

$$60\% \times n = 24$$

$$0.60 \times n = 24$$

$$\frac{0.60 \times n}{0.60} = \frac{24}{0.60}$$

$$n = 40$$

The swim team has qualified for the finals for 40 years.

55. Find 12% of 30% of $1600.

$$n = 12\% \times 30\% \times 1600$$

$$n = 0.12 \times 0.30 \times 1600$$

$$n = 57.60$$

$57.60

Cumulative Review

57.
$$
\begin{array}{r}
1.36 \\
\times\ 1.8 \\
\hline
1\ 088 \\
1\ 36\ \ \ \\
\hline
2.448
\end{array}
$$

58.
$$
\begin{array}{r}
5.06 \\
\times\ 0.82 \\
\hline
1012 \\
4\ 048\ \ \ \\
\hline
4.1492
\end{array}
$$

59.
$$
\begin{array}{r}
2834 \\
0.06_{\wedge}\overline{)170.04_{\wedge}} \\
\underline{12\ \ \ \ \ \ \ } \\
50\ \ \ \ \ \\
\underline{48\ \ \ \ \ } \\
2\ 0\ \ \ \\
\underline{1\ 8\ \ \ } \\
24 \\
\underline{24} \\
0
\end{array}
$$

60.
$$
\begin{array}{r}
2.36 \\
0.9_{\wedge}\overline{)2.1_{\wedge}24} \\
\underline{1\ 8\ \ \ \ } \\
3\ 2\ \ \\
\underline{2\ 7\ \ } \\
54 \\
\underline{54} \\
0
\end{array}
$$

Quick Quiz 5.3A

1. What is 152% of 84?

$n = 152\% \times 84$

$n = 1.52 \times 84$

$n = 127.68$

2. 72 is 0.8% of what number?

$72 = 0.8\% \times n$

$72 = 0.008 \times n$

$$\frac{72}{0.008} = \frac{0.008 \times n}{0.008}$$

$9000 = n$

3. 68 is what percent of 400?

$68 = n \times 400$

$$\frac{68}{400} = \frac{n \times 400}{400}$$

$0.17 = n$

17%

4. Answers may vary. Possible solution:

Find "What is 85% of 120?"

$n = 85\% \times 120$

$n = 0.85 \times 120$

$n = 102$

102 people were previous Mustang owners.

5.3B Exercises

		p	b	a
1.	75% of 660 is 495.	75	660	495
3.	What is 22% of 60?	22	60	a
5.	49% of what is 2450?	49	b	2450
7.	30 is what percent of 50?	p	50	30

9. 40% of 70 is what?

$$\frac{a}{70} = \frac{40}{100}$$

$100a = 70 \times 40$

$$\frac{100a}{100} = \frac{2800}{100}$$

$a = 28$

11. Find 210% of 40.

$$\frac{a}{40} = \frac{210}{100}$$

$100a = 40 \times 210$

$$\frac{100a}{100} = \frac{8400}{100}$$

$a = 84$

13. 0.7% of 8000 is what?

$$\frac{a}{8000} = \frac{0.7}{100}$$

$100a = 8000 \times 0.7$

$$\frac{100a}{100} = \frac{5600}{100}$$

$a = 56$

15. 20 is 25% of what?
$$\frac{20}{b} = \frac{25}{100}$$
$$20 \times 100 = 25b$$
$$\frac{2000}{25} = \frac{25b}{25}$$
$$80 = b$$

17. 250% of what is 200?
$$\frac{200}{b} = \frac{250}{100}$$
$$200 \times 100 = 250b$$
$$\frac{20,000}{250} = \frac{250b}{250}$$
$$80 = b$$

19. 3000 is 0.5% of what?
$$\frac{3000}{b} = \frac{0.5}{100}$$
$$3000 \times 100 = 0.5b$$
$$\frac{300,000}{0.5} = \frac{0.5b}{0.5}$$
$$600,000 = b$$

21. 56 is what percent of 280?
$$\frac{p}{100} = \frac{56}{280}$$
$$280p = 56 \times 100$$
$$\frac{280p}{280} = \frac{5600}{280}$$
$$p = 20$$
20%

23. What percent of 90 is 18?
$$\frac{18}{90} = \frac{p}{100}$$
$$18 \times 100 = 90 \times p$$
$$\frac{1800}{90} = \frac{90 \times p}{90}$$
$$20 = p$$
20%

25. 25% of 88 is what?
$$\frac{25}{100} = \frac{a}{88}$$
$$25 \times 88 = 100a$$
$$\frac{2200}{100} = \frac{100a}{100}$$
$$22 = a$$

27. 300% of what is 120?
$$\frac{300}{100} = \frac{120}{b}$$
$$300b = 100 \times 120$$
$$\frac{300b}{300} = \frac{12,000}{300}$$
$$b = 40$$

29. 82 is what percent of 500?
$$\frac{p}{100} = \frac{82}{500}$$
$$500p = 100 \times 82$$
$$\frac{500p}{500} = \frac{8200}{500}$$
$$p = 16.4$$
16.4%

31. Find 0.7% of 520.
$$\frac{a}{520} = \frac{0.7}{100}$$
$$100a = 520 \times 0.7$$
$$\frac{100a}{100} = \frac{364}{100}$$
$$a = 3.64$$

33. What percent of 66 is 16.5?
$$\frac{p}{100} = \frac{16.5}{66}$$
$$66p = 100 \times 16.5$$
$$\frac{66p}{66} = \frac{1650}{66}$$
$$p = 25$$
25%

35. 68 is 40% of what?
$$\frac{40}{100} = \frac{68}{b}$$
$$40b = 100 \times 68$$
$$\frac{40b}{40} = \frac{6800}{40}$$
$$b = 170$$

37. 5% of what is $48?
$$\frac{5}{100} = \frac{48}{b}$$
$$5b = 100 \times 48$$
$$\frac{5b}{5} = \frac{4800}{5}$$
$$b = 960$$
Lowell's paycheck was $960.

39. $3.90 is what percent of $26.00?

$$\frac{p}{100} = \frac{3.90}{26.00}$$

$$26.00p = 100 \times 3.90$$

$$\frac{26.00p}{26.00} = \frac{390}{26.00}$$

$$p = 15$$

15% of the check was the tip.

41. What is 15% of 120?

$$\frac{a}{120} = \frac{15}{100}$$

$$100a = 120 \times 15$$

$$\frac{100a}{100} = \frac{1800}{100}$$

$$a = 18$$

18 gallons had passed the expiration date.

43. What is 18% of 10,500?

$$\frac{a}{10,500} = \frac{18}{100}$$

$$100a = 10,500 \times 18$$

$$\frac{100a}{100} = \frac{189,000}{100}$$

$$a = 1890$$

Her down payment was $1890.

45.
$$\begin{array}{r} 3\ 870 \\ 2\ 213 \\ 1\ 757 \\ 727 \\ 1\ 010 \\ 1\ 653 \\ +\ \ 1\ 098 \\ \hline 12,328 \end{array}$$

What percent of $12,328 is $2213?

$$\frac{2213}{12,328} = \frac{p}{100}$$

$$2213 \times 100 = 12,328p$$

$$\frac{221,300}{12,328} = \frac{12,328p}{12,328}$$

$$17.95 \approx p$$

About 18% was spent on food.

47. Housing: What is 125% of $3870?

$$\frac{a}{3870} = \frac{125}{100}$$

$$100a = 125 \times 3870$$

$$\frac{100a}{100} = \frac{483,750}{100}$$

$$a = 4837.5$$

The new amount is $4837.5.

Child care/education: What is 110% of $1653?

$$\frac{a}{1653} = \frac{110}{100}$$

$$100a = 1653 \times 110$$

$$\frac{100a}{100} = \frac{181,830}{100}$$

$$a = 1818.3$$

The new amount is $1818.3.

Total is:
$$\begin{array}{r} 4\ 837.5 \\ 2\ 213.0 \\ 1\ 757.0 \\ 727.0 \\ 1\ 010.0 \\ 1\ 818.3 \\ +\ 1\ 098.0 \\ \hline 13,460.8 \end{array}$$

What percent of 13,460.8 is 2213?

$$\frac{2213}{13,460.8} = \frac{p}{100}$$

$$2213 \times 100 = 13,460.8p$$

$$\frac{221,300}{13,460.8} = \frac{13,460.8p}{13,460.8}$$

$$16.4 \approx p$$

About 16.4% was used for food.

Cumulative Review

49. $\dfrac{4}{5} + \dfrac{8}{9} = \dfrac{4}{5} \times \dfrac{9}{9} + \dfrac{8}{9} \times \dfrac{5}{5}$

$$= \frac{36}{45} + \frac{40}{45}$$

$$= \frac{76}{45} \text{ or } 1\frac{31}{45}$$

50. $\dfrac{7}{13} - \dfrac{1}{2} = \dfrac{14}{26} - \dfrac{13}{26} = \dfrac{1}{26}$

51. $\left(2\dfrac{4}{5}\right)\left(1\dfrac{1}{2}\right) = \dfrac{14}{5} \times \dfrac{3}{2} = \dfrac{42}{10} = \dfrac{21}{5} \text{ or } 4\dfrac{1}{5}$

52. $1\dfrac{2}{5} \div \dfrac{3}{4} = \dfrac{7}{5} \div \dfrac{3}{4} = \dfrac{7}{5} \times \dfrac{4}{3} = \dfrac{28}{15} \text{ or } 1\dfrac{13}{15}$

Quick Quiz 5.3B

1. What is 0.09% of 17,000?
$$\frac{a}{17,000} = \frac{0.09}{100}$$
$$a \times 100 = 17,000 \times 0.09$$
$$\frac{100a}{100} = \frac{1530}{100}$$
$$a = 15.3$$

2. 64.8 is 54% of what number?
$$\frac{64.8}{b} = \frac{54}{100}$$
$$64.8 \times 100 = b \times 54$$
$$\frac{6480}{54} = \frac{54b}{54}$$
$$120 = b$$

3. 132 is what percent of 600?
$$\frac{132}{600} = \frac{p}{100}$$
$$132 \times 100 = 600 \times p$$
$$\frac{13,200}{600} = \frac{600p}{600}$$
$$22 = p$$
22%

4. Answers may vary. Possible solution:
Find "0.7% of what number is $140?"
Let $a = 140$, $b = $ unknown, and $p = 0.7$.
Substitute in formula and solve for b.
$$\frac{a}{b} = \frac{p}{100}$$
$$\frac{140}{b} = \frac{0.7}{100}$$
$$140 \times 100 = 0.7b$$
$$\frac{14,000}{0.7} = \frac{0.7b}{0.7}$$
$$20,000 = b$$
The stock's value was $20,000.

How Am I Doing? Sections 5.1–5.3

1. $0.17 = 17\%$

2. $0.387 = 38.7\%$

3. $7.95 = 795\%$

4. $12.25 = 1225\%$

5. $0.006 = 0.6\%$

6. $0.0004 = 0.04\%$

7. $\frac{17}{100} = 17\%$

8. $\frac{89}{100} = 89\%$

9. $\frac{13.4}{100} = 13.4\%$

10. $\frac{19.8}{100} = 19.8\%$

11. $\frac{6\frac{1}{2}}{100} = 6\frac{1}{2}\%$

12. $\frac{3\frac{5}{8}}{100} = 3\frac{5}{8}\%$

13. $10\overline{)8.0}$ with quotient 0.8
$$\frac{8}{10} = 0.8 = 80\%$$

14. $30\overline{)15.00}$
$$\begin{array}{r} .50 \\ 30\overline{)15.00} \\ \underline{15\ 0} \\ 00 \end{array}$$
$$\frac{15}{30} = 0.50 = 50\%$$

15. $$\begin{array}{r} 2.6 \\ 20\overline{)52.0} \\ \underline{40} \\ 12\ 0 \\ \underline{12\ 0} \\ 0 \end{array}$$
$$\frac{52}{20} = 2.6 = 260\%$$

16.

$$
\begin{array}{r}
1.0625 \\
16\overline{)17.0000} \\
\underline{16} \\
1\ 00 \\
\underline{96} \\
40 \\
\underline{32} \\
80 \\
\underline{80} \\
0
\end{array}
$$

$$\frac{17}{16} = 1.0625 = 106.25\%$$

17.

$$
\begin{array}{r}
.71428 \\
7\overline{)5.000} \\
\underline{4\ 9} \\
10 \\
\underline{7} \\
30 \\
\underline{28} \\
20 \\
\underline{14} \\
60 \\
\underline{56} \\
4
\end{array}
$$

$$\frac{5}{7} \approx 0.7143 = 71.43\%$$

18.

$$
\begin{array}{r}
.28571 \\
7\overline{)2.00000} \\
\underline{1\ 4} \\
60 \\
\underline{56} \\
40 \\
\underline{35} \\
50 \\
\underline{49} \\
10
\end{array}
$$

$$\frac{2}{7} \approx 0.2857 = 28.57\%$$

19.

$$
\begin{array}{r}
.75 \\
24\overline{)18.00} \\
\underline{16\ 8} \\
1\ 20 \\
\underline{1\ 20} \\
0
\end{array}
$$

$$\frac{18}{24} = 0.75 = 75\%$$

20.

$$
\begin{array}{r}
.25 \\
36\overline{)9.00} \\
\underline{7\ 2} \\
1\ 80 \\
\underline{1\ 80} \\
0
\end{array}
$$

$$\frac{9}{36} = 0.25 = 25\%$$

21. $4\dfrac{2}{5} = \dfrac{22}{5} = 4.4 = 440\%$

22. $2\dfrac{3}{4} = \dfrac{11}{4} = 2.75 = 275\%$

23. $\dfrac{1}{300} = \dfrac{1}{3}\left(\dfrac{1}{100}\right) = \dfrac{1}{3}\% \approx 0.33\%$

24. $\dfrac{1}{400} = \dfrac{1}{4}\left(\dfrac{1}{100}\right) = \dfrac{1}{4}\% = 0.25\%$

25. $22\% = \dfrac{22}{100} = \dfrac{11}{50}$

26. $53\% = \dfrac{53}{100}$

27. $150\% = \dfrac{150}{100} = \dfrac{3}{2} = 1\dfrac{1}{2}$

28. $160\% = \dfrac{160}{100} = \dfrac{8}{5} = 1\dfrac{3}{5}$

29. $6\dfrac{1}{3}\% = \dfrac{6\frac{1}{3}}{100} = \dfrac{\frac{19}{3}}{100} = \dfrac{19}{300}$

30. $3\dfrac{1}{8}\% = \dfrac{3\frac{1}{8}}{100} = \dfrac{\frac{25}{8}}{100} = \dfrac{25}{800} = \dfrac{1}{32}$

31. $51\frac{1}{4}\% = \frac{\frac{205}{4}}{100} = \frac{205}{400} = \frac{41}{80}$

32. $43\frac{3}{4}\% = \frac{\frac{175}{4}}{100} = \frac{175}{400} = \frac{7}{16}$

33. What is 70% of 60?
$n = 70\% \times 60$
$n = 0.70(60)$
$n = 42$

34. Find 15% of 140.
$n = 15\% \times 140$
$n = 0.15(140)$
$n = 21$

35. 68 is what percent of 72?
$\frac{p}{100} = \frac{68}{72}$
$72p = 68 \times 100$
$\frac{72p}{72} = \frac{6800}{72}$
$p \approx 94.44$
94.44%

36. What percent of 80 is 64?
$\frac{p}{100} = \frac{64}{80}$
$80p = 64 \times 100$
$\frac{80p}{80} = \frac{6400}{80}$
$p = 80$
80%

37. 8% of what number is 240?
$\frac{8}{100} = \frac{240}{b}$
$8b = 240 \times 100$
$\frac{8b}{8} = \frac{24,000}{8}$
$b = 3000$

38. 354 is 40% of what number?
$\frac{40}{100} = \frac{354}{b}$
$40b = 354 \times 100$
$\frac{40b}{40} = \frac{35,400}{40}$
$b = 885$

5.4 Exercises

1. $n \times 2.5\% = 4500$
$\frac{n \times 0.025}{0.025} = \frac{4500}{0.025}$
$n = 180,000$
180,000 pencils were in the order.

3. $63 = 140\% \times n$
$63 = 1.4n$
$\frac{63}{1.4} = \frac{1.4n}{1.4}$
$45 = n$
His average monthly bill was $45.

5. $432 = n \times 2100$
$\frac{432}{2100} = \frac{2100n}{2100}$
$0.2057 \approx n$
The basement accounts for 20.57%.

7. $n = 6\% \times 65$
$n = 0.06 \times 65$
$n = 3.9$
$3.90 is the tax she paid.

9. $\$38.5 = 7\% \times n$
$38.5 = 0.07n$
$\frac{38.5}{0.07} = \frac{0.07n}{0.07}$
$550 = n$
The price of the mountain bike was $550.

11. $\$1254 = n \times \$4180$
$\frac{1254}{4180} = \frac{4180n}{4180}$
$0.3 = n$
30% of their income goes toward paying the mortgage.

13. $n \times 75\% = 7,200,000$
$\frac{0.75n}{0.75} = \frac{7,200,000}{0.75}$
$n = 9,600,000$
$9,600,000 was the charity's total income last year.

15. $n \times 10,001 = 248$
$\frac{10,001n}{10,001} = \frac{248}{10,001}$
$n \approx 0.0248$
Ted Williams hit 2.48%.

17. $n = 0.6\% \times 6{,}862{,}600{,}000$
$n = 0.006 \times 6{,}862{,}600{,}000$
$n = 41{,}175{,}600$
41,175,600 people lived in Kenya.

19. $105\% \times n = 800$
$$\frac{1.05n}{1.05} = \frac{800}{1.05}$$
$n \approx 761.90$
He can afford to spend $761.90.

21. $100\% \times n + 9\% \times n = 163{,}500$
$109\% \times n = 163{,}500$
$1.09n = 163{,}500$
$$\frac{1.09n}{1.09} = \frac{163{,}500}{1.09}$$
$n = 150{,}000$
The price was $150,000.

23. $3\% + 8\% = 11\%$
$n = 11\% \times 20{,}000$
$n = 0.11 \times 20{,}000$
$n = 2200$
2200 pounds are thrown away.

25. a. $15\% + 12\% + 10\% = 37\%$
$n = 37\% \times 33{,}000{,}000$
$n = 0.37 \times 33{,}000{,}000$
$n = 12{,}210{,}000$
$12,210,000 will go for personnel, food and decorations.

   b. $33{,}000{,}000$
   $\underline{-\ 12{,}210{,}000}$
   $20{,}790{,}000$
   $20,790,000 will go for security, facility rental and all other expenses.

27. $n = 35\% \times 190$
$n = 0.35 \times 190$
$n = 66.5$
Discount is $65.50
Sales price is $190 − $65.50 = $123.50.

29. a. $n = 15\% \times \$8800$
$n = 0.15 \times 8800$
$n = 1320$
Discount is $1320.

   b. Cost of snowmobile is
   $8800 − $1320 = $7480.

Cumulative Review

31. 1,698,4̲81 rounds to 1,698,000.

32. 2,452,3̲9̲9 rounds to 2,452,400.

33. 1.634̲74 rounds to 1.63.

34. 0.799̲5 rounds to 0.800.

35. 0.0556̲13 rounds to 0.0556.

36. 0.07915̲2 rounds to 0.0792.

Quick Quiz 5.4

1. a. $n = 28\% \times 596$
$n = 0.28 \times 596$
$n = 166.88$
The discount was $166.88.

   b. Chris paid $596 − $166.88 = $429.12.

2. $56 = n \times 87$
$$\frac{56}{87} = \frac{87n}{87}$$
$0.6436 \approx n$
64.4% was spam.

3. $4500 = 30\% \times n$
$4500 = 0.3n$
$$\frac{4500}{0.3} = \frac{0.3n}{0.3}$$
$15{,}000 = n$
15,000 people live in the city.

4. Answers may vary. Possible solution: Add the three percents ($23\% + 14\% + 17\% = 54\%$). Then find 54% of $80,000 ($0.54 \times \$80{,}000 = \$43{,}200$). Since $43,200 is less than $48,000, he did not stay within his budget.

5.5 Exercises

1. Commission $= 2\% \times 170{,}000$
$= 0.02 \times 170{,}000$
$= 3400$
He earned (in commission) $3400.

3. Total income $= 300 + 4\% \times 96{,}000$
$= 300 + 0.04 \times 96{,}000$
$= 300 + 3840$
$= 4140$
Her total income was $4140.

5. Increase = 330 − 275 = 55

$n \times 275 = 55$

$\dfrac{n \times 275}{275} = \dfrac{55}{275}$

$n = 0.2$

This is a 20% increase.

7. Decrease = 10.9 − 9.2 = 1.7

$n \times 10.9 = 1.7$

$\dfrac{n \times 10.9}{10.9} = \dfrac{1.7}{10.9}$

$n \approx 0.15596$

This is a 15.6% decrease.

9. $I = P \times R \times T$

$ = 2000 \times 7\% \times 1$

$ = 2000 \times 0.07 \times 1$

$ = 140$

Phil will earn $140.

11. $I = P \times R \times T$

$ = 500 \times 1.5\% \times 1$

$ = 500 \times 0.015 \times 1$

$ = 7.5$

She was charged $7.50.

13. 4 months $= \dfrac{4}{12} = \dfrac{1}{3}$ year

$I = P \times R \times T$

$ = 26,000 \times 12\% \times \dfrac{1}{3}$

$ = 26,000 \times 0.12 \times \dfrac{1}{3}$

$ = 3120 \times \dfrac{1}{3}$

$ = 1040$

He needs to pay $1040.

15. Rate $= \dfrac{72,000}{12,000,000} = 0.006 = 0.6\%$

His commission rate was 0.6%.

17. Sales total $= \dfrac{48,000}{3\%} = \dfrac{48,000}{0.03} = 1,600,000$

The sales total was $1,600,000.

19. Spending $= 15\% \times 265 = 0.15 \times 265 = 39.75$

Each week he can spend $39.75.

21. Thin mints $= 25\% \times 156 = 0.25 \times 156 = 39$

39 boxes are Thin Mints.

23. Increase = 40.2 − 28 = 12.2

$n \times 28 = 12.2$

$\dfrac{n \times 28}{28} = \dfrac{12.2}{28}$

$n \approx 0.4357$

This is a 43.57% increase.

25. Increase = 72 − 40 = 32

$n \times 40 = 32$

$\dfrac{n \times 40}{40} = \dfrac{32}{40}$

$n = 0.8$

This is an 80% increase.

27. a. $n = 2.3\% \times 3700$

$n = 0.023 \times 3700$

$n = 85.1$

$85.10 is the interest he earned.

b.
$$\begin{array}{r} 3700 \\ +\ \ 85.10 \\ \hline 3785.10 \end{array}$$

He withdrew $3785.10.

29. a. Purchases = 52 + 38 + 26 = 116

Tax = 6% × 116 = 0.06 × 116 = $6.96

The total sales tax was $6.96.

b. Total cost = $116 + $6.96 = $122.96

The total cost was $122.96.

31. $n = 114\% \times 9500$

$n = 1.14 \times 9500$

$n = 10,830$

The total amount that will pay off the debt is $10,830.

33. a. $n = 8\% \times 349,000$

$n = 0.08 \times 349,000$

$n = 27,920$

Their down payment was $27,920.

b.
$$\begin{array}{r} 349,000 \\ -\ \ 27,920 \\ \hline 321,080 \end{array}$$

Their mortgage was $321,080.

35. $n \times 840 = 814$

$\dfrac{n \times 840}{840} = \dfrac{814}{840}$

$n \approx 0.969$

96.9% is used to pay off the interest charge.

37. $\text{Tax} = 4.6\% \times 20,456.82$
$= 0.046 \times 20,456.82$
≈ 941.01
The sales tax would be $941.01.

Cumulative Review

39. $3(12-6) - 4(12 \div 3) = 3(6) - 4(4) = 18 - 16 = 2$

40. $7 + 4^3 \times 2 - 15 = 7 + 64 \times 2 - 15$
$= 7 + 128 - 15$
$= 135 - 15$
$= 120$

41. $\left(\dfrac{5}{2}\right)\left(\dfrac{1}{3}\right) - \left(\dfrac{2}{3} - \dfrac{1}{3}\right)^2 = \left(\dfrac{5}{2}\right)\left(\dfrac{1}{3}\right) - \left(\dfrac{1}{3}\right)^2$
$= \left(\dfrac{5}{2}\right)\left(\dfrac{1}{3}\right) - \dfrac{1}{9}$
$= \dfrac{5}{6} - \dfrac{1}{9}$
$= \dfrac{15}{18} - \dfrac{2}{18}$
$= \dfrac{13}{18}$

42. $(6.8 - 6.6)^2 + 2(1.8) = (0.2)^2 + 2(1.8)$
$= 0.04 + 3.6$
$= 3.64$

Quick Quiz 5.5

1. $n = 8\% \times 325,000$
$n = 0.08 \times 325,000$
$n = 26,000$
His commission is $26,000.

2. $\text{Increase} = 275 - 160 = 115$
$n \times 160 = 115$
$\dfrac{n \times 160}{160} = \dfrac{115}{160}$
$n = 0.71875$
The percent of increase is 71.875%.

3. $\text{Six months} = \dfrac{6}{12} = \dfrac{1}{2}$ year
$I = P \times R \times T$
$= 4600 \times 13\% \times \dfrac{1}{2}$
$= 4600 \times 0.13 \times \dfrac{1}{2}$
$= 598 \times \dfrac{1}{2}$
$= 299$
The simple interest is $299.

4. Answers will vary. Possible solution:
Use the formula $I = P \times R \times T$.

Let $P = 5800$, $R = 16\%$ of 0.16, and $T = \dfrac{3}{12} = \dfrac{1}{4}$

(since T must be in years).
$I = 5800 \times 0.16 \times \dfrac{1}{4} = 232$
The simple interest is $232.

Use Math to Save Money

1. $1000 + 36 \times 388.06 = 1000 + 13,970.16$
$= 14,970.16$
Louvy would pay $14,970.16 over the entire length of the lease.

2. $1000 + 36 \times 669.28 = 1000 + 24,094.08$
$= 25,094.08$
Louvy would pay $25,094.08 over the entire length of the loan.

3. $14,970.16 + 11,000.00 = 25,970.16$
The total cost would be $25,970.16.

4. $25,970.16 - 25,094.08 = 876.08$
Louvy saves $876.08 if he buys the car instead of leasing it.

5. $669.28 - 388.06 = 281.22$
He will save $281.22 each month in car payments.

6. To get the best overall price, Louvy should buy the car since he will save $876.08 on the total price. To get a lower monthly payment, Louvy should lease the car since he will save $281.22 each month in car payments.

7. Answers will vary.

8. Answers will vary.

9. Answers will vary.

You Try It

1. a. $0.12 = 12\%$

   b. $2.35 = 235\%$

   c. $0.0071 = 0.71\%$

   d. $6.5 = 650\%$

   e. $1.005 = 100.5\%$

2. a. $\dfrac{78}{100} = 78\%$

   b. $\dfrac{125}{100} = 125\%$

   c. $\dfrac{9}{100} = 9\%$

   d. $\dfrac{3\frac{1}{4}}{100} = 3\frac{1}{4}\%$

   e. $\dfrac{18.5}{100} = 18.5\%$

3. a. $\dfrac{34}{50} = 0.68 = 68\%$

   b. $\dfrac{8}{25} = 0.32 = 32\%$

   c. $\dfrac{4}{500} = 0.008 = 0.8\%$

   d. $\dfrac{480}{300} = 1.6 = 160\%$

   e. $3\dfrac{1}{4} = 3.25 = 325\%$

4. a. $86\% = 0.86$

   b. $1\% = 0.01$

   c. $1.8\% = 0.018$

   d. $335\% = 3.35$

   e. $7.54\% = 0.0754$

5. a. $20\% = \dfrac{20}{100} = \dfrac{1}{5}$

   b. $45\% = \dfrac{45}{100} = \dfrac{9}{20}$

   c. $260\% = \dfrac{260}{100} = \dfrac{13}{5} = 2\dfrac{3}{5}$

   d. $1.5\% = 0.015 = \dfrac{15}{1000} = \dfrac{3}{200}$

   e. $5.84\% = 0.0584 = \dfrac{584}{10,000} = \dfrac{73}{1250}$

   f. $9\dfrac{2}{3}\% = \dfrac{\frac{29}{3}}{100} = \dfrac{29}{300}$

6. a. $360\% = 3.60 = 3\dfrac{60}{100} = 3\dfrac{3}{5}$

   b. $185\% = 1.85 = 1\dfrac{85}{100} = 1\dfrac{17}{20}$

   c. $208\% = 2.08 = 2\dfrac{8}{100} = 2\dfrac{2}{25}$

7. a. What is 12% of 300?
$$n = 12\% \times 300$$
$$n = (0.12)(300)$$
$$n = 36$$

   b. 22% of what number is 165?
$$22\% \times n = 165$$
$$0.22n = 165$$
$$\dfrac{0.22n}{0.22} = \dfrac{165}{0.22}$$
$$n = 750$$

   c. What percent of 90 is 32?
$$n \times 90 = 32$$
$$90n = 32$$
$$\dfrac{90n}{90} = \dfrac{32}{90}$$
$$n = 0.3555...$$
$$n \approx 35.56\%$$

8. a. What is 32% of 180?

$$\frac{a}{180} = \frac{32}{100}$$

$$\frac{a}{180} = \frac{8}{25}$$

$$25a = (8)(180)$$

$$\frac{25a}{25} = \frac{1440}{25}$$

$$a = 57.6$$

b. 48% of what number is 15?

$$\frac{15}{b} = \frac{48}{100}$$

$$\frac{15}{b} = \frac{12}{25}$$

$$(15)(25) = 12b$$

$$\frac{375}{12} = \frac{12b}{12}$$

$$31.25 = b$$

c. What percent of 300 is 120?

$$\frac{120}{300} = \frac{p}{100}$$

$$\frac{2}{5} = \frac{p}{100}$$

$$(2)(100) = 5p$$

$$\frac{200}{5} = \frac{5p}{5}$$

$$40\% = p$$

9. a. Discount = (0.15)($25) = $33.75

b. 225 − 33.75 = 191.25
He paid $191.25 for the DVD player.

10. Commission = (0.08)(15,500) = 1240
She earns a commission of $1240.

11. $I = (8000)(0.06)(3) = (480)(3) = 1440$
Olivia owed $1440 in interest.

12. Increase = 310,000 − 270,000 = 40,000

$$\frac{40,000}{270,000} \approx 0.148$$

This is approximately a 14.8% increase.

Chapter 5 Review Problems

1. 0.62 = 62%

2. 0.43 = 43%

3. 0.372 = 37.2%

4. 2.2 = 220%

5. 2.52 = 252%

6. 1.036 = 103.6%

7. 0.006 = 0.6%

8. $\dfrac{62.5}{100} = 62.5\%$

9. $\dfrac{4\frac{1}{12}}{100} = 4\frac{1}{12}\%$

10. $\dfrac{317}{100} = 317\%$

11. 32% = 0.32

12. 15.75% = 0.1575

13. 236% = 2.36

14. $32\frac{1}{8} = 32.125\% = 0.32125$

15. $\dfrac{19}{25} = 0.76 = 76\%$

16. $\dfrac{11}{20} = 0.55 = 55\%$

17. $\dfrac{9}{40} = 0.225 = 22.5\%$

18. $\dfrac{7}{12} \approx 0.5833 = 58.33\%$

19. $\dfrac{14}{15} \approx 0.9333 = 93.33\%$

20. $2\frac{1}{4} = 2.25 = 225\%$

21. $3\frac{3}{4} = 3.75 = 375\%$

22. $2\frac{7}{9} \approx 2.7778 = 277.78\%$

23. $\dfrac{200}{80} = 2.5 = 250\%$

24. $\dfrac{5}{800} = 0.00625 \approx 0.63\%$

25. $72\% = \dfrac{72}{100} = \dfrac{72 \div 4}{100 \div 4} = \dfrac{18}{25}$

26. $175\% = \dfrac{175}{100} = \dfrac{175 \div 25}{100 \div 25} = \dfrac{7}{4} = 1\dfrac{3}{4}$

27. $16.4\% = 0.164 = \dfrac{164}{1000} = \dfrac{164 \div 4}{1000 \div 4} = \dfrac{41}{250}$

28. $13\dfrac{1}{4}\% = \dfrac{31\frac{1}{4}}{100}$

$\qquad = 31\dfrac{1}{4} \div 100$

$\qquad = \dfrac{125}{4} \times \dfrac{1}{100}$

$\qquad = \dfrac{125}{400}$

$\qquad = \dfrac{125 \div 25}{400 \div 25}$

$\qquad = \dfrac{5}{16}$

29. $0.08\% = 0.0008 = \dfrac{8}{10,000} = \dfrac{8 \div 8}{10,000 \div 8} = \dfrac{1}{1250}$

30. $0.04\% = 0.0004 = \dfrac{4}{10,000} = \dfrac{4 \div 4}{10,000 \div 4} = \dfrac{1}{2500}$

31. $\begin{array}{r} 0.6 \\ 5\overline{)3.0} \\ \underline{3\ 0} \\ 0 \end{array}$

$0.6 = 60\%$

$\dfrac{3}{5};\ 0.6;\ 60\%$

32. $\dfrac{7}{10} = 0.7 = 70\%$

$\dfrac{7}{10};\ 0.7;\ 70\%$

33. $37.5\% = 0.375 = \dfrac{375}{1000} = \dfrac{3}{8}$

$\dfrac{3}{8};\ 0.375;\ 37.5\%$

34. $56.25\% = 0.5625 = \dfrac{5625}{10,000} = \dfrac{9}{16}$

$\dfrac{9}{16};\ 0.5625;\ 56.25\%$

35. $0.008 = 0.8\%$

$0.008 = \dfrac{8}{1000} = \dfrac{1}{125}$

$\dfrac{1}{125};\ 0.008;\ 0.8\%$

36. $0.45 = 45\%$

$0.45 = \dfrac{45}{100} = \dfrac{9}{20}$

$\dfrac{9}{20};\ 0.45;\ 45\%$

37. What is 20% of 85?
$n = 20\% \times 85$
$n = 0.2 \times 85$
$n = 17$

38. What is 25% of 92?
$n = 25\% \times 92$
$n = 0.25 \times 92$
$n = 23$

39. Find 162% of 60.
$n = 162\% \times 60$
$n = 162 \times 60$
$n = 97.2$

40. 15 is 25% of what number?
$25\% \times n = 15$
$0.25n = 15$
$\dfrac{0.25n}{0.25} = \dfrac{15}{0.25}$
$n = 60$

41. 30 is 75% of what number?
$75\% \times n = 30$
$0.75n = 30$
$\dfrac{0.75n}{0.75} = \dfrac{30}{0.75}$
$n = 40$

42. 92% of what number is 147.2?

$$92\% \times n = 147.2$$
$$0.92n = 147.2$$
$$\frac{0.92n}{0.92} = \frac{147.2}{0.92}$$
$$n = 160$$

43. 50 is what percent of 130?

$$50 = n \times 130$$
$$\frac{50}{130} = \frac{n \times 130}{130}$$
$$0.3846 \approx n$$
$$38.46\%$$

44. 70 is what percent of 180?

$$70 = n \times 180$$
$$\frac{70}{180} = \frac{n \times 180}{180}$$
$$0.3889 \approx n$$
$$38.89\%$$

45. What percent of 70 is 14?

$$n \times 70 = 14$$
$$\frac{n \times 70}{70} = \frac{14}{70}$$
$$n = 0.2$$
$$20\%$$

46. $n = 35\% \times 140$
$n = 0.35 \times 140$
$n = 49$
49 students are sophomores.

47. $n = 64\% \times 150$
$n = 0.64 \times 150$
$n = 96$
96 trucks had four-wheel drive.

48. $n \times 61\% = 6832$

$$0.61n = 6832$$
$$\frac{0.61n}{0.61} = \frac{6832}{0.61}$$
$$n = 11,200$$

Two years ago, it was worth $11,200.

49. $n \times 12\% = 9624$

$$0.12n = 9624$$
$$\frac{0.12n}{0.12} = \frac{9624}{0.12}$$
$$n = 80,200$$

The total budget was $80,200.

50. Days = 29 + 31 + 30 = 90
Rain days = 20 + 18 + 16 = 54

$$90 \times n = 54$$
$$\frac{90 \times n}{90} = \frac{54}{90}$$
$$n = 0.6$$

It rained 60% of the time.

51. $600 \times n = 45$

$$\frac{600 \times n}{600} = \frac{45}{600}$$
$$n = 0.075$$

7.5% of the applicants obtained a job.

52. What is 5% of 3670?

$n = 5\% \times 3670$
$n = 0.05 \times 3670$
$n = 183.5$
He paid $183.50.

53. What is 6% of 12,600?

$n = 6\% \times 12,600$
$n = 0.06 \times 12,600$
$n = 756$
They paid $756.

54. Commission $= 6\% \times 83,500$
$= 0.06 \times 83,500$
$= 5010$
His commission was $5010.

55. Commission $= 7.5\% \times 16,000$
$= 0.075 \times 16,000$
$= 1200$
She earned $1200.

56. a. Discount $= 25\% \times \$1450$
$= 0.25 \times 1450$
$= 362.5$
$362.50 was the discount.

b. Total pay $= \$1450 - \$362.50 = 1087.5$
$1087.50 was the price for the set.

57. Increase = 18,400 − 18,040 = 360

$$\frac{p}{100} = \frac{360}{18,040}$$
$$18,040p = 100 \times 360$$
$$\frac{18,040p}{18,040} = \frac{36,000}{18,040}$$
$$p \approx 1.995$$

This was a 2% increase.

58. Increase = 6585 − 5000 = 1585

$$\frac{p}{100} = \frac{1585}{5000}$$
$$5000p = 100 \times 1585$$
$$\frac{5000p}{5000} = \frac{158,500}{5000}$$
$$p = 31.7$$

32% was the percent of increase.

59. a. Discount = 14% of 24,000
$$= 0.14 \times 24,000$$
$$= \$3360$$
The discount is $3360.

   b. Cost = 24,000 − 3360 = $20,640.

60. $I = PRT$

   a. $I = 6000(0.11)(0.5) = \$330$
   In six months, she will earn $330.

   b. $I = 6000(0.11)(2) = \$1320$
   In two years, she will earn $1320.

61. a. $3 \text{ months} = \frac{3}{12} = \frac{1}{4} \text{ year}$
$$I = P \times R \times T$$
$$= 3000 \times 8\% \times \frac{1}{4}$$
$$= 3000 \times 0.08 \times \frac{1}{4}$$
$$= 240 \times \frac{1}{4}$$
$$= 60$$
In three months $60 will be due.

   b. $I = P \times R \times T$
$$= 3000 \times 8\% \times 3$$
$$= 3000 \times 0.08 \times 3$$
$$= 240 \times 3$$
$$= 720$$
In three years $720 will be due.

How Am I Doing? Chapter 5 Test

   1. $0.57 = 57\%$

   2. $0.01 = 1\%$

   3. $0.008 = 0.8\%$

   4. $12.8 = 1280\%$

   5. $3.56 = 356\%$

6. $\dfrac{71}{100} = 71\%$

7. $\dfrac{1.8}{100} = 1.8\%$

8. $\dfrac{3\frac{1}{7}}{100} = 3\frac{1}{7}\%$

9.
$$\begin{array}{r} 0.475 \\ 40\overline{)19.000} \\ \underline{16\ 0} \\ 3\ 00 \\ \underline{2\ 80} \\ 200 \\ \underline{200} \\ 0 \end{array}$$

$\dfrac{19}{40} = 0.475 = 47.5\%$

10.
$$\begin{array}{r} 0.75 \\ 36\overline{)27.00} \\ \underline{25\ 2} \\ 1\ 80 \\ \underline{1\ 80} \\ 0 \end{array}$$

$\dfrac{27}{36} = 75\%$

11.
$$\begin{array}{r} 3 \\ 75\overline{)225} \\ \underline{225} \\ 0 \end{array}$$

$\dfrac{225}{75} = 3 = 300\%$

12.
$$\begin{array}{r} 0.75 \\ 4\overline{)3.00} \\ \underline{2\ 8} \\ 20 \\ \underline{20} \\ 0 \end{array}$$

$1\dfrac{3}{4} = 1.75 = 175\%$

13. $0.0825 = 8.25\%$

14. $3.024 = 302.4\%$

15. $152\% = \dfrac{152}{100} = \dfrac{38}{25} = 1\dfrac{13}{25}$

16. $7\dfrac{3}{4}\% = \dfrac{7\frac{3}{4}}{100} = 7\dfrac{3}{4} \div 100 = \dfrac{31}{4} \times \dfrac{1}{100} = \dfrac{31}{400}$

17. $n = 40\% \times 50$
$n = 0.4 \times 50$
$n = 20$

18. $33.8 = 26\% \times n$
$33.8 = 0.26n$
$\dfrac{33.8}{0.26} = \dfrac{0.26n}{0.26}$
$130 = n$

19. $n \times 72 = 40$
$\dfrac{n \times 72}{72} = \dfrac{40}{72}$
$n \approx 0.5556 = 55.56\%$

20. $n = 0.8\% \times 25{,}000$
$n = 0.008 \times 25{,}000$
$n = 200$

21. $16\% \times n = 800$
$0.16n = 800$
$\dfrac{0.16n}{0.16} = \dfrac{800}{0.16}$
$n = 5000$

22. $92 = n \times 200$
$\dfrac{92}{200} = \dfrac{n \times 200}{200}$
$n = 0.46$
$n = 46\%$

23. $132\% \times 530 = n$
$1.32 \times 530 = n$
$699.6 = n$

24. $p \times 75 = 15$
$\dfrac{75p}{75} = \dfrac{15}{75}$
$p = 0.2 = 20\%$

25. $n = 4\% \times 152{,}300$
$n = 0.04 \times 152{,}300$
$n = 6092$
$6092 is her commission.

26. a. $33\% \times 457 = n$
$0.33 \times 457 = n$
$150.81 = n$
$150.81 is the discount.

  b. $457 - 150.81 = 306.19$
$306.19 is the amount they paid.

27. $75 = n \times 84$
$\dfrac{75}{84} = \dfrac{n \times 84}{84}$
$n \approx 0.8929$
89.29% of the parts were not defective.

28. Increase $= 228 - 185 = 43$
$\dfrac{p}{100} = \dfrac{43}{185}$
$185p = 100 \times 43$
$\dfrac{185p}{185} = \dfrac{4300}{185}$
$p \approx 23.24$
23.24% is the percent of increase.

29. $5160 = 43\% \times n$
$5160 = 0.43n$
$\dfrac{5160}{0.43} = \dfrac{0.43n}{0.43}$
$12{,}000 = n$
There are 12,000 registered voters.

30. a. $I = P \times R \times T = 3000 \times 0.16 \times \dfrac{6}{12} = 240$
In six months she paid $240.

  b. $I = P \times R \times T = 3000 \times 0.16 \times 2 = 960$
In two years she paid $960.

Chapter 6

1. We know that each mile has 5280 feet. Each foot has 12 inches. Therefore, we know that one mile is $5280 \times 12 = 63,360$ inches. The unit fraction is $\dfrac{63,360 \text{ inches}}{1 \text{ mile}}$. So we multiply

   $23 \text{ miles} \times \dfrac{63,360 \text{ inches}}{1 \text{ mile}}$. The mile units divide out. We obtain 1,457,280 inches. Thus, 23 miles = 1,457,280 inches.

3. 1760 yards = 1 mile

5. 1 ton = 2000 pounds

7. 4 quarts = 1 gallon

9. 1 quart = 2 pints

11. $21 \text{ feet} \times \dfrac{1 \text{ yard}}{3 \text{ feet}} = 7 \text{ yards}$

13. $108 \text{ inches} \times \dfrac{1 \text{ foot}}{12 \text{ inches}} = 9 \text{ feet}$

15. $9 \text{ feet} \times \dfrac{12 \text{ inches}}{1 \text{ foot}} = 108 \text{ inches}$

17. $10,560 \text{ feet} \times \dfrac{1 \text{ mile}}{5280 \text{ feet}} = 2 \text{ miles}$

19. $7 \text{ miles} \times \dfrac{1760 \text{ yards}}{1 \text{ mile}} = 12,320 \text{ yards}$

21. $12 \text{ gallons} \times \dfrac{4 \text{ quarts}}{1 \text{ gallon}} = 48 \text{ quarts}$

23. $48 \text{ quarts} \times \dfrac{1 \text{ gallon}}{4 \text{ quarts}} = 12 \text{ gallons}$

25. $16 \text{ cups} \times \dfrac{8 \text{ fluid ounces}}{1 \text{ cup}} = 128 \text{ fluid ounces}$

27. $8\dfrac{1}{2} = \dfrac{17}{2} \text{ gallons} \times \dfrac{8 \text{ pints}}{1 \text{ gallon}} = 68 \text{ pints}$

29. $77 \text{ days} \times \dfrac{1 \text{ week}}{7 \text{ days}} = 11 \text{ weeks}$

31. $960 \text{ seconds} \times \dfrac{1 \text{ minute}}{60 \text{ seconds}} = 16 \text{ minutes}$

33. $8 \text{ ounces} \times \dfrac{1 \text{ pound}}{16 \text{ ounces}} = 0.5 \text{ pound}$

35. $12,500 \text{ pounds} \times \dfrac{1 \text{ ton}}{2000 \text{ pounds}} = 6.25 \text{ tons}$

37. $15 \text{ pints} \times \dfrac{2 \text{ cups}}{1 \text{ pint}} = 30 \text{ cups}$

39. $2.25 \text{ pounds} \times \dfrac{16 \text{ ounces}}{1 \text{ pound}} = 36 \text{ ounces}$

41. $66 \text{ inches} \times \dfrac{1 \text{ foot}}{12 \text{ inches}} = 5.5 \text{ feet}$

43. $26.2 \text{ miles} \times \dfrac{5280 \text{ feet}}{1 \text{ mile}} = 138,336 \text{ feet}$
    He traveled 138,336 feet.

45. $35,840 \text{ feet} \times \dfrac{1 \text{ mile}}{5280 \text{ feet}} = 6.79 \text{ miles}$
    The depth of the trench is 6.79 miles.

47. $26 \text{ ounces} \times \dfrac{1 \text{ pound}}{16 \text{ ounces}} \times \dfrac{\$6.00}{1 \text{ pound}} = \$9.75$
    The mushrooms cost \$9.75.

49. **a.** $3 \text{ feet} \times \dfrac{12 \text{ inches}}{1 \text{ foot}} = 36 \text{ inches}$
    3 feet 6 inches = 36 + 6 = 42 inches
    $2 \text{ feet} \times \dfrac{12 \text{ inches}}{1 \text{ foot}} = 24 \text{ inches}$
    2 feet 5 inches = 24 + 5 = 29 inches
    perimeter = 2(42) + 2(29)
    $\qquad\quad = 84 + 58$
    $\qquad\quad = 142 \text{ inches}$
    The perimeter is 142 inches.

    b. cost = \$0.60 × 142 = \$85.20
    It will cost \$85.20.

51. $7200 \text{ quarts} \times \dfrac{4 \text{ cups}}{1 \text{ quart}} = 28,800 \text{ cups}$

Your heart pumps 28,800 cups of blood through your body.

53. 1760 rounds to 2000

$6 \text{ miles} \times \dfrac{2000 \text{ yards}}{1 \text{ mile}} \approx 12,000 \text{ yards}$

The distance is approximately 12,000 yards.

55. 33,000 rounds to 30,000
5280 rounds to 5000

$30,000 \text{ feet} \times \dfrac{1 \text{ mile}}{5000 \text{ feet}} \approx 6 \text{ miles}$

The plane is about 6 miles high.

Cumulative Review

57. $560 - 515 = \$45 \text{ per month}$
$\$45 \times 12 \times 20 = \$10,800$
They will save a total of \$10,800.

58. $\text{part used} = 3(1.5) + 2(2.25) = 4.5 + 4.5 = 9 \text{ GB}$

$\text{part not used} = 60 - 9 = 51 \text{ GB}$

$\text{percent} = \dfrac{51}{60} = 0.85 = 85\%$

85% of her drive is still free for storage.

59. $\dfrac{n}{115} = \dfrac{7}{5}$
$n \times 5 = 7 \times 115$
$\dfrac{n \times 5}{5} = \dfrac{805}{5}$
$n = 161 \text{ miles}$
They are likely to cover 161 miles.

60. $\dfrac{n}{1300} = \dfrac{12}{150}$
$n \times 150 = 1300 \times 12$
$\dfrac{n \times 150}{150} = \dfrac{15,600}{150}$
$n = 104$
104 students dropped the course during the first two weeks of the semester.

Quick Quiz 6.1

1. $3.5 \text{ tons} \times \dfrac{2000 \text{ pounds}}{1 \text{ ton}} = 7000 \text{ pounds}$

2. $4.5 \text{ yards} \times \dfrac{3 \text{ feet}}{1 \text{ yard}} = 13.5 \text{ feet}$

3. $24 \text{ ounces} \times \dfrac{1 \text{ pound}}{16 \text{ ounces}} = 1.5 \text{ pounds}$

4. Answers may vary. Possible solution: There are 2 pints in 1 quart. Use the unit fraction $\dfrac{1 \text{ quart}}{2 \text{ pints}}$ to convert.

$250 \text{ pints} \times \dfrac{1 \text{ quart}}{2 \text{ pints}} = 125 \text{ quarts}$

6.2 Exercises

1. hecto- is the prefix for hundred.

3. deci- is the prefix for tenth.

5. kilo- is the prefix for thousand.

7. 46 centimeters = 460 millimeters

9. 2.61 kilometers = 2610 meters

11. 12,500 millimeters = 12.5 meters

13. 7.32 centimeters = 0.0732 meters

15. 2 kilometers = 200,000 centimeters

17. 78,000 millimeters = 0.078 kilometer

19. 35 mm = 3.5 cm = 0.035 m

21. 4.5 km = 4500 m = 450,000 cm

23. (b), 24 km would be the most reasonable measurement.

25. (c), 80.5 mm would be the most reasonable measurement.

27. (a), 45 cm would be the most reasonable measurement.

29. (a), 0.5 km would be the most reasonable measurement.

31. (b), 11.9 cm would be the most reasonable measurement.

33. 390 decimeters = 39 meters

35. 800 dekameters = 8000 meters

37. 48.2 meters = 0.482 hectometer

Copyright © 2012 Pearson Education, Inc.

39. 243 m + 2.7 km + 312 m
= 243 m + 2700 m + 312 m
= 3255 m

41. 225 mm + 12.7 cm + 148 cm
= 22.5 cm + 12.7 cm + 148 cm
= 183.2 cm

43. 15 mm + 2 dm + 42 cm
= 1.5 cm + 20 cm + 42 cm
= 63.5 cm

45. 0.95 cm + 1.35 cm + 2.464 mm
= 0.95 cm + 1.35 cm + 0.2464 cm
= 2.5464 cm or 25.464 mm
The stereo casing is 2.5464 cm or 25.464 mm
thick.

47. 65 cm + 80 mm + 2.5 m
= 0.65 m + 0.08 m + 2.5 m
= 3.23 m

49. 46 m + 986 cm + 0.884 km
= 46 m + 9.86 m + 884 m
= 939.86 m

51. 96.4 centimeters = 0.964 meter

53. False; 1 kilometer = 1000 meters

55. True

57. True

59. False; a meter is longer than a yard.

61. a. 5072 meters = 507,200 centimeters

    b. 5072 meters = 5.072 kilometers
       The track is 5.072 kilometers high.

63. 0.000000254 centimeter
= 0.00000000254 meter
The diameter is 0.00000000254 meter.

65. $\begin{array}{r} 420,000 \\ -\,314,000 \\ \hline 106,000 \end{array}$

The subway system in Shanghai is
106,000 meters longer than that in Seoul.

67. 415 kilometers $\times \dfrac{1}{1000}$ = 0.415 megameter

69. $420 \times 0.62 = 260.4$
Shanghai's subway system is about 260.4 miles
long.

Cumulative Review

70. 57% of what number is 2850?
$$\frac{57}{100} = \frac{2850}{n}$$
$$57 \times n = 100 \times 2850$$
$$\frac{57 \times n}{57} = \frac{285,000}{57}$$
$$n = 5000$$

71. $n = 0.03\% \times 5900$
$n = 0.0003 \times 5900$
$n = 1.77$

72. Discount = (0.20)(660) = 132
Sale price = 660 − 132 = 528
Chloe paid $528

73. Commission = 0.08(27,000) = 2160
Frank's commission is $2160.

Quick Quiz 6.2

1. 45.9 meters = 4590 centimeters

2. 0.0283 centimeters = 0.283 millimeter

3. 5160 meters = 5.16 kilometers

4. Answers may vary. Possible solution: We are
converting from a smaller unit, cm, to a larger
unit, km. Therefore, there will be fewer
kilometers than centimeters. Move the decimal
place 5 units to the left.
5643 cm = 0.05643 km

6.3 Exercises

1. 1 kL

3. 1 mg

5. 1 g

7. 9 kL = 9000 L

9. 12 L = 12,000 mL

11. 18.9 mL = 0.0189 L

13. 752 L = 0.752 kL

15. 5.652 kL = 5,652,000 mL

17. 82 mL = 82 cm^3 = 82 cc

19. 24,418 mL = 0.024418 kL

21. 74 L = 74,000 mL = 74,000 cm^3

23. 216 g = 0.216 kg

25. 35 mg = 0.035 g

27. 6328 mg = 6.328 g

29. 2.92 kg = 2920 g

31. 2.4 t = 2400 kg

33. 7 mL = 0.007 L = 0.000007 kL

35. 84 cm^3 = 0.084 L = 0.000084 kL

37. 0.033 kg = 33 g = 33,000 mg

39. 2.58 metric tons = 2580 kg = 2,580,000 g

41. (b), 0.32 L would be the most reasonable measurement.

43. (a), 100 t would be the most reasonable measurement.

45. 83 L + 822 mL + 30.1 L
= 83 L + 0.822 L + 30.1 L
= 113.922 L or 113,922 mL

47. 20 g + 52 mg + 1.5 kg
= 20 g + 0.052 g + 1500 g
= 1520.052 g or 1,520,052 mg

49. True

51. False; it can be purchased in liter containers.

53. False; a reasonable weight would be 100,000 grams.

55. True

57. $\dfrac{n}{3.624 \text{ kg}} = \dfrac{\$8.99}{0.453 \text{ kg}}$

$n = \dfrac{3.624 \text{ kg} \times \$8.99}{0.453 \text{ kg}}$

$n = \$71.92$

She spent $71.92 on coffee.

59. 0.4 L = 400 mL

$\dfrac{\$850}{\text{mL}} \times 400 \text{ mL} = \$340,000$

It will cost the company $340,000.

61.   12,900,000,000
  − 10,400,000,000
  ‾‾‾‾‾‾‾‾‾‾‾‾‾‾
   2,500,000,000

It increased by 2,500,000,000 metric tons.

63. 17,200,000,000 × 1000 = 17,200,000,000,000

It is estimated to increase to 17,200,000,000,000 kg.

65. Increase = 12.9 − 11.6 = 1.3 billion metric tons

Percent of increase: $\dfrac{1.3}{11.6} = \dfrac{p}{100}$

$11.6p = 130$

$\dfrac{11.6p}{11.6} = \dfrac{130}{11.6}$

$p = 11.21\%$

$15.6 \times 11.21\% = 0.156 \times 0.1121 \approx 1.7$

$15.6 + 1.7 = 17.3$

The emissions will be about 17,300,000,000 metric tons.

Cumulative Review

67. $\text{percent} = \dfrac{14}{70} = 0.20 = 20\%$

68. $n = 23\%$ of 250
$n = 0.23 \times 250$
$n = 57.5$

69. $n = 10\%$ of 4800
$n = 0.1 \times 4800$
$n = 480$
Price = 4800 − 480 = 4320
$n = 105\% \times 4320$
$n = 1.05 \times 4320$
$n = 4536$
Marilyn paid $4536.

70. Commission = 8% × 8960
= 0.08 × 8960
= 716.80
She earned a commission of $716.80.

Quick Quiz 6.3

1. 671 grams = 0.671 kilogram

Copyright © 2012 Pearson Education, Inc.

2. 8.52 liters = 8520 milliliters

3. 45.62 milligrams = 0.04562 gram

4. Answers may vary. Possible solution: We are converting from a larger unit, kg, to a smaller unit, mg. Therefore, there will be more mg than km. Move the decimal place 6 units to the right. 54 kg = 54,000,000 mg

How Am I Doing? Sections 6.1–6.3

1. $48 \text{ feet} \times \dfrac{1 \text{ yard}}{3 \text{ feet}} = 16 \text{ yards}$

2. $24 \text{ quarts} \times \dfrac{1 \text{ gallon}}{4 \text{ quarts}} = 6 \text{ gallons}$

3. $3 \text{ miles} \times \dfrac{1760 \text{ yards}}{\text{miles}} = 5280 \text{ yards}$

4. $4.5 \text{ tons} \times \dfrac{2000 \text{ pounds}}{1 \text{ ton}} = 9000 \text{ pounds}$

5. $22 \text{ minutes} \times \dfrac{60 \text{ seconds}}{1 \text{ minute}} = 1320 \text{ seconds}$

6. $48 \text{ pints} \times \dfrac{1 \text{ gallon}}{8 \text{ pints}} = 6 \text{ gallons}$

7. 2 pounds = 32 ounces
32 + 4 = 36 ounces

$36 \text{ ounces} \times \dfrac{1 \text{ pound}}{16 \text{ ounces}} = 2.25 \text{ pounds}$

$2.25 \text{ pounds} \times \dfrac{\$6.80}{1 \text{ pound}} = \$15.30$

Isabel spent $15.30.

8. 6.75 km = 6750 m

9. 73.9 m = 7390 cm

10. 34 cm = 340 mm

11. 27 mm = 0.027 m

12. 5296 mm = 529.6 cm

13. 482 m = 0.482 km

14. 1.2 km = 1200 m

$$\begin{array}{r} 1200 \text{ m} \\ 192 \text{ m} \\ +\ \ 984 \text{ m} \\ \hline 2376 \text{ m} \end{array}$$

15. 305 cm + 82.5 m + 6150 mm
= 3.05 m + 82.5 m + 6.150 m
= 91.7 m

16.
$$\begin{array}{r} 78 \text{ cm} = 0.78 \text{ m} \\ +\ 128 \text{ cm} = 1.28 \text{ m} \\ \hline 2.06 \text{ m} \end{array}$$
$$\begin{array}{r} 3.40 \\ -2.06 \\ \hline 1.34 \text{ m or } 134 \text{ cm} \end{array}$$
The length that has single insulation is 1.34 m or 134 cm.

17. 5.66 L = 5660 mL

18. 535 g = 0.535 kg

19. 56.3 kg = 0.0563 t

20. 4.8 kL = 4800 L

21. 568 mg = 0.568 g

22. $8.9 \text{ L} = 8900 \text{ cm}^3$

23. 75 kg = 75,000 g
$$\dfrac{x}{75,000} = \dfrac{7.75}{5000}$$
$$5000x = 7.75(75,000)$$
$$\dfrac{5000x}{5000} = \dfrac{581,250}{5000}$$
$$x = 116.25$$
The mission will spend $116.25.

24. 35 kg = 35,000 g
$$\dfrac{x}{35,000} = \dfrac{32.50}{5000}$$
$$5000x = 32.50(35,000)$$
$$\dfrac{5000x}{5000} = \dfrac{1,137,500}{5000}$$
$$x = 227.5$$
It will cost $227.50.

25. 200 mL = 0.2 L
$$0.2 \text{ L} \times \dfrac{36}{1 \text{ L}} = \$7.20$$
They paid $7.20.

26. $0.5 \text{ kg} = 500 \text{ g}$

$500 \text{ g} \times \dfrac{\$38.40}{1 \text{ g}} = \$19,200$

She paid $19,200.

6.4 Exercises

1. A meter is approximately the same length as a yard. The meter is slightly longer.

3. An inch is approximately twice the length of a centimeter.

5. $8 \text{ ft} \times \dfrac{0.305 \text{ m}}{1 \text{ ft}} \approx 2.44 \text{ m}$

7. $9 \text{ in.} \times \dfrac{2.54 \text{ cm}}{1 \text{ in.}} \approx 22.86 \text{ cm}$

9. $32 \text{ m} \times \dfrac{1.09 \text{ yd}}{1 \text{ m}} \approx 34.88 \text{ yd}$

11. $30.8 \text{ yd} \times \dfrac{0.914 \text{ m}}{1 \text{ yd}} \approx 28.15 \text{ m}$

13. $82 \text{ mi} \times \dfrac{1.61 \text{ km}}{1 \text{ mi}} = 132.02 \text{ km}$

15. $9.25 \text{ m} \times \dfrac{1.09 \text{ yd}}{1 \text{ m}} \approx 10.08 \text{ yd}$

17. $17.5 \text{ cm} \times \dfrac{0.394 \text{ in.}}{1 \text{ cm}} \approx 6.90 \text{ in.}$

19. $200 \text{ m} \times \dfrac{3.28 \text{ ft}}{1 \text{ m}} \approx 656 \text{ ft}$

21. $5 \text{ km} \times \dfrac{0.62 \text{ mi}}{1 \text{ km}} = 3.1 \text{ mi}$

23. $48 \text{ gal} \times \dfrac{3.79 \text{ L}}{1 \text{ gal}} \approx 181.92 \text{ L}$

25. $23 \text{ qt} \times \dfrac{0.946 \text{ L}}{1 \text{ qt}} \approx 21.76 \text{ L}$

27. $19 \text{ L} \times \dfrac{0.264 \text{ gal}}{1 \text{ L}} \approx 5.02 \text{ gal}$

29. $4.5 \text{ L} \times \dfrac{1.06 \text{ qt}}{1 \text{ L}} = 4.77 \text{ qt}$

31. $82 \text{ kg} \times \dfrac{2.2 \text{ lb}}{1 \text{ kg}} = 180.4 \text{ lb}$

33. $130 \text{ lb} \times \dfrac{0.454 \text{ kg}}{1 \text{ lb}} = 59.02 \text{ kg}$

35. $26 \text{ oz} \times \dfrac{28.35 \text{ g}}{1 \text{ oz}} = 737.1 \text{ g}$

37. $152 \text{ kg} \times \dfrac{2.2 \text{ lb}}{\text{kg}} = 334.4 \text{ lb}$

39. $158 \text{ g} \times \dfrac{0.0353 \text{ oz}}{1 \text{ g}} \approx 5.58 \text{ oz}$

41. $35 \text{ ft} \times \dfrac{12 \text{ in.}}{1 \text{ ft}} \times \dfrac{2.54 \text{ cm}}{1 \text{ in.}} = 1066.8 \text{ cm}$

43. $\dfrac{55 \text{ km}}{\text{hr}} \times \dfrac{0.62 \text{ mi}}{1 \text{ km}} = 34.1 \text{ mi/hr}$

45. $\dfrac{400 \text{ ft}}{1 \text{ sec}} \times \dfrac{3600 \text{ sec}}{1 \text{ hr}} \times \dfrac{1 \text{ mi}}{5280 \text{ ft}} \approx 273 \text{ mi/hr}$

47. $13 \text{ mm} \times \dfrac{1 \text{ cm}}{10 \text{ mm}} \times \dfrac{0.394 \text{ in.}}{1 \text{ cm}} \approx 0.51 \text{ in.}$

49. $F = 1.8 \times C + 32 = 1.8 \times 85 + 32 = 185$

$185°F$

51. $F = 1.8 \times C + 32 = 1.8 \times 12 + 32 = 53.6$

$53.6°F$

53. $C = \dfrac{5 \times F - 160}{9} = \dfrac{5 \times 140 - 160}{9} = \dfrac{540}{9} = 60$

$60°C$

55. $C = \dfrac{5 \times F - 160}{9} = \dfrac{5 \times 95 - 160}{9} = \dfrac{315}{9} = 35$

$35°C$

57. $\dfrac{65 \text{ miles}}{\text{hr}} \times \dfrac{1.61 \text{ km}}{\text{mile}} = 104.65 \text{ km/hr}$

Yes, she is speeding.

59. $15 \text{ gal} \times \dfrac{3.79 \text{ L}}{1 \text{ gal}} = 56.85 \text{ L}$

Difference $= 56.85 - 38 = 18.85 \text{ L}$
There were 18.85 liters left in the tank.

61. $540 \text{ kg} \times \dfrac{2.2 \text{ lb}}{1 \text{ kg}} = 1188 \text{ lb}$

Her weight was 1188 pounds.

63. $2.72 \text{ m} \times \dfrac{3.28 \text{ ft}}{1 \text{ m}} \approx 8.92 \text{ ft}$

His height is 8.92 feet.

65. $F = 1.8 \times C + 32$
4 A.M.: $F = 1.8 \times 19 + 32 = 66.2$
7 A.M.: $F = 1.8 \times 45 + 32 = 113$
It is 66.2°F at 4 A.M.
It may reach 113°F after 7 A.M.

67. $96,550 \text{ km} \times \dfrac{0.62 \text{ mi}}{1 \text{ km}} = 59,861 \text{ mi}$

This is equivalent to 59,861 miles of blood vessels.

69. $28 \times 2.54 \times 2.54 = 180.6448 \text{ sq cm}$

71. American: $8 \times 4 \times 28 = \$896$

German: $8 \text{ yd} \times \dfrac{0.914 \text{ m}}{1 \text{ yd}} = 7.312 \text{ m}$

$4 \text{ yd} \times \dfrac{0.914 \text{ m}}{1 \text{ yd}} = 3.656 \text{ m}$

$7.312 \times 3.656 \times 30 \approx \$802$

German carpet is cheaper by $896 - 802 = \$94$.

Cumulative Review

73. $3^4 \times 2 - 5 + 12 = 81 \times 2 - 5 + 12$
$ = 162 - 5 + 12$
$ = 169$

74. $96 + 24 \div 4 \times 3 = 96 + 6 \times 3 = 96 + 18 = 114$

75. $\dfrac{6}{11} \overset{?}{=} \dfrac{18}{21}$
$6 \times 21 \overset{?}{=} 11 \times 18$
$126 \neq 198$
It is not a proportion.

76. $\dfrac{1\frac{1}{2}}{12} \overset{?}{=} \dfrac{2}{16}$

$1\frac{1}{2} \times 16 \overset{?}{=} 12 \times 2$

$\dfrac{3}{2} \times 16 \overset{?}{=} 24$

$24 = 24$
It is a proportion.

Quick Quiz 6.4

1. $5 \text{ ounces} \times \dfrac{28.35 \text{ g}}{1 \text{ oz}} = 141.75 \text{ grams}$

2. $24 \text{ kilometers} \times \dfrac{0.62 \text{ mi}}{1 \text{ km}} = 14.88 \text{ miles}$

3. $6 \text{ liters} \times \dfrac{1.06 \text{ qt}}{1 \text{ L}} = 6.36 \text{ quarts}$

4. Answers may vary. Possible solution:

Multiply by the unit fraction $\dfrac{1 \text{ km}}{0.62 \text{ mi}}$.

$\dfrac{65 \text{ mi}}{1 \text{ hr}} \cdot \dfrac{1 \text{ km}}{0.62 \text{ mi}} \approx 104.8 \text{ km/hr}$

6.5 Exercises

1. $14\frac{1}{3} + 8\frac{2}{3} + 13 = 36 \text{ in.}$

$36 \text{ in.} \times \dfrac{1 \text{ ft}}{12 \text{ in.}} = 3 \text{ ft}$
The perimeter is 3 feet.

3. 480 feet = 160 yards
$42 + 65 = 107$

$\begin{array}{r} 160 \\ -\ 107 \\ \hline 53 \end{array}$ yards

There is 53 yards of fencing left for the third side.

5. Length $= 200 + 90 + 200 = 490 \text{ cm} = 4.9 \text{ m}$

Cost $= 4.9 \text{ m} \times \dfrac{\$6.00}{1 \text{ m}} = \$29.40$

It will cost \$29.40 to weatherstrip the door.

7. 1.863 km = 1863 m

$\dfrac{1863 \text{ m}}{230} = 8.1 \text{ m/space}$

Each space is 8.1 meters long.

9. $880 \text{ yd} \times \dfrac{0.914 \text{ m}}{1 \text{ yd}} = 804.32 \text{ m}$

880 yd is 4.32 m longer than 800 m.

11. $\dfrac{\$0.89}{\text{L}} \times \dfrac{3.79 \text{ L}}{1 \text{ gal}} = \$3.37/\text{gal}$

Gasoline is more expensive in Mexico.

13. $F = 1.8 \times C + 32$
$= 1.8 \times 25 + 32$
$= 45 + 32$
$= 77°F$
$86 - 77 = 9$
The high temperature in Scotland was 77°F.
The difference is 9°F. The temperature in Boston was 9°F greater.

15. $F = 1.8 \times 180 + 32 = 324 + 32 = 356°F$
$356 - 350 = 6$
The difference is 6°F. 180°C is hotter.

17. a. $\dfrac{520 \text{ mi}}{8 \text{ hr}} = 65 \text{ mi/hr}$

$\dfrac{65 \text{ mi}}{1 \text{ hr}} \times \dfrac{1.61 \text{ km}}{1 \text{ mi}} \approx 105 \text{ km/hr}$
They averaged 105 km/hr.

b. Probably not; we cannot be sure, since they could speed for a short time, but we have no evidence to indicate that they broke the speed limit.

19. $2 \times \dfrac{1 \text{ pt}}{1 \text{ min}} \times \dfrac{1 \text{ qt}}{2 \text{ pt}} \times \dfrac{1 \text{ gal}}{4 \text{ qt}} \times \dfrac{60 \text{ min}}{1 \text{ hr}} = 15 \text{ gal/hr}$

21. $1.8 \text{ tons} \times \dfrac{2000 \text{ lb}}{1 \text{ ton}} \times \dfrac{\$0.05}{1 \text{ lb}} = \$180$
His annual tax is $180.

23. $\dfrac{708 \text{ g}}{12} = 59 \text{ g}$

$59 \text{ g} \times \dfrac{0.0353 \text{ oz}}{1 \text{ g}} \approx 2.08 \text{ oz}$
There are 2.08 oz per serving.

25. a. $11 \text{ L} \times \dfrac{1.06 \text{ qt}}{1 \text{ L}} = 11.66 \text{ qt}$
$16 - 11.66 = 4.34 \text{ qt}$
He has 4.34 qt extra.

b. $11.66 \text{ qt} \times \dfrac{\$2.89}{1 \text{ qt}} \approx \$33.70$
It will cost him $33.70.

27. a. $392 \text{ km} \times \dfrac{1 \text{ L}}{56 \text{ km}} \times \dfrac{\$1.09}{1 \text{ L}} = \$7.63$
It will cost $5.46 for the trip.

b. $\dfrac{56 \text{ km}}{1 \text{ L}} \times \dfrac{0.62 \text{ mi}}{1 \text{ km}} \times \dfrac{1 \text{ L}}{0.264 \text{ gal}} \approx 132 \text{ mi/gal}$
His motorcycle gets about 132 mi/gal.

29. $\dfrac{240,000 \text{ gal}}{1 \text{ hr}} \times \dfrac{8 \text{ pt}}{1 \text{ gal}} \times \dfrac{1 \text{ hr}}{3,600 \text{ sec}} = 533\dfrac{1}{3} \text{ pt/sec}$
Yes; 240,000 gal/hr is equivalent to $533\dfrac{1}{3}$ pt/sec.

Cumulative Review

31. $6 \text{ in.} \times \dfrac{7.75 \text{ mi}}{3 \text{ in.}} = 15.5 \text{ mi}$
The train must go 15.5 miles.

32. $\dfrac{2}{7.5} = \dfrac{11}{n}$
$2 \times n = 7.5 \times 11$
$\dfrac{2 \times n}{2} = \dfrac{82.5}{2}$
$n = 41.25 \text{ yards}$
The famous fishing schooner is 41.25 yards.

Quick Quiz 6.5

1. $F = 1.8 \times 2 + 32$
$F = 3.6 + 32$
$F = 35.6$
$35.6 - 28 = 7.6$
The prediction was 7.6°F cooler.

2. $P = 2(4 \text{ cm}) + 2(8 \text{ cm})$
$P = 8 \text{ cm} + 16 \text{ cm}$
$P = 24 \text{ cm}$
$24 \text{ cm} \times \dfrac{0.394 \text{ in.}}{1 \text{ cm}} \approx 9.5 \text{ in.}$
The perimeter is 9.5 inches.

3. $\dfrac{100 \text{ km}}{1 \text{ hr}} \times \dfrac{0.62 \text{ mi}}{1 \text{ km}} \times \dfrac{1 \text{ hr}}{60 \text{ min}} = 1.0\overline{3} \text{ mi/min}$

$45 \text{ mi} \times \dfrac{1 \text{ min}}{1.0\overline{3} \text{ mi}} \approx 43.5 \text{ min}$

It will take her 43.5 minutes.

4. Answers may vary. Possible solution:
Set up a ratio. Cross-multiply and solve for the unknown.

$\dfrac{3 \text{ in.}}{6.5 \text{ mi}} = \dfrac{5 \text{ in.}}{x \text{ mi}}$

$3x = 6.5(5)$

$\dfrac{3x}{3} = \dfrac{32.5}{3}$

$x \approx 10.8$

The two cities are 10.8 miles apart.

Use Math to Save Money

1.   200.00
     150.50
     120.25
      50.00
  + 25.00
     545.75
The total amount of his deposits is $545.75.

2.   238.50
      75.00
     200.00
      28.56
  + 36.00
     578.06
The total amount of his checks is $578.06.

3. Since $578.06 > $545.75, he spent more than he deposited, but the $300.50 already in the bank would help him to cover his expenses.

4. 300.50 + 545.75 − 578.06 = 268.19
His balance would be $268.19.

5. Eventually he will be in debt.

6. Answers will vary.

7. Answers will vary.

You Try It

1. a. $45 \text{ yd} \times \dfrac{3 \text{ ft}}{1 \text{ yd}} = 135 \text{ ft}$

b. $88 \text{ oz} \times \dfrac{1 \text{ lb}}{16 \text{ oz}} = 5.5 \text{ lb}$

2. a. 1500 km = 1,500,000 m

b. 25 cm = 250 mm

c. 12,500 mL = 12.5 L

3. a. $6.5 \text{ yd} \times \dfrac{0.914 \text{ m}}{1 \text{ yd}} = 5.941 \text{ m}$

b. $8 \text{ mi} \times \dfrac{1.61 \text{ km}}{1 \text{ mi}} = 12.88 \text{ km}$

4. a. $50 \text{ m} \times \dfrac{3.28 \text{ ft}}{1 \text{ m}} = 164 \text{ ft}$

b. $\dfrac{120 \text{ km}}{\text{hr}} \times \dfrac{0.62 \text{ mi}}{1 \text{ km}} = 74.4 \text{ mi/hr}$

5. $F = 1.8 \times C + 32 = 1.8 \times 40 + 32 = 104$
104°F

6. $C = \dfrac{5 \times F - 160}{9} = \dfrac{5 \times 85 - 160}{9} = \dfrac{265}{9} \approx 25.44$
25.44°C

Chapter 6 Review Problems

1. $33 \text{ ft} \times \dfrac{1 \text{ yd}}{3 \text{ ft}} = 11 \text{ yd}$

2. $5 \text{ mi} \times \dfrac{1760 \text{ yd}}{1 \text{ mi}} = 8800 \text{ yd}$

3. $126 \text{ in.} \times \dfrac{1 \text{ ft}}{12 \text{ in.}} = 10.5 \text{ ft}$

4. $2.5 \text{ mi} \times \dfrac{5280 \text{ ft}}{1 \text{ mi}} = 13,200 \text{ ft}$

5. $14 \text{ ft} \times \dfrac{12 \text{ in.}}{1 \text{ ft}} = 168 \text{ in.}$

6. $7 \text{ tons} \times \dfrac{2000 \text{ lb}}{1 \text{ ton}} = 14,000 \text{ lb}$

7. $8 \text{ oz} \times \dfrac{1 \text{ lb}}{16 \text{ oz}} = 0.5 \text{ lb}$

8. $3.5 \text{ lb} \times \dfrac{16 \text{ oz}}{1 \text{ lb}} = 56 \text{ oz}$

9. $15 \text{ gal} \times \dfrac{4 \text{ qt}}{1 \text{ gal}} = 60 \text{ qt}$

10. $31 \text{ pt} \times \dfrac{1 \text{ qt}}{2 \text{ pt}} = 15.5 \text{ qt}$

11. $56 \text{ cm} = 560 \text{ mm}$

12. $1763 \text{ mm} = 176.3 \text{ cm}$

13. $13.25 \text{ m} = 1325 \text{ cm}$

14. $10{,}000 \text{ m} = 10 \text{ km}$

15. $9.2 \text{ km} = 9200 \text{ m}$

16. $2400 \text{ cm} = 24 \text{ m}$

17. $6.2 \text{ m} + 121 \text{ cm} + 0.52 \text{ m}$
$= 6.2 \text{ m} + 1.21 \text{ m} + 0.52 \text{ m}$
$= 7.93 \text{ m}$

18. $0.024 \text{ km} + 1.8 \text{ m} + 983 \text{ cm}$
$= 24 \text{ m} + 1.8 \text{ m} + 9.83 \text{ m}$
$= 35.63 \text{ m}$

19. $17 \text{ kL} = 17{,}000 \text{ L}$

20. $8000 \text{ L} = 8 \text{ kL}$

21. $196 \text{ kg} = 196{,}000 \text{ g}$

22. $95 \text{ mg} = 0.095 \text{ g}$

23. $3500 \text{ g} = 3.5 \text{ kg}$

24. $15.1 \text{ g} = 15{,}100 \text{ mg}$

25. $765 \text{ cc} = 765 \text{ mL}$

26. $423 \text{ cm}^3 = 423 \text{ mL}$

27. $0.256 \text{ L} = 256 \text{ mL} = 256 \text{ cm}^3$

28. $42 \text{ kg} \times \dfrac{2.2 \text{ lb}}{1 \text{ kg}} = 92.4 \text{ lb}$

29. $9 \text{ ft} \times \dfrac{0.305 \text{ m}}{1 \text{ ft}} \approx 2.75 \text{ m}$

30. $45 \text{ mi} \times \dfrac{1.61 \text{ km}}{1 \text{ mi}} = 72.45 \text{ km}$

31. $14 \text{ cm} \times \dfrac{0.394 \text{ in.}}{1 \text{ cm}} \approx 5.52 \text{ in.}$

32. $20 \text{ lb} \times \dfrac{0.454 \text{ kg}}{1 \text{ lb}} = 9.08 \text{ kg}$

33. $50 \text{ yd} \times \dfrac{0.914 \text{ m}}{1 \text{ yd}} = 45.7 \text{ m}$

34. $\dfrac{80 \text{ km}}{1 \text{ hr}} \times \dfrac{0.62 \text{ mi}}{1 \text{ km}} = 49.6 \text{ mi/hr}$

35. $F = 1.8 \times C + 32$
$ = 1.8 \times 12 + 32$
$ = 21.6 + 32$
$ = 53.6$
$53.6°\text{F}$

36. $F = 1.8 \times 32 + 32 = 57.6 + 32 = 89.6°\text{F}$

37. $C = \dfrac{5 \times F - 160}{9} = \dfrac{5 \times 221 - 160}{9} = 105$
$105°\text{C}$

38. $C = \dfrac{5 \times F - 160}{9} = \dfrac{5 \times 32 - 160}{9} = 0$
$0°\text{C}$

39. $13 \text{ L} \times \dfrac{0.264 \text{ gal}}{1 \text{ L}} \approx 3.43 \text{ gal}$

40. $27 \text{ qt} \times \dfrac{0.946 \text{ L}}{1 \text{ qt}} \approx 25.54 \text{ L}$

41. a. $16 \text{ m} + 84 \text{ m} + 16 \text{ m} + 84 \text{ m} = 200 \text{ m}$
The perimeter is 200 meters.

b. $200 \text{ m} = 0.2 \text{ km}$
The perimeter is 0.2 kilometer.

42. a. $7\dfrac{2}{3} \text{ ft} \times 4\dfrac{1}{3} \text{ ft} + 5 \text{ ft} = 16\dfrac{3}{3} \text{ ft} = 17 \text{ feet}$
The perimeter is 17 feet.

b. $17 \text{ feet} \times \dfrac{12 \text{ in.}}{1 \text{ foot}} = 204 \text{ in.}$
The perimeter is 204 inches.

43. $5 \text{ yd} \times \dfrac{3 \text{ ft}}{\text{yd}} = 15 \text{ ft}$

$A = lw$
$A = (18)(15)$
$A = 270 \text{ sq ft}$

$270 \text{ sq ft} \times \dfrac{1 \text{ sq yd}}{9 \text{ sq ft}} = 30 \text{ sq yd}$

The area is 270 square feet or 30 square yards.

44. $510 \text{ g} \times \dfrac{0.0353 \text{ oz}}{1 \text{ g}} \approx 18 \text{ oz}$

$\dfrac{\$0.16}{1 \text{ oz}} \times 18 \text{ oz} \approx \$2.88$

The cereal costs \$2.88.

45. $70 \dfrac{\text{mi}}{\text{hr}} \times \dfrac{1.61 \text{ km}}{1 \text{ mi}} = 112.7 \dfrac{\text{km}}{\text{hr}}$

Yes; she was driving 112.7 km/hr.

46. $F = 1.8 \times C + 32 = 1.8 \times 185 + 32 = 365°$
$390° - 365° = 25°F$ too hot

47. $19 \text{ m} = 1900 \text{ cm}$

Bottom part $= 1900 \times \dfrac{1}{5} = 380 \text{ cm}$

The length of the pole that has the water seal is 380 cm.

48. $\dfrac{1 \text{ mi}}{91.41 \text{ sec}} \times \dfrac{60 \text{ min}}{1 \text{ hr}} \times \dfrac{60 \text{ sec}}{1 \text{ min}} = \dfrac{3600}{91.41}$
$\approx 39.38 \text{ mi/hr}$

The horse's speed was 39.38 mi/hr.

49. $\dfrac{\$1.05}{1 \text{ L}} \times \dfrac{1 \text{ L}}{0.264 \text{ gal}} \approx \$3.98/\text{gal}$

The gas costs \$3.98/gal.

50. $4 \text{ m} \times \dfrac{3.28 \text{ ft}}{1 \text{ m}} = 13.12 \text{ ft}$

$12 \text{ m} \times \dfrac{3.28 \text{ ft}}{1 \text{ m}} = 39.36 \text{ ft}$

Area $= 13.12 \times 39.36 \approx 516.4 \text{ sq ft}$
He needs approximately 516.4 square feet of sealer.

How Am I Doing? Chapter 6 Test

1. $1.6 \text{ tons} \times \dfrac{2000 \text{ lb}}{1 \text{ ton}} = 3200 \text{ lb}$

2. $19 \text{ ft} \times \dfrac{12 \text{ in.}}{1 \text{ ft}} = 228 \text{ in.}$

3. $21 \text{ gal} \times \dfrac{4 \text{ qt}}{1 \text{ gal}} = 84 \text{ qt}$

4. $36,960 \text{ ft} \times \dfrac{1 \text{ mi}}{5280 \text{ ft}} = 7 \text{ mi}$

5. $1800 \text{ sec} \times \dfrac{1 \text{ min}}{60 \text{ sec}} = 30 \text{ min}$

6. $3 \text{ cups} \times \dfrac{1 \text{ qt}}{4 \text{ cups}} = 0.75 \text{ qt}$

7. $8 \text{ oz} \times \dfrac{1 \text{ lb}}{16 \text{ oz}} = 0.5 \text{ lb}$

8. $5.5 \text{ yd} \times \dfrac{3 \text{ ft}}{1 \text{ yd}} = 16.5 \text{ ft}$

9. $9.2 \text{ km} = 9200 \text{ m}$

10. $9.88 \text{ cm} = 0.0988 \text{ m}$

11. $46 \text{ mm} = 4.6 \text{ cm}$

12. $12.7 \text{ m} = 1270 \text{ cm}$

13. $0.936 \text{ cm} = 9.36 \text{ mm}$

14. $46 \text{ L} = 0.046 \text{ kL}$

15. $28.9 \text{ mg} = 0.0289 \text{ g}$

16. $983 \text{ g} = 0.983 \text{ kg}$

17. $0.92 \text{ L} = 920 \text{ mL}$

18. $9.42 \text{ g} = 9420 \text{ mg}$

19. $42 \text{ mi} \times \dfrac{1.61 \text{ km}}{1 \text{ mi}} = 67.62 \text{ km}$

20. $1.78 \text{ yd} \times \dfrac{0.914 \text{ m}}{1 \text{ yd}} \approx 1.63 \text{ m}$

21. $9 \text{ cm} \times \dfrac{0.394 \text{ in.}}{1 \text{ cm}} \approx 3.55 \text{ in.}$

22. $30 \text{ km} \times \dfrac{0.62 \text{ mi}}{1 \text{ km}} = 18.6 \text{ mi}$

23. $7.3 \text{ kg} \times \dfrac{2.2 \text{ lb}}{1 \text{ kg}} = 16.06 \text{ lb}$

24. $3 \text{ oz} \times \dfrac{28.35 \text{ g}}{1 \text{ oz}} = 85.05 \text{ g}$

25. $15 \text{ gal} \times \dfrac{3.79 \text{ L}}{1 \text{ gal}} = 56.85 \text{ L}$

26. $3 \text{ L} \times \dfrac{1.06 \text{ qt}}{1 \text{ L}} = 3.18 \text{ qt}$

27. a. $3 \text{ m} + 7 \text{ m} + 3 \text{ m} + 7 \text{ m} = 20 \text{ m}$
The perimeter is 20 meters.

    b. $20 \text{ m} \times \dfrac{1.09 \text{ yd}}{1 \text{ m}} = 21.8 \text{ yd}$
The perimeter is 21.8 yards.

28. a. $F = 1.8 \times C + 32 = 1.8 \times 35 + 32 = 95$
$95°F - 80°F = 15°F$
There are 15°F between the two temperatures.

    b. Yes; $80°F < 95°F$

29. $\dfrac{5.5 \text{ qt}}{1 \text{ min}} \times \dfrac{1 \text{ gal}}{4 \text{ qt}} \times \dfrac{60 \text{ min}}{1 \text{ hr}} = 82.5 \text{ gal/hr}$
The pump is running at 82.5 gal/hr.

30. a. $100 \times 3 = 300$
He can travel 300 km.

    b. $300 \text{ km} \times \dfrac{0.62 \text{ mi}}{1 \text{ km}} = 186 \text{ mi}$
$200 - 186 = 14$
He will need to travel 14 miles farther.

31. $1\dfrac{6}{16} + 2\dfrac{2}{16} + 1\dfrac{12}{16} = 4\dfrac{20}{16} = 5\dfrac{4}{16} = 5\dfrac{1}{4}$

He bought a total of $5\dfrac{1}{4}$ pounds of fruit.

32. $F = 1.8 \times C + 32 = 1.8 \times 40 + 32 = 104$
The temperature was 104°F.

Cumulative Test for Chapters 1–6

1.
$$\begin{array}{r} 12,655 \\ +\ 8,915 \\ \hline 21,570 \end{array}$$

2.
$$\begin{array}{r} 30,075 \\ -\ 19,328 \\ \hline 10,747 \end{array}$$

3.
$$\begin{array}{r} 4256 \\ \times\ \ \ \ 534 \\ \hline 17\ 024 \\ 127\ 68\ \ \\ 2\ 128\ 0\ \ \ \\ \hline 2,272,704 \end{array}$$

4.
$$\begin{array}{r} 2\ 667 \\ 7\overline{)18,669} \\ \underline{14}\ \ \ \ \ \ \ \\ 4\ 6\ \ \ \ \\ \underline{4\ 2}\ \ \ \ \\ 46\ \ \\ \underline{42}\ \ \\ 49 \\ \underline{49} \\ 0 \end{array}$$

5. $2 \times (6-3)^3 \div 9 + 15 = 2 \times 3^3 \div 9 + 15$
$$\begin{aligned} &= 2 \times 27 \div 9 + 15 \\ &= 54 \div 9 + 15 \\ &= 6 + 15 \\ &= 21 \end{aligned}$$

6. $322\overline{)27,551}$

$$\begin{array}{r} 100 \\ 300\overline{)30,000} \\ \underline{30\ 0}\ \ \ \\ 000 \end{array}$$

$27,551 \div 322 \approx 100$

7. $\dfrac{24}{60} = \dfrac{24 \div 12}{60 \div 12} = \dfrac{2}{5}$

8.
$$\begin{array}{r} 8 \\ 10\overline{)85} \\ \underline{80} \\ 5 \end{array}$$
$\qquad \dfrac{85}{10} = 8\dfrac{5}{10} = 8\dfrac{1}{2}$

9. $2\dfrac{1}{3} \times 3\dfrac{3}{4} = \dfrac{7}{3} \times \dfrac{15}{4} = \dfrac{7 \times 3 \times 5}{3 \times 4} = \dfrac{35}{4} \text{ or } 8\dfrac{3}{4}$

10. $\dfrac{9}{10} \div \dfrac{3}{5} = \dfrac{9}{10} \times \dfrac{5}{3} = \dfrac{3 \times 3 \times 5}{2 \times 5 \times 3} = \dfrac{3}{2} \text{ or } 1\dfrac{1}{2}$

11.

$$4\frac{2}{3}$$
$$+5\frac{1}{2}$$

$$4\frac{4}{6}$$
$$+5\frac{3}{6}$$
$$9\frac{7}{6} = 9 + 1\frac{1}{6} = 10\frac{1}{6}$$

12.

$$\left(\frac{1}{2}\right)^2 + \frac{7}{8} \times \frac{1}{3} = \frac{1}{4} + \frac{7}{8} \times \frac{1}{3}$$
$$= \frac{1}{4} + \frac{7}{24}$$
$$= \frac{6}{24} + \frac{7}{24}$$
$$= \frac{13}{24}$$

13. $0.48 = \dfrac{48}{100} = \dfrac{12}{25}$

14. 1.86374 rounds to 1.864.
Find the thousandths place: 1.86374. The digit to the right of the thousandths place is greater than 5. We round up to 1.864 and drop the digits to the right.

15.

$$\begin{array}{r} 32.15 \\ 8.50 \\ + 320.42 \\ \hline 361.07 \end{array}$$

16.

$$\begin{array}{r} 0.35 \\ 28\overline{)9.80} \\ 8\,4 \\ \hline 1\,40 \\ 1\,40 \\ \hline 0 \end{array}$$

17.

$$\begin{array}{r} 0.65 \\ 20\overline{)13.00} \\ 12\,0 \\ \hline 1\,00 \\ 1\,00 \\ \hline 0 \end{array}$$

$$\frac{13}{20} = 0.65$$

18. $0.75 \text{ lb} \times \dfrac{\$9.80}{1 \text{ lb}} = \$7.35$
He paid $7.35.

19. $\dfrac{384 \text{ points}}{12 \text{ games}} = 32$ points/game

20.

$$\frac{5.5}{8} \overset{?}{=} \frac{22}{32}$$
$$5.5 \times 32 \overset{?}{=} 8 \times 22$$
$$176 = 176$$
Yes, it is a proportion.

21.

$$\frac{0.4}{n} = \frac{2}{30}$$
$$0.4 \times 30 = n \times 2$$
$$\frac{12}{2} = \frac{n \times 2}{2}$$
$$n = 6$$

22.

$$\frac{6.5 \text{ cm}}{68 \text{ g}} = \frac{20 \text{ cm}}{n \text{ g}}$$
$$6.5 \times n = 68 \times 20$$
$$6.5 \times n = 1360$$
$$\frac{6.5 \times n}{6.5} = \frac{1360}{6.5}$$
$$n \approx 209.23 \text{ g}$$
The 20-cm wire will weigh 209.23 g.

23. $85\% = \dfrac{85}{100} = \dfrac{17}{20}$

24. $\dfrac{3}{40} = 0.075 = 7.5\%$

25. $n = 15\% \times 800$
$n = 0.15 \times 800$
$n = 120$

26. What percent of 74 is 148?
$n \times 74 = 148$
$$\frac{n \times 74}{74} = \frac{148}{74}$$
$$n = 2 = 200\%$$

27. 0.5% of what number is 100?
$0.5\% \times n = 100$
$$\frac{0.005 \times n}{0.005} = \frac{100}{0.005}$$
$$n = 20,000$$

28. Discount = (0.15)(1220) = 183
$1220 - 183 = 1037$
She will pay $1037.

29. $38 \text{ qt} \times \dfrac{1 \text{ gal}}{4 \text{ qt}} = 9.5 \text{ gal}$

30. $3.7 \text{ kL} = 3700 \text{ L}$

31. $5 \text{ cm} = 0.05 \text{ m}$

32. $42 \text{ lb} \times \dfrac{16 \text{ oz}}{1 \text{ lb}} = 672 \text{ oz}$

33. $6 \text{ yd} + 4 \text{ yd} + 3 \text{ yd} = 13 \text{ yd}$

$13 \text{ yd} \times \dfrac{0.914 \text{ m}}{1 \text{ yd}} \approx 11.88 \text{ m}$

The perimeter is 11.88 meters.

34. $F = 1.8 \times C + 32 = 1.8 \times 15 + 32 = 59°\text{F}$
$59° - 15° = 44°\text{F}$
The difference is 44°F; the 15°C temperature is higher.

Chapter 7

7.1 Exercises

1. An acute angle is an angle whose measure is between 0° and 90°.

3. Complementary angles are two angles whose measures have a sum of 90°.

5. When two lines intersect, two angles that are opposite each other are called vertical angles.

7. A transversal is a line that intersects two or more other lines at different points.

9. $\angle ABD$, $\angle CBE$

11. $\angle ABD$ and $\angle CBE$
    $\angle DBC$ and $\angle ABE$

13. There are no complementary angles.

15. $\angle LOJ = 180° - 90° = 90°$

17. $\angle JON = 90° - 65° = 25°$

19. $\angle JOM = 90° + 20° = 110°$

21. $\angle NOK = 65° + 90° = 155°$

23. $90° - 31° = 59°$

25. $180° - 127° = 53°$

27. $\angle a = 180° - 146° = 34°$

29. $\angle a = 123° - 88° = 35°$

31. $\angle a = 85° - 18° - 42° = 25°$

33. $\angle b = 102°$
    $\angle a = \angle c = 180° - 102° = 78°$

35. $\angle b = 38°$
    $\angle a = \angle c = 180° - 38° = 142°$

37. $\angle a = \angle c = 48°$
    $\angle b = 180° - 48° = 132°$

39. $\angle e = \angle d = \angle a = 123°$
    $\angle b = \angle c = \angle f = \angle g = 180° - 123° = 57°$

41. $\angle x = 90° - 86° = 4°$
    The tower deviates by 4°.

43. New angle is $62° - 9° = 53°$ north of east.

Cumulative Review

45. $204.5 + 179 + 184.5 + 210.5 = 778.5$ km
    $778.5 \text{ km} \times \dfrac{0.62 \text{ mi}}{1 \text{ km}} = 482.67 \text{ mi}$
    The total distance is 778.5 km or 482.67 mi.

46. $58 \text{ km} \times \dfrac{0.62 \text{ mi}}{1 \text{ km}} \approx 36.0 \text{ mi}$
    He needs to drive about 36.0 miles farther.

47. Percent miles $= 100 + 5 = 105\%$
    Miles $= 105\%$ of $24 = 1.05 \times 24 = 25.2$ miles
    He should jog at most 25.2 miles.

48. Increase $= 27 - 24 = 3$
    3 is what percent of 24?
    $p \times 24 = 3$
    $p = \dfrac{3}{24}$
    $p = 0.125$
    12.5% is the percent of increase.

Quick Quiz 7.1

1. $\angle a = \angle c = 124°$

2. $\angle b = 180° - 124° = 56°$

3. $\angle e = \angle b = 56°$

4. Answers may vary. Possible solution:
   $\angle e$ and $\angle a$ are supplementary angles.
   measure of $\angle a = 180° -$ measure of $\angle e$

7.2 Exercises

1. A rectangle has two properties: (1) any two adjoining sides are <u>perpendicular</u> and (2) the lengths of opposite sides are <u>equal</u>.

3. To find the area of a rectangle, we <u>multiply</u> the length by the width.

5. $P = 2l + 2w$
   $= 2(5.5 \text{ mi}) + 2(2 \text{ mi})$
   $= 11 \text{ mi} + 4 \text{ mi}$
   $= 15 \text{ mi}$

7. $P = 2l + 2w$
$= 2(9.3 \text{ ft}) + 2(2.5 \text{ ft})$
$= 18.6 \text{ ft} + 5 \text{ ft}$
$= 23.6 \text{ ft}$

9. $P = 4s = 4(4.3 \text{ in.}) = 17.2 \text{ in.}$

11. $P = 2l + 2w$
$= 2(0.84 \text{ mm}) + 2(0.12 \text{ mm})$
$= 1.68 \text{ mm} + 0.24 \text{ mm}$
$= 1.92 \text{ mm}$

13. $P = 4s = 4(4.28 \text{ km}) = 17.12 \text{ km}$

15. $3.2 \text{ ft} \times \dfrac{12 \text{ in.}}{1 \text{ ft}} = 38.4 \text{ in.}$
$P = 2l + 2w$
$= 2(38.4 \text{ in.}) + 2(48 \text{ in.})$
$= 76.8 \text{ in.} + 96 \text{ in.}$
$= 172.8 \text{ in.}$
or
$48 \text{ in.} \times \dfrac{1 \text{ ft}}{12 \text{ in.}} = 4 \text{ ft}$
$P = 2l + 2w$
$= 2(4 \text{ ft}) + 2(3.2 \text{ ft})$
$= 8 \text{ ft} + 6.4 \text{ ft}$
$= 14.4 \text{ ft}$

17. $P = 4s = 4(0.068 \text{ mm}) = 0.272 \text{ mm}$

19. $P = 4s = 4\left(3\dfrac{1}{2} \text{ cm}\right) = 4\left(\dfrac{7}{2} \text{ cm}\right) = 14 \text{ cm}$

21.
$\begin{array}{r}
10\frac{1}{2} \text{ cm} \\
7 \text{ cm} \\
6 \text{ cm} \\
5 \text{ cm} \\
4\frac{1}{2} \text{ cm} \\
+ \quad 2 \text{ cm} \\
\hline
P = 35 \text{ cm}
\end{array}$

23.
$\begin{array}{r}
9 \text{ cm} \\
13 \text{ cm} \\
11 \text{ cm} \\
13 \text{ cm} \\
16 \text{ cm} \\
41 \text{ cm} \\
36 \text{ cm} \\
+ 41 \text{ cm} \\
\hline
P = 180 \text{ cm}
\end{array}$

25. $A = s^2 = (2.5 \text{ ft})^2 = 6.25 \text{ ft}^2$

27. $A = lw = 8 \text{ mi} \times 1.5 \text{ mi} = 12 \text{ mi}^2$

29. $39 \text{ yd} \times \dfrac{3 \text{ ft}}{1 \text{ yd}} = 117 \text{ ft}$
$A = lw = (117 \text{ ft})(9 \text{ ft}) = 1053 \text{ ft}^2$
or
$9 \text{ ft} \times \dfrac{1 \text{ yd}}{3 \text{ ft}} = 3 \text{ yd}$
$A = lw = (39 \text{ yd})(3 \text{ yd}) = 117 \text{ yd}^2$

31. a. $A = 21 \text{ m} \times 12 \text{ m} + 6 \text{ m} \times 7 \text{ m}$
$= 252 \text{ m}^2 + 42 \text{ m}^2$
$= 2894 \text{ m}^2$
The area is 2894 m^2.

b.
$\begin{array}{r}
21 \text{ m} \\
12 \text{ m} \\
11 \text{ m} \\
6 \text{ m} \\
7 \text{ m} \\
6 \text{ m} \\
3 \text{ m} \\
+ 12 \text{ m} \\
\hline
78 \text{ m}
\end{array}$
The perimeter is 78 meters.

33. $A = 40 \text{ ft} \times 55 \text{ ft} = 2200 \text{ ft}^2$
$\text{Cost} = 2200 \text{ ft}^2 \times \dfrac{\$15.00}{\text{ft}^2} = \$33,000$
It will cost \$33,000.

35. a. $\text{length} = \dfrac{84 \text{ in.}}{1} \cdot \dfrac{1 \text{ ft}}{12 \text{ in.}} = 7 \text{ ft}$
$\text{new width} = \dfrac{72 \text{ in.}}{1} \cdot \dfrac{1 \text{ ft}}{12 \text{ in.}} + 1 \text{ ft} = 7 \text{ ft}$
$A = 7 \text{ ft} \cdot 7 \text{ ft} = 49 \text{ ft}^2$
The blanket will be 49 ft^2.

b. $\text{Border} = 2(\text{length}) + 2(\text{width})$
$= 2(7 \text{ ft}) + 2(7 \text{ ft})$
$= 28 \text{ ft}$
She should buy 28 feet.

37. a. $1 \times 7, 2 \times 6, 3 \times 5, 4 \times 4$
There are four possible shapes.

b. $1 \times 7 = 7 \text{ ft}^2$, $2 \times 6 = 12 \text{ ft}^2$, $3 \times 5 = 15 \text{ ft}^2$,

$4 \times 4 = 16 \text{ ft}^2$

c. The square garden measuring 4 ft on a side has the largest area.

39. $A = lw + lw$

$= (24 \text{ ft})(12 \text{ ft}) + (7 \text{ ft})(8 \text{ ft})$

$= 288 \text{ ft}^2 + 56 \text{ ft}^2$

$= 344 \text{ ft}^2$

Cost of carpet $= 344 \text{ ft}^2 \times \dfrac{\$14.50}{\text{yd}^2} \times \dfrac{1 \text{ yd}^2}{9 \text{ ft}^2}$

$\approx \$554.24$

$P = 12 \text{ ft} + 17 \text{ ft} + 8 \text{ ft} + 7 \text{ ft} + 20 \text{ ft} + 24 \text{ ft}$

$= 88 \text{ ft}$

Cost of binding $= 88 \text{ ft} \times \dfrac{\$1.50}{\text{yd}} \times \dfrac{1 \text{ yd}}{3 \text{ ft}} = \$44$

Total cost $= \$544.22 + \$44 = \$598.22$

Cumulative Review

41.
$$\begin{array}{r} 156.8 \\ 27.2 \\ + \ 39.3 \\ \hline 223.3 \end{array}$$

42.
$$\begin{array}{r} 200.57 \\ - 193.39 \\ \hline 7.18 \end{array}$$

43.
$$\begin{array}{r} 1076 \\ \times \quad 20.3 \\ \hline 322 \ 8 \\ 21 \ 520 \ \ \\ \hline 21,842.8 \end{array}$$

44.
$$\begin{array}{r} 1.57593 \\ 12.3 \overline{\smash{\big)}\, 19.3 \wedge 84000} \\ \underline{12\ 3} \ \ \\ 7\ 0\ 8 \\ \underline{6\ 1\ 5} \\ 9\ 34 \\ \underline{8\ 61} \\ 730 \\ \underline{615} \\ 1150 \\ \underline{1107} \\ 430 \\ \underline{369} \end{array}$$

≈ 1.5759

Quick Quiz 7.2

1. $P = 2l + 2w$

$= 2(2.3 \text{ cm}) + 2(1.5 \text{ cm})$

$= 4.6 \text{ cm} + 3 \text{ cm}$

$= 7.6 \text{ cm}$

2. $A = s^2 = (11 \text{ mi})^2 = 121 \text{ mi}^2$

3. $A = lw = (11 \text{ ft})(16 \text{ ft}) = 176 \text{ ft}^2$

Cost $= \dfrac{176 \text{ ft}^2}{1} \cdot \dfrac{\$6}{1 \text{ ft}^2} = \$1056$

4. Answers may vary. Possible solution: Find the area of the rectangle. Find the area of the square. Then add the results.

Area $= 12 \text{ ft} \times 15 \text{ ft} + (15 \text{ ft})^2$

$= 180 \text{ ft}^2 + 225 \text{ ft}^2$

$= 405 \text{ ft}^2$

7.3 Exercises

1. The perimeter of a parallelogram is found by <u>adding</u> the lengths of all the sides of the figure.

3. The height of a parallelogram is a line segment that is <u>perpendicular</u> to the base.

5. $P = 2(2.8 \text{ m}) + 2(17.3 \text{ m})$

$= 5.6 \text{ m} + 34.6 \text{ m}$

$= 40.2 \text{ m}$

7. $P = 2(9.2 \text{ in.}) + 2(15.6 \text{ in.})$

$= 18.4 \text{ in.} + 31.2 \text{ in.}$

$= 49.6 \text{ in.}$

9. $A = bh = (17.6 \text{ m})(20.15 \text{ m}) = 354.64 \text{ m}^2$

11. $A = bh = (28 \text{ yd})(21.5 \text{ yd}) = 602 \text{ yd}^2$

13. $P = 4(12 \text{ m}) = 48 \text{ m}$
$A = bh = (12 \text{ m})(6 \text{ m}) = 72 \text{ m}^2$

15. $P = 4(2.4 \text{ ft}) = 9.6 \text{ ft}$
$A = bh = (2.4 \text{ ft})(1.5 \text{ ft}) = 3.6 \text{ ft}^2$

17. $P = 13 \text{ m} + 20 \text{ m} + 15 \text{ m} + 34 \text{ m} = 82 \text{ m}$

19. $P = 55 \text{ ft} + 135 \text{ ft} + 80.5 \text{ ft} + 75.5 \text{ ft} = 346 \text{ ft}$

21. $A = \dfrac{h(b+B)}{2}$
$= \dfrac{(12 \text{ yd})(9.6 \text{ yd} + 10.2 \text{ yd})}{2}$
$= \dfrac{(12 \text{ yd})(19.8 \text{ yd})}{2}$
$= 118.8 \text{ yd}^2$

23. $A = \dfrac{h(b+B)}{2}$
$= \dfrac{(265 \text{ m})(280 \text{ m} + 300 \text{ m})}{2}$
$= \dfrac{(256 \text{ m})(580 \text{ m})}{2}$
$= 76,850 \text{ m}^2$

25. a. $A = 28 \text{ m} \times 16 \text{ m} + \dfrac{1}{2} \times 9 \text{ m} \times (28 \text{ m} + 32 \text{ m})$
$= 718 \text{ m}^2$

b. The orange object is a rectangle.

c. The yellow object is a trapezoid.

27. a. $A = (12 \text{ ft} \times 5 \text{ ft}) + \dfrac{1}{2} \times 18 \text{ ft} \times (12 \text{ ft} + 21 \text{ ft})$
$= 357 \text{ ft}^2$

b. The orange object is a parallelogram.

c. The yellow object is a trapezoid.

29. Top area $= bh = (46 \text{ yd})(49 \text{ yd}) = 2254 \text{ yd}^2$
Bottom area $= (46 \text{ yd})(31 \text{ yd}) = 1426 \text{ yd}^2$
Total area $= 2254 \text{ yd}^2 + 1426 \text{ yd}^2 = 3680 \text{ yd}^2$
Cost $= 3680 \text{ yd}^2 \times \dfrac{\$22}{\text{yd}^2} = \$80,960$
The carpet will cost a total of \$80,960.

Cumulative Review

31. $40 \text{ qt} \times \dfrac{1 \text{ gal}}{4 \text{ qt}} = 10 \text{ gal}$

32. $500 \text{ cm} = 5 \text{ m}$

33. $4 \text{ yd} \times \dfrac{36 \text{ in.}}{1 \text{ yd}} = 144 \text{ in.}$

34. $8.2 \text{ kg} = 8200 \text{ g}$

Quick Quiz 7.3

1. $P = 9 \text{ yd} + 15 \text{ yd} + 9 \text{ yd} + 17 \text{ yd} = 50 \text{ yd}$

2. $A = \dfrac{h(b+B)}{2}$
$= \dfrac{9 \text{ m}(30 \text{ m} + 34 \text{ m})}{2}$
$= \dfrac{9 \text{ m}(64 \text{ m})}{2}$
$= 288 \text{ m}^2$

3. $A = bh = (4.8 \text{ cm})(2.5 \text{ cm}) = 12 \text{ cm}^2$

4. Answers may vary. Possible solution: The area would increase by $\dfrac{16}{9}$.
$A = \dfrac{16(30+34)}{2} = 512 \text{ m}^2$; $288 \times \dfrac{16}{9} = 512 \text{ m}^2$

7.4 Exercises

1. A 90° angle is called a <u>right</u> angle.

3. Add the measures of the two known angles and subtract that value from 180°.

5. You could have concluded that the lengths of all three sides of the triangle are equal.

7. True

9. True

11. False; all angles in an equilateral triangle are exactly 60°.

13. False; to find the area of a triangle, multiply its base by its height and divide by 2.

15. $180° - (36° + 74°) = 180° - 110° = 70°$

17. $180° - (44.6° + 52.5°) = 180° - 97.1° = 82.9°$

19. $P = 18 \text{ m} + 45 \text{ m} + 55 \text{ m} = 118 \text{ m}$

21. $P = 45.25 \text{ in.} + 35.75 \text{ in.} + 35.75 \text{ in.} = 116.75 \text{ in.}$

23. $P = 3l = 3\left(3\dfrac{1}{3} \text{ mi}\right) = 3\left(\dfrac{10}{3} \text{ mi}\right) = 10 \text{ mi}$

25. $A = \dfrac{bh}{2} = \dfrac{(9 \text{ in.})(12.5 \text{ in.})}{2} = 56.25 \text{ in.}^2$

27. $A = \dfrac{bh}{2} = \dfrac{(17.5 \text{ cm})(9.5 \text{ cm})}{2} = 83.125 \text{ cm}^2$

29. $A = \dfrac{1}{2}bh$

$= \dfrac{1}{2}\left(3\dfrac{1}{2} \text{ yd}\right)\left(4\dfrac{1}{3} \text{ yd}\right)$

$= \dfrac{1}{2}\left(\dfrac{7}{2}\right)\left(\dfrac{13}{3}\right)$

$= \dfrac{91}{12} \text{ yd}^2$

$= 7\dfrac{7}{12} \text{ yd}^2$

31. Large triangle area $= \dfrac{bh}{2}$

$\qquad\qquad = \dfrac{(16 \text{ cm})(18 \text{ cm})}{2}$

$\qquad\qquad = 144 \text{ cm}^2$

Small triangle area $= \dfrac{bh}{2}$

$\qquad\qquad = \dfrac{(5 \text{ cm})(7 \text{ cm})}{2}$

$\qquad\qquad = 17.5 \text{ cm}^2$

Shaded area $= 144 \text{ cm}^2 - 17.5 \text{ cm}^2 = 126.5 \text{ cm}^2$

33. Top area $= \dfrac{bh}{2} = \dfrac{(16 \text{ yd})(4.5 \text{ yd})}{2} = 36 \text{ yd}^2$

Bottom area $= lw = (16 \text{ yd})(9.5 \text{ yd}) = 152 \text{ yd}^2$

Total area $= 36 \text{ yd}^2 + 152 \text{ yd}^2 = 188 \text{ yd}^2$

35. Area of side $= lw = (30 \text{ ft})(15 \text{ ft}) = 450 \text{ ft}^2$

Area of front (or back)

$= lw + \dfrac{bh}{2}$

$= (20 \text{ ft})(15 \text{ ft}) + \dfrac{(20 \text{ ft})(12 \text{ ft})}{2}$

$= 300 \text{ ft}^2 + 120 \text{ ft}^2$

$= 420 \text{ ft}^2$

Area of 4 sides (total area)

$= 2(450 \text{ ft}^2) + 2(420 \text{ ft}^2)$

$= 900 \text{ ft}^2 + 840 \text{ ft}^2$

$= 1740 \text{ ft}^2$

The total area of all four vertical sides of the building is 1740 ft^2.

37. Draw a vertical line through the middle of the wing to produce two identical triangles. For each triangle, $b = 18 \text{ yd}$ and $h = \dfrac{26 \text{ yd}}{2} = 13 \text{ yd}$.

$A = \dfrac{bh}{2} = \dfrac{(18 \text{ yd})(13 \text{ yd})}{2} = 117 \text{ yd}^2$

Total area $= 2(117 \text{ yd}^2) = 234 \text{ yd}^2$

Cost $= 234 \text{ yd}^2 \times \dfrac{\$90}{\text{yd}^2} = \$21,060$

The total cost is $21,060.

39. Largest triangle area $= \dfrac{bh}{2} = \dfrac{20h}{2} = 10h \text{ m}$

Smallest triangle area $= \dfrac{bh}{2}$

$\qquad\qquad = \dfrac{5(0.25h)}{2}$

$\qquad\qquad = \dfrac{1.25h}{2}$

$\qquad\qquad = 0.625h \text{ m}$

Percent $= \dfrac{0.625h}{10h} = 0.0625 = 6.25\%$

The smallest triangle area is 6.25% of the largest triangle area.

Cumulative Review

41. $\dfrac{5}{n} = \dfrac{7.5}{18}$

$5 \times 18 = 7.5 \times n$

$\dfrac{90}{7.5} = \dfrac{7.5n}{7.5}$

$12 = n$

42. $\dfrac{n}{\frac{3}{4}} = \dfrac{7}{\frac{1}{8}}$

$\dfrac{1}{8} \times n = \dfrac{3}{4} \times 7$

$\dfrac{1}{8}n = \dfrac{21}{4}$

$\dfrac{1}{8} \times \dfrac{8}{1} \times n = \dfrac{21}{4} \times \dfrac{8}{1}$

$n = 21 \times 2 = 42$

43. $\dfrac{2800}{124} = \dfrac{3500}{n}$

$2800 \times n = 124 \times 3500$

$\dfrac{2800n}{2800} = \dfrac{434,000}{2800}$

$n = 155$ tons

155 tons of trash could be collected.

$\dfrac{124}{122} = \dfrac{155}{n}$

$124 \times n = 122 \times 155$

$\dfrac{124n}{124} = \dfrac{18,910}{124}$

$n = 152.5$ miles

152.5 miles of coastline could be cleaned up.

44. $\dfrac{n}{425} = \dfrac{68}{300}$

$300 \times n = 425 \times 68$

$\dfrac{300 \times n}{300} = \dfrac{28,900}{300}$

$n = 96.\overline{3}$

$n \approx 96$

About 96 magazines would be taken.

Quick Quiz 7.4

1. $P = 22.8 \text{ m} + 21.9 \text{ m} + 36.7 \text{ m} = 81.4$ meters

2. $A = \dfrac{bh}{2} = \dfrac{(17 \text{ in.})(12 \text{ in.})}{2} = 102 \text{ in.}^2$

3. $180° - (75.4° + 53.7°) = 180° - 129.1° = 50.9°$

4.   Answers may vary. Possible solution: Find the area of each separately. Then add the results.

Triangle area: $A = \dfrac{1}{2}(20 \text{ yd})(20 \text{ yd}) = 200 \text{ yd}^2$

Rectangle area: $A = 20 \text{ yd} \times 15 \text{ yd} = 300 \text{ yd}^2$

Combined area $= 200 \text{ yd}^2 + 300 \text{ yd}^2 = 500 \text{ yd}^2$

7.5 Exercises

1. $\sqrt{25} = 5$ since $(5)(5) = 25$.

3.   25 is a perfect square because its square root, 5, is a <u>whole</u> number.

5.   To approximate the square root of a number that is not a perfect square, use a square root table or a calculator.

7. $\sqrt{9} = 3$

9. $\sqrt{64} = 8$

11. $\sqrt{144} = 12$

13. $\sqrt{0} = 0$

15. $\sqrt{169} = 13$

17. $\sqrt{100} = 10$

19. $\sqrt{49} + \sqrt{9} = 7 + 3 = 10$

21. $\sqrt{81} + \sqrt{1} = 9 + 1 = 10$

23. $\sqrt{225} - \sqrt{144} = 15 - 12 = 3$

25. $\sqrt{169} - \sqrt{121} + \sqrt{36} = 13 - 11 + 6 = 2 + 6 = 8$

27. $\sqrt{4} \times \sqrt{121} = 2 \times 11 = 22$

29. a.   Yes, because $16 \times 16 = 256$.

b. $\sqrt{256} = 16$

31. $\sqrt{18} \approx 4.243$

33. $\sqrt{76} \approx 8.718$

35. $\sqrt{200} \approx 14.142$

37. $\sqrt{34 \text{ m}^2} \approx 5.831 \text{ m}$

39. $\sqrt{136 \text{ m}^2} \approx 11.662 \text{ m}$

41. $\sqrt{36} + \sqrt{20} \approx 6 + 4.472 = 10.472$

43. $\sqrt{198} - \sqrt{49} \approx 14.071 - 7 = 7.071$

45. $\sqrt{10,964} \text{ ft} \approx 104.7 \text{ ft}$
The diagonal measures about 104.7 ft.

47. $\sqrt{16,200} \text{ ft} \approx 127.3 \text{ ft}$
This distance is about 127.3 ft.

49. $\sqrt{456} + \sqrt{322} \approx 21.3542 + 17.9443$
$= 39.2985$ which rounds to 39.299

Cumulative Review

51. $A = lw = (60 \text{ in.})(80 \text{ in.}) = 4800 \text{ in.}^2$
The area is 4800 in.^{2}

52. $19.68 \text{ km} \times \dfrac{0.62 \text{ mi}}{1 \text{ km}} \approx 12.2 \text{ mi}$
This is about 12.2 miles.

53. $30 \text{ km} \times \dfrac{1 \text{ mi}}{1.61 \text{ km}} \approx 18.6 \text{ mi}$
The race is about 18.6 miles.

54. $17 \text{ cm} = \dfrac{0.394 \text{ in.}}{1 \text{ cm}} \approx 6.7 \text{ in.}$
It can reach a length of about 6.7 in.

Quick Quiz 7.5

1. $\sqrt{64} = 8$

2. $\sqrt{36} + \sqrt{144} = 6 + 12 = 18$

3.
$$A = s^2$$
$$196 \text{ ft}^2 = s^2$$
$$\sqrt{196 \text{ ft}^2} = \sqrt{s^2}$$
$$14 \text{ ft} = s$$

4. Answers may vary. Possible solution: Since the square root of 81 is 9 $\left(\sqrt{81} = 9\right)$, then the square root of 0.81 is 0.9 $\left(\sqrt{0.81} = 0.9\right)$.

How Am I Doing? Sections 7.1–7.5

1. $90° - 72° = 18°$

2. $180° - 63° = 117°$

3. $\angle a = 180° - 136° = 44°$
$\angle b = 136°$
$\angle a = \angle c = 44°$

4. $P = 2(6.5 \text{ m}) + 2(2.5 \text{ m}) = 13 \text{ m} + 5 \text{ m} = 18 \text{ m}$

5. $P = 4(3.5 \text{ m}) = 14 \text{ m}$

6. $A = (4.8 \text{ cm})^2 = 23.04 \text{ cm}^2$

7. $A = 5.8 \text{ yd} \times 3.9 \text{ yd} = 22.62 \text{ yd}^2$

8. $P = 2(9.2 \text{ yd}) + 2(3.6 \text{ yd})$
$= 18.4 \text{ yd} + 7.2 \text{ yd}$
$= 25.6 \text{ yd}$

9. $P = 17 \text{ ft} + 15 \text{ ft} + 25\dfrac{1}{2} \text{ ft} + 21\dfrac{1}{2} \text{ ft} = 79 \text{ ft}$

10. $A = 27 \text{ in.} \times 13 \text{ in.} = 351 \text{ in.}^2$

11. $A = \dfrac{1}{2}h(b + B)$
$= \dfrac{1}{2}(9 \text{ in.})(16 \text{ in.} + 22 \text{ in.})$
$= \dfrac{1}{2}(9 \text{ in.})(38 \text{ in.})$
$= 171 \text{ in.}^2$

12. $A = 7 \text{ m} \times 9 \text{ m} + \dfrac{1}{2}(10 \text{ m} + 7 \text{ m})(4 \text{ m})$
$= 63 \text{ m}^2 + \dfrac{1}{2}(17 \text{ m})(4 \text{ m})$
$= 97 \text{ m}^2$

13. $180° - (22.5° + 54.5°) = 180° - 77° = 103°$

14. $P = 9\dfrac{1}{2} \text{ in.} + 4 \text{ in.} + 6\dfrac{1}{2} \text{ in.}$
$= 13\dfrac{1}{2} \text{ in.} + 6\dfrac{1}{2} \text{ in.}$
$= 20 \text{ in.}$

15. $A = \dfrac{1}{2}(21 \text{ m})(10 \text{ m}) = 105 \text{ m}^2$

16. **a.** $A = \frac{1}{2}(44 \text{ ft} + 30 \text{ ft})(16 \text{ ft})$

$= \frac{1}{2}(74 \text{ ft})(16 \text{ ft})$

$= 592 \text{ ft}^2$

592 ft^2 of paint is needed for the sign.

b. $P = 20 \text{ ft} + 30 \text{ ft} + 20 \text{ ft} + 44 \text{ ft} = 114 \text{ ft}$
114 ft of trim is needed for the sign.

17. $\sqrt{64} = 8$

18. $\sqrt{225} + \sqrt{16} = 15 + 4 = 19$

19. $\sqrt{169} = 13$

20. $\sqrt{256} = 16$

21. $\sqrt{46} \approx 6.782$

7.6 Exercises

1. Square the length of each leg and add these two results. Then take the square root of the result.

3. $h = \sqrt{12^2 + 9^2} = \sqrt{144 + 81} = \sqrt{225} = 15 \text{ yd}$

5. $\text{leg} = \sqrt{16^2 - 5^2} = \sqrt{256 - 25} = \sqrt{231} \approx 15.199 \text{ ft}$

7. $h = \sqrt{11^2 + 3^2} = \sqrt{121 + 9} = \sqrt{130} \approx 11.402 \text{ m}$

9. $h = \sqrt{10^2 + 10^2}$

$= \sqrt{100 + 100}$

$= \sqrt{200}$

$\approx 14.142 \text{ m}$

11. $\text{leg} = \sqrt{11^2 - 4^2}$

$= \sqrt{121 - 16}$

$= \sqrt{105}$

$\approx 10.247 \text{ ft}$

13. $\text{leg} = \sqrt{14^2 - 10^2}$

$= \sqrt{196 - 100}$

$= \sqrt{96}$

$\approx 9.798 \text{ yd}$

15. $h = \sqrt{12^2 + 9^2} = \sqrt{144 + 81} = \sqrt{225} = 15 \text{ m}$

17. $\text{leg} = \sqrt{16^2 - 11^2}$

$= \sqrt{256 - 121}$

$= \sqrt{135}$

$\approx 11.619 \text{ ft}$

19. $\text{hypotenuse} = \sqrt{12^2 + 5^2}$

$= \sqrt{144 + 25}$

$= \sqrt{169}$

$= 13 \text{ ft}$

The distance is 13 ft.

21. $\text{hypotenuse} = \sqrt{9^2 + 4^2}$

$= \sqrt{81 + 16}$

$= \sqrt{97}$

$\approx 9.8 \text{ cm}$

The distance is 9.8 cm.

23. $\text{leg} = \sqrt{32^2 - 30^2}$

$= \sqrt{1024 - 900}$

$= \sqrt{124}$

$\approx 11.1 \text{ yd}$

The kite is 11.1 yards above the ground.

25. Side opposite $30° = \frac{1}{2}(8) = 4 \text{ in.}$

The other leg $= \sqrt{8^2 - 4^2}$

$= \sqrt{64 - 16}$

$= \sqrt{48}$

$\approx 6.9 \text{ in.}$

27. The other leg = 6 m

$\text{hypotenuse} = \sqrt{2} \times \text{leg}$

$= \sqrt{2} \times 6$

$\approx 1.414 \times 6$

$\approx 8.5 \text{ m}$

29. The other leg = 18 cm

$\text{hypotenuse} = \sqrt{2} \times \text{leg}$

$= \sqrt{2} \times 18$

$\approx 1.414 \times 18$

$\approx 25.5 \text{ cm}$

31. $\text{leg} = \sqrt{10^2 - 7^2} = \sqrt{100 - 49} = \sqrt{51} \approx 7.1 \text{ in.}$

The rectangular base will be 7.1 inches tall.

33. $h = \sqrt{(0.25)^2 + (0.4)^2}$
$= \sqrt{0.0625 + 0.16}$
$= \sqrt{0.2225}$
≈ 0.47 mi
The length of the walkway is 0.47 mile.

35. $h = \sqrt{14^2 + 5^2} = \sqrt{196 + 25} = \sqrt{221} \approx 14.866$ cm

Cumulative Review

37. $A = \dfrac{bh}{2} = \dfrac{(31 \text{ m})(22 \text{ m})}{2} = 341$ m^2
The area of the land is 341 m^2.

38. $A = lw = (20.5 \text{ ft})(14.5 \text{ ft}) = 297.25$ ft^2
The area of the garden is 297.25 ft^2.

39. $A = s^2 = (21 \text{ in.})^2 = 441$ in.^{2}
The area of the window is 441 in.^{2}.

40. $A = bh = (88 \text{ yd})(48 \text{ yd}) = 4224$ yd^2
The area of the roof is 4224 yd^2.

Quick Quiz 7.6

1. $h = \sqrt{10^2 + 5^2} = \sqrt{100 + 25} = \sqrt{125} \approx 11.18$ feet

2. $\text{leg} = \sqrt{26^2 - 24^2}$
$= \sqrt{676 - 576}$
$= \sqrt{100}$
$= 10$ centimeters

3. $h = \sqrt{3^2 + 8^2} = \sqrt{9 + 64} = \sqrt{73} \approx 8.54$ miles
She is 8.54 miles from where she started.

4. Answers may vary. Possible solution:
The hypotenuse is 2 mi and one leg is 1.5 mi.
Find the distance to the supplies by using the Pythagorean Theorem.
$\text{leg} = \sqrt{2^2 - 1.5^2}$
$= \sqrt{4 - 2.25}$
$= \sqrt{1.75}$
≈ 1.32 miles

7.7 Exercises

1. The distance around a circle is called the circumference.

3. The diameter is two times the radius of the circle.

5. You need to multiply the radius by 2 and then use the formula $C = \pi d$.

7. $d = 2r = 2(29 \text{ in.}) = 58$ in.

9. $d = 2r = 2\left(8\dfrac{1}{2} \text{ mm}\right) = 17$ mm

11. $r = \dfrac{d}{2} = \dfrac{45 \text{ yd}}{2} = 22.5$ yd

13. $r = \dfrac{d}{2} = \dfrac{32.18 \text{ ft}}{2} = 16.09$ ft

15. $C = \pi d = 3.14(32 \text{ cm}) = 100.48$ cm

17. $C = 2\pi r = 2(3.14)(18.5 \text{ in.}) = 116.18$ in.

19. $C = \pi d = (3.14)(32 \text{ in.}) = 100.48$ in.
$\text{Distance} = (100.48 \text{ in.})(5 \text{ rev}) \times \dfrac{1 \text{ ft}}{12 \text{ in.}} \approx 41.9$ ft

21. $A = \pi r^2$
$= 3.14(5 \text{ yd})^2$
$= 3.14(25 \text{ yd}^2)$
$= 78.5$ yd^2

23. $A = \pi r^2$
$= 3.14(8.5)^2$
$= 3.14(72.25)$
≈ 226.87 in.^{2}

25. $r = \dfrac{d}{2} = \dfrac{32 \text{ cm}}{2} = 16$ cm
$A = \pi r^2$
$= 3.14(16 \text{ cm})^2$
$= 3.14(256 \text{ cm}^2)$
$= 803.84$ cm^2

27. $A = \pi r^2$

$\quad = 3.14(12 \text{ ft})^2$

$\quad = 3.14(144 \text{ ft}^2)$

$\quad = 452.16 \text{ ft}^2$

29. $r = \dfrac{90 \text{ mi}}{2} = 45 \text{ mi}$

$A = \pi r^2$

$\quad = 3.14(45 \text{ mi})^2$

$\quad = 3.14(2025 \text{ mi}^2)$

$\quad = 6358.5 \text{ mi}^2$

31. $A = \pi r^2 - \pi r^2$

$\quad = 3.14(14 \text{ m})^2 - 3.14(12 \text{ m})^2$

$\quad = 3.14(196 \text{ m}^2) - 3.14(144 \text{ m}^2)$

$\quad = 615.44 \text{ m}^2 - 452.16 \text{ m}^2$

$\quad = 163.28 \text{ m}^2$

33. $A = lw - \pi r^2$

$\quad = (12 \text{ m})(12 \text{ m}) - (3.14)(6 \text{ m})^2$

$\quad = 144 \text{ m}^2 - (3.14)(36 \text{ m}^2)$

$\quad = 144 \text{ m}^2 - 113.04 \text{ m}^3$

$\quad = 30.96 \text{ m}^2$

35. $r = \dfrac{d}{2} = \dfrac{10 \text{ m}}{2} = 5 \text{ m}$

$A = \dfrac{1}{2}\pi r^2 + lw$

$\quad = \dfrac{1}{2}(3.14)(5 \text{ m})^2 + (15 \text{ m})(10 \text{ m})$

$\quad = \dfrac{1}{2}(3.14)(25 \text{ m}^2) + 150 \text{ m}^2$

$\quad = 39.25 \text{ m}^2 + 150 \text{ m}^2$

$\quad = 189.25 \text{ m}^2$

37. $r = \dfrac{d}{2} = \dfrac{40 \text{ yd}}{2} = 20 \text{ yd}$

$A = \pi r^2 + lw$

$\quad = (3.14)(20 \text{ yd})^2 + (120 \text{ yd})(40 \text{ yd})$

$\quad = 1256 \text{ yd}^2 + 4800 \text{ yd}^2$

$\quad = 6056 \text{ yd}^2$

$\text{Cost} = 6056 \text{ yd}^2 \times \dfrac{\$0.20}{\text{yd}^2} = \$1211.20$

39. $C = \pi d = 3.14(3 \text{ ft}) = 9.42 \text{ ft}$

The length of the strip is 9.42 feet.

41. $C = 2\pi r = 2(3.14)(30 \text{ in.}) = 188.4 \text{ in.}$

$\text{Distance} = (188.4)(9 \text{ rev}) \times \dfrac{1 \text{ ft}}{12 \text{ in.}} = 141.3 \text{ ft}$

The truck travels 141.3 feet.

43. $1 \text{ mi} = 5280 \text{ ft} \times \dfrac{12 \text{ in.}}{1 \text{ ft}} = 63,360 \text{ in.}$

$C = 2\pi r = 2(3.14)(7.5 \text{ in.}) = 47.1 \text{ in.}$

$\text{rev} = \dfrac{63,360 \text{ in.}}{47.1 \text{ in.}} \approx 1345.22$

Her wheels make 1345.22 revolutions.

45. a. $C = \pi d = 3.14(8) = 25.12 \text{ ft}$

The tape will be 25.12 feet long.

b. $r = \dfrac{d}{2} = \dfrac{8}{2} = 4 \text{ ft}$

$A = \pi r^2 = (3.14)(4)^2 = 50.24 \text{ ft}^2$

The area is 50.24 ft^2.

47. $A = \pi r^2$

$\quad = 3.14(200)^2$

$\quad = 3.14(40,000)$

$\quad = 125,600 \text{ mi}^2$

The receiving range is $125,600 \text{ mi}^2$.

49. a. $\text{Cost} = \dfrac{12.00}{8} = 1.5 = \$1.50$

$r = \dfrac{d}{2} = \dfrac{16 \text{ in.}}{2} = 8 \text{ in.}$

$\dfrac{A}{8} = \dfrac{\pi r^2}{8} = \dfrac{3.14(8 \text{ in.})^2}{8} = 25.12 \text{ in.}^2$

The cost is \$1.50 per slice and the area is 25.12 in.^2.

b. $\text{Cost} = \dfrac{8.00}{6} \approx 1.33 = \$1.33$

$r = \dfrac{d}{2} = \dfrac{12}{2} = 6 \text{ in.}$

$\dfrac{A}{6} = \dfrac{\pi r^2}{6} = \dfrac{3.14(6 \text{ in.})^2}{6} = 18.84 \text{ in.}^2$

The cost is \$1.33 per slice and the area is 18.84 in.^2.

c. For the 12-in. pizza, it is about

$\dfrac{\$1.33}{18.84 \text{ in.}^2} \approx \$0.07$ per in.^{2}; for the 16-in.

pizza, it is about

$\dfrac{\$1.50}{25.12 \text{ in.}^2} \approx \$0.06$ per in.^{2}; the 16-in. pizza

is a better value.

Cumulative Review

51. $n = 25\% \times 120$
$n = 0.25 \times 120$
$n = 30$

52. $n = 0.5\% \times 60$
$n = 0.005 \times 60$
$n = 0.3$

53. $10\% \times n = 7$
$0.10n = 7$
$\dfrac{0.10n}{0.10} = \dfrac{7}{0.10}$
$n = 70$

54. $19\% \times n = 570$
$0.19n = 570$
$\dfrac{0.19n}{l0.19} = \dfrac{570}{0.19}$
$n = 3000$

Quick Quiz 7.7

1. $C = \pi d \approx 3.14(9 \text{ in.}) = 28.26$ in.

2. $A = \pi r^2 \approx 3.14(11)^2 = 379.94$ m^2

3. rectangle $A = (3.5 \text{ cm})(3.1 \text{ cm}) = 10.85$ cm^2

circle $A = \pi r^2 \approx 3.14(1.5 \text{ cm})^2 = 7.065$ cm^2

Area after drilled hole $= 10.85 \text{ cm}^2 - 7.065 \text{ cm}^2$
$\qquad = 3.785$ cm^2

4. Answers may vary. Possible solution: Find the area of a circle with a radius of 3 feet. Then divide the result by 2.

$\dfrac{\pi r^2}{2} = \dfrac{3.14(3 \text{ ft})^2}{2} = 14.13$ ft^2

7.8 Exercises

1. a. sphere

b. $V = \dfrac{4\pi r^3}{3}$

3. a. cylinder

b. $V = \pi r^2 h$

5. a. cone

b. $V = \dfrac{\pi r^2 h}{3}$

7. $V = lwh$
$\quad = (30 \text{ mm})(12 \text{ mm})(1.5 \text{ mm})$
$\quad = 540$ mm^3

9. $V = \pi r^2 h$
$\quad = 3.14(3 \text{ m})^2(8 \text{ m})$
$\quad = 3.14(9 \text{ m}^2)(8 \text{ m})$
$\quad \approx 226.1$ m^2

11. $r = \dfrac{d}{2} = \dfrac{22 \text{ m}}{2} = 11$ m
$V = \pi r^2 h$
$\quad = 3.14(11 \text{ m})^2(17 \text{ m})$
$\quad \ 3.14(121 \text{ m}^2)(17 \text{ m})$
$\quad \approx 6459.0$ m^3

13. $V = \dfrac{4\pi r^3}{3}$
$\quad = \dfrac{4(3.14)(9 \text{ yd})^3}{3}$
$\quad = \dfrac{4(3.14)(729 \text{ yd}^3)}{3}$
$\quad \approx 3052.1$ yd^3

15. $V = \dfrac{Bh}{3} = \dfrac{(18 \text{ ft}^2)(35 \text{ ft})}{3} = 210$ ft^3

17. $V = s^3 = (0.6 \text{ cm})^3 = 0.216$ cm^3

19. $V = \dfrac{\pi r^2 h}{3}$

$\qquad = \dfrac{3.14(3 \text{ yd})^2(7 \text{ yd})}{3}$

$\qquad = \dfrac{3.14(9 \text{ yd}^2)(7 \text{ yd})}{3}$

$\qquad = 65.94 \text{ yd}^3$

21. $V = \dfrac{1}{2} \times \dfrac{4\pi r^3}{3}$

$\qquad = \dfrac{1}{2} \times \dfrac{4(3.14)(7 \text{ m})^3}{3}$

$\qquad = \dfrac{1}{2} \times \dfrac{4(3.14)(343 \text{ m}^3)}{3}$

$\qquad \approx \dfrac{1}{2} \times 1436.027 \text{ m}^3$

$\qquad \approx 718.0 \text{ m}^3$

23. $V = \dfrac{\pi r^2 h}{3}$

$\qquad = \dfrac{3.14(8 \text{ cm})^2(14 \text{ cm})}{3}$

$\qquad = \dfrac{3.14(64 \text{ cm}^2)(14 \text{ cm})}{3}$

$\qquad \approx 937.8 \text{ cm}^3$

25. $V = \dfrac{\pi r^2 h}{3}$

$\qquad = \dfrac{3.14(7 \text{ ft})^2(12.5 \text{ ft})}{3}$

$\qquad = \dfrac{3.14(49 \text{ ft}^2)(12.5 \text{ ft})}{3}$

$\qquad \approx 641.1 \text{ ft}^3$

27. $B = (7 \text{ m})(7 \text{ m}) = 49 \text{ m}^2$

$\qquad V = \dfrac{Bh}{3} = \dfrac{(49 \text{ m}^2)(10 \text{ m})}{3} = 163.3 \text{ m}^3$

29. $B = (8 \text{ m})(14 \text{ m}) = 112 \text{ m}^2$

$\qquad V = \dfrac{Bh}{3} = \dfrac{(112 \text{ m}^2)(10 \text{ m})}{3} \approx 373.3 \text{ m}^3$

31. $3 \text{ in.} \times \dfrac{1 \text{ ft}}{12 \text{ in.}} = \dfrac{1}{4} \text{ ft}$

$\quad V = lwh$

$\qquad = (9 \text{ ft})(16 \text{ ft})\left(\dfrac{1}{4} \text{ ft}\right)$

$\qquad = (144 \text{ ft}^2)\left(\dfrac{1}{4} \text{ ft}\right)$

$\qquad = 36 \text{ ft}^3$

$\quad \dfrac{36 \text{ ft}^3}{3 \text{ ft}^3} = 12 \text{ bags}$

She should purchase 12 bags.

33. $\text{Outer} = \pi r^2 h$

$\qquad = 3.14(5 \text{ in.})^2(20 \text{ in.})$

$\qquad = 3.14(25 \text{ in.}^2)(20 \text{ in.})$

$\qquad = 1570 \text{ in.}^3$

$\quad \text{Inner} = \pi r^2 h$

$\qquad = 3.14(3 \text{ in.})^2(20 \text{ in.})$

$\qquad = 3.14(9 \text{ in.}^2)(20 \text{ in.})$

$\qquad = 565.2 \text{ in.}^3$

$\text{Difference} = 1570 \text{ in.}^3 - 565.2 \text{ in.}^3 = 1004.8 \text{ in.}^3$

The volume is 1004.8 in.^3.

35. Jupiter: $V = \dfrac{4\pi r^3}{3}$

$\qquad = \dfrac{4(3.14)(45{,}000 \text{ mi})^3}{3}$

$\qquad = 381{,}510{,}000{,}000{,}000 \text{ mi}^3$

$\quad$ Earth: $V = \dfrac{4\pi r^3}{3}$

$\qquad = \dfrac{4(3.14)(3950 \text{ mi})^3}{3}$

$\qquad \approx 258{,}023{,}743{,}333 \text{ mi}^3$

$\quad$ Difference: $\quad 381{,}510{,}000{,}000{,}000 \text{ mi}^3$

$\qquad \underline{- \qquad 258{,}023{,}743{,}333 \text{ mi}^3}$

$\qquad\qquad 381{,}251{,}976{,}256{,}667 \text{ mi}^3$

37. Smaller: $V = lwh$
$$= (18 \text{ in.})(6 \text{ in.})(12 \text{ in.})$$
$$= (108 \text{ in.}^2)(12 \text{ in.})$$
$$= 1296 \text{ in.}^3$$
Larger: $V = lwh$
$$= (22 \text{ in.})(12 \text{ in.})(16 \text{ in.})$$
$$= (264 \text{ in.}^2)(16 \text{ in.})$$
$$= 4224 \text{ in.}^3$$
Difference $= 4224 \text{ in.}^3 - 1296 \text{ in.}^3 = 2928 \text{ in.}^3$
There are 2928 in.^3 of Styrofoam peanuts.

39. $V = \dfrac{\pi r^2 h}{3}$
$$= \frac{3.14(1.5 \text{ in.})^2 (3 \text{ in.})}{3}$$
$$= \frac{3.14(2.25 \text{ in.}^2)(3 \text{ in.})}{3}$$
$$\approx 7.1 \text{ in}^3$$
$$\frac{1150 \text{ in.}^3}{7.1 \text{ in.}^3} \approx 162$$
Each cup holds about 7.1 in.^3, and the dispenser fills about 162 cups.

41. $h = 13.5 - 1.4 = 12.1 \text{ cm}$
$$r = \frac{d}{2} = \frac{6.6}{2} = 3.3 \text{ cm}$$
$$V = \pi r^2 h$$
$$\approx 3.14(3.3 \text{ cm})^2 \times 12.1 \text{ cm}$$
$$\approx 3.14(10.89 \text{ cm}^2)(12.1 \text{ cm})$$
$$\approx 413.8 \text{ cm}^3$$
The volume of the new can is 413.8 cm^3.

43. $B = (87 \text{ yd})(130 \text{ yd}) = 11{,}310 \text{ yd}^2$
$$V = \frac{Bh}{3} = \frac{(11{,}310 \text{ yd}^2)(70 \text{ yd})}{3} = 263{,}900 \text{ yd}^3$$
The volume is $263{,}900 \text{ yd}^3$.

Cumulative Review

45.
$$7\frac{1}{3} \qquad 7\frac{4}{12}$$
$$+2\frac{1}{4} \qquad +2\frac{3}{12}$$
$$\overline{\qquad\qquad \quad 9\frac{7}{12}}$$

46. $9\dfrac{1}{8} - 2\dfrac{3}{4} = 8\dfrac{9}{8} - 2\dfrac{6}{8} = 6\dfrac{3}{8}$

47. $2\dfrac{1}{4} \times 3\dfrac{3}{4} = \dfrac{9}{4} \times \dfrac{15}{4} = \dfrac{135}{16} = 8\dfrac{7}{16}$

48. $7\dfrac{1}{2} \div 4\dfrac{1}{5} = \dfrac{15}{2} \div \dfrac{21}{5} = \dfrac{15}{2} \times \dfrac{5}{21} = \dfrac{25}{14} = 1\dfrac{11}{14}$

49. $\left(\dfrac{5}{8} - \dfrac{1}{4}\right)^2 + \dfrac{7}{32} = \left(\dfrac{5}{8} - \dfrac{2}{8}\right)^2 + \dfrac{7}{32}$
$$= \left(\frac{3}{8}\right)^2 + \frac{7}{32}$$
$$= \frac{9}{64} + \frac{14}{64}$$
$$= \frac{23}{64}$$

50. $\left(6\dfrac{5}{6} + 2\dfrac{3}{4}\right) \times \dfrac{2}{3} = \left(6\dfrac{10}{12} + 2\dfrac{9}{12}\right) \times \dfrac{2}{3}$
$$= \left(8\frac{19}{12}\right) \times \frac{2}{3}$$
$$= \frac{115}{12} \times \frac{2}{3}$$
$$= \frac{115}{18}$$
$$= 6\frac{7}{18}$$

Quick Quiz 7.8

1. $V = \dfrac{4\pi r^3}{3} = \dfrac{4(3.14)(4 \text{ cm})^3}{3} \approx 267.95 \text{ cm}^3$

2. $V = \dfrac{Bh}{3} = \dfrac{(7 \text{ yd} \times 6 \text{ yd})(8 \text{ yd})}{3} = 112 \text{ yd}^3$

3. $V = \pi r^2 h = 3.14(3 \text{ m})^2 (13 \text{ m}) = 367.38 \text{ m}^3$

4. Answers may vary. Possible solution:
The volume is larger by $\dfrac{4^2}{3^2} = \dfrac{16}{9}$.

7.9 Exercises

1. Similar figures may be different in <u>size</u> but they are alike in <u>shape</u>.

3. The perimeter of similar figures have the same ratio as their corresponding <u>sides</u>.

5. $\dfrac{n}{2} = \dfrac{12}{3}$

$3n = (2)(12)$

$3n = 24$

$\dfrac{3n}{3} = \dfrac{24}{3}$

$n = 8$

8 m

7. $\dfrac{n}{6} = \dfrac{9}{21}$

$21n = (6)(9)$

$21n = 54$

$\dfrac{21n}{21} = \dfrac{54}{21}$

$n \approx 2.6$

2.6 ft

9. $\dfrac{n}{8.5} = \dfrac{2}{5}$

$5n = 8.5(2)$

$\dfrac{5n}{5} = \dfrac{17}{5}$

$n = 3.4$ yd

11. a corresponds to f.

b corresponds to e.

c corresponds to d.

13. $\dfrac{n}{8} = \dfrac{10.5}{25}$

$25n = (8)(10.5)$

$25n = 84$

$\dfrac{25n}{25} = \dfrac{84}{25}$

$n \approx 3.4$

The shortest side will be 3.4 m.

15. $\dfrac{6}{39} = \dfrac{4}{n}$

$6n = 39(4)$

$\dfrac{6n}{6} = \dfrac{156}{6}$

$n = 26$

The width of the poster is 26 inches.

17. $\dfrac{n}{2} = \dfrac{5.5}{5}$

$5n = 2 \times 5.5$

$\dfrac{5n}{5} = \dfrac{11}{5}$

$n = 2.2$

The tub will be 2.2 ft.

19. $\dfrac{n}{6} = \dfrac{24}{4}$

$4n = (6)(24)$

$4n = 144$

$\dfrac{4n}{4} = \dfrac{144}{4}$

$n = 36$

The flagpole is 36 ft.

21. $\dfrac{n}{144} = \dfrac{5.5}{18}$

$18n = (144)(5.5)$

$18n = 792$

$\dfrac{18n}{18} = \dfrac{792}{18}$

$n = 44$

The wall is 44 feet tall.

23. $\dfrac{n}{15} = \dfrac{5}{9}$

$9n = (15)(5)$

$9n = 75$

$\dfrac{9n}{9} = \dfrac{75}{9}$

$n \approx 8.3$ ft

25. $\dfrac{n}{9} = \dfrac{8}{6}$

$6n = 9 \times 8$

$\dfrac{6n}{6} = \dfrac{72}{6}$

$n = 12$ cm

Cumulative Review

27. $2 \times 3^2 + 4 - 2 \times 5 = 2 \times 9 + 4 - 2 \times 5$

$\qquad\qquad\qquad\qquad = 18 + 4 - 10$

$\qquad\qquad\qquad\qquad = 22 - 10$

$\qquad\qquad\qquad\qquad = 12$

28. $100 \div (8-3)^2 \times 2^3 = 100 \div (5^2)(2^3)$
$= 100 \div (25)(8)$
$= 4 \times 8$
$= 32$

29. $\dfrac{4}{5} \times \dfrac{5}{3} - \dfrac{1}{3} = \dfrac{4}{3} - \dfrac{1}{3} = \dfrac{3}{3} = 1$

30. $\dfrac{8}{5} \div 3 - \dfrac{1}{3} = \dfrac{8}{5} \cdot \dfrac{1}{3} - \dfrac{1}{3} = \dfrac{8}{15} - \dfrac{1}{3} = \dfrac{8}{15} - \dfrac{5}{15} = \dfrac{3}{15} = \dfrac{1}{5}$

Quick Quiz 7.9

1. $\dfrac{6}{n} = \dfrac{7}{100}$
$6(100) = 7n$
$\dfrac{600}{7} = n$
$85.71 \text{ ft} \approx n$
The cliff is 85.71 feet tall.

2. $\dfrac{15}{x} = \dfrac{14}{3}$
$15(3) = x(14)$
$45 = 14x$
$\dfrac{45}{14} = \dfrac{14x}{14}$
$3.21 \text{ m} \approx x$

3. $\dfrac{7}{x} = \dfrac{12}{36}$
$7(36) = x(12)$
$\dfrac{252}{12} = \dfrac{12x}{12}$
$21 \text{ ft} = x$
The new house will be 21 feet wide.

4. Answers may vary. Possible solution: Draw a picture. Set up a proportion and solve for the unknown.
$\dfrac{5}{6} = \dfrac{9}{n}$
$5n = 6(9)$
$5n = 54$
$\dfrac{5n}{5} = \dfrac{54}{5}$
$n = 10.8$
The largest side in the drawing is 10.8 inches.

7.10 Exercises

1. a. Trip = 13 mi + 17 mi = 30 mi
Speed $= \dfrac{30 \text{ mi}}{0.6 \text{ hr}} = 50 \text{ mi/hr}$

b. Trip = 12 mi + 15 mi + 11 mi = 38 mi
Speed $= \dfrac{38 \text{ mi}}{0.7 \text{ hr}} \approx 54.3 \text{ mi/hr}$

c. Through Woodville and Palermo is the more rapid rate route.

3. $A = (7 \text{ ft})(10 \text{ ft}) = 70 \text{ ft}^2$
$A = (7 \text{ ft})(14 \text{ ft}) = 98 \text{ ft}^2$
$A = (6 \text{ ft})(10 \text{ ft}) = 60 \text{ ft}^2$
$A = (6 \text{ ft})(8 \text{ ft}) = 48 \text{ ft}^2$
Total $A = 70 \text{ ft}^2 + 98 \text{ ft}^2 + 60 \text{ ft}^2 + 48 \text{ ft}^2$
$= 276 \text{ ft}^2$
Time $= 276 \text{ ft}^2 \times \dfrac{35 \text{ min}}{120 \text{ ft}^2} = 80.5 \text{ min}$
It will take her 80.5 minutes or 1 hour 20.5 minutes to wallpaper all four walls.

5. Front and back:
$2(55 \text{ ft} \times 24 \text{ ft}) = 2(1320 \text{ ft}^2) = 2640 \text{ ft}^2$
Sides: $2(32 \text{ ft} \times 24 \text{ ft}) = 2(768 \text{ ft}^2) = 1536 \text{ ft}^2$
Windows: $16(4 \text{ ft} \times 2 \text{ ft}) = 16(8 \text{ ft}^2) = 128 \text{ ft}^2$
Doors: $2(7 \text{ ft} \times 3 \text{ ft}) = 2(21 \text{ ft}^2) = 42 \text{ ft}^2$
Total:
$2640 \text{ ft}^2 + 1536 \text{ ft}^2 - 128 \text{ ft}^2 - 42 \text{ ft}^2 = 4006 \text{ ft}^2$
The total area to be painted is 4006 ft^2.
$4006 \text{ ft}^2 \times \dfrac{1 \text{ gal}}{350 \text{ ft}^2} \approx 11.4 \text{ gal}$
They need 12 gallons of paint.

7. $A = lw - \dfrac{bh}{2}$
$= (21 \text{ ft})(15 \text{ ft}) - \dfrac{(3 \text{ ft})(6 \text{ ft})}{2}$
$= 315 \text{ ft}^2 - 9 \text{ ft}^2$
$= 306 \text{ ft}^2$
Cost $= 306 \text{ ft}^2 \times \dfrac{1 \text{ yd}^2}{9 \text{ ft}^2} \times \dfrac{\$15}{\text{yd}^2} = \$510$
It will cost $510 to carpet the room.

9. $V = \dfrac{1}{2} \times \dfrac{4}{3} \times 3.14 \times (1 \text{ mm})^3$

$\qquad\qquad\qquad + 3.14 \times 2 \text{ mm} \times (1 \text{ mm})^2$

$V \approx 8.37 \text{ mm}^3$

$\text{Cost} = 8.37 \text{ mm}^3 \times \dfrac{\$95}{\text{mm}^3} = \$795.15$

The gold for the filling cost \$795.15.

11. a. $C = 2\pi r = 2(3.14)(6500 \text{ km}) = 40{,}820 \text{ km}$
    The orbit is 40,820 km long.

b. $S = \dfrac{40{,}820 \text{ km}}{2 \text{ hr}} = 20{,}410 \text{ km/hr}$
    The speed is 20,410 km/hr.

13. $V = 400 \times \pi r^2 h$

$\qquad = 400 \times (3.14)(2 \text{ in.})^2 (10 \text{ in.})$

$\qquad = 400 \times (3.14)(4 \text{ in.}^2)(10 \text{ in.})$

$\qquad = 400 \times 125.6 \text{ in.}^3$

$\qquad \approx 50{,}240 \text{ in.}^3$

The total volume is $50{,}240 \text{ in.}^3$.

Cumulative Review

15.
$$
\begin{array}{r}
128 \\
16\overline{)2048} \\
\underline{16} \\
44 \\
\underline{32} \\
128 \\
\underline{128} \\
0
\end{array}
$$

16.
$$
\begin{array}{r}
308 \\
42\overline{)12{,}936} \\
\underline{12\,6} \\
336 \\
\underline{336} \\
0
\end{array}
$$

17.
$$
\begin{array}{r}
0.25 \\
1.3_\wedge\overline{)0.3_\wedge 25} \\
\underline{2\,6} \\
65 \\
\underline{65} \\
0
\end{array}
$$

18.
$$
\begin{array}{r}
4.87 \\
0.52_\wedge\overline{)2.53_\wedge 24} \\
\underline{2\,08} \\
45\,2 \\
\underline{41\,6} \\
3\,64 \\
\underline{3\,64} \\
0
\end{array}
$$

Quick Quiz 7.10

1. $A = \text{rectangle area} + \text{circle area (2 halves)}$

$\qquad = lw + \pi r^2$

$\qquad = (140 \text{ yd})(50 \text{ yd}) + 3.14(25 \text{ yd})^2$

$\qquad = 8962.5 \text{ yd}^2$

The area of the new field is 8962.5 yd^2.

2. $A = \text{rectangle area} - 7(\text{window area})$

$\qquad = (20 \text{ ft} \times 34 \text{ ft}) - 7(3 \text{ ft} \times 2 \text{ ft})$

$\qquad = 638 \text{ ft}^2$

He will need to paint 638 ft^2.

3. Total cost

$= (\text{cost per m}^2)(\text{Area off triangle in m}^2)$

$= \$0.05/\text{m} \left(\dfrac{1}{2} \times 200 \text{ m} \times 140 \text{ m} \right)$

$= \$700$

The weed killer will cost \$700.

4. Answers may vary. Possible solution: Find the area of the new field. Multiply the area by \$0.20 to find the cost. Then subtract the result found in problem 3 from this result.

$\dfrac{1}{2}(300)(140) = 21{,}000 \text{ m}^2$

$21{,}000 \times 0.20 = \$4200$

The difference in cost is $\$4200 - \$700 = \$3500$.

Use Math to Save Money

1. A: $15{,}000 \text{ mi} \times \dfrac{1 \text{ gal}}{20 \text{ mi}} = 750 \text{ gal}$

B: $15{,}000 \text{ mi} \times \dfrac{1 \text{ gal}}{32 \text{ mi}} = 468.75 \text{ gal}$

C: $15{,}000 \text{ mi} \times \dfrac{1 \text{ gal}}{60 \text{ mi}} = 250 \text{ gal}$

2. A: $750 \text{ gal} \times \dfrac{\$3.60}{1 \text{ gal}} = \$2700$

B: $468.75 \text{ gal} \times \dfrac{\$3.60}{1 \text{ gal}} = \$1687.50$

C: $250 \text{ gal} \times \dfrac{\$3.60}{1 \text{ gal}} = \$900$

3. $\begin{array}{r} 2700.00 \\ -\ 1687.50 \\ \hline 1012.50 \end{array}$

Compared to Option A, Option B would save $1012.50 in one year.

4. $\$6610 \times \dfrac{1 \text{ yr}}{\$1012.50} \approx 6.5 \text{ yr}$

It would take about six and a half years.

5. $\dfrac{\$1012.50}{1 \text{ yr}} \times 5 \text{ yr} = \$5062.50$

$\dfrac{\$1012.50}{1 \text{ yr}} \times 10 \text{ yr} = \$10{,}125.00$

Michele would save $5062.50 over five years and $10,125.00 over ten years.

6. $\begin{array}{r} 2700 \\ -\ 900 \\ \hline 1800 \end{array}$

Compared to Option A, Option C would save $1800 in one year.

7. $\$2230 \times \dfrac{1 \text{ yr}}{\$1800} \approx 1.24 \text{ yr}$

It would take about one year and three months.

8. $\dfrac{\$1800}{1 \text{ yr}} \times 5 \text{ yr} = \$9000$

$\dfrac{\$1800}{1 \text{ yr}} \times 10 \text{ yr} = \$18{,}000$

Michele would save $9000 over five years and $18,000 over ten years.

You Try It

1. $P = 2(12 \text{ ft}) + 2(5 \text{ ft}) = 24 \text{ ft} + 10 \text{ ft} = 34 \text{ ft}$

2. $P = 4(8.5 \text{ in.}) = 34 \text{ in.}$

3. $A = (10 \text{ m})(3 \text{ m}) = 30 \text{ m}^2$

4. $A = s^2 = (9 \text{ ft})^2 = 81 \text{ ft}^2$

5. $7 \text{ in.} + 21 \text{ in.} + 15 \text{ in.} = 43 \text{ in.}$

6. $A = bh = (13 \text{ cm})(6 \text{ cm}) = 78 \text{ cm}^2$

7. $\begin{aligned} A &= \frac{h(b+B)}{2} = \frac{(8 \text{ ft})(3.5 \text{ ft} + 12.5 \text{ ft})}{2} \\ &= \frac{(8 \text{ ft})(16 \text{ ft})}{2} \\ &= \frac{128 \text{ ft}^2}{2} \\ &= 64 \text{ ft}^2 \end{aligned}$

8. $43° + 58° = 101°$

$\begin{array}{r} 180° \\ -\ 101° \\ \hline 79° \end{array}$

The missing angle is 79°.

9. $A = \dfrac{bh}{2} = \dfrac{(10 \text{ cm})(6.5 \text{ cm})}{2} = \dfrac{65 \text{ cm}^2}{2} = 32.5 \text{ cm}^2$

10. a. $\sqrt{1} = 1$ because $(1)(1) = 1$.

b. $\sqrt{9} = 3$ because $(3)(3) = 9$.

c. $\sqrt{144} = 12$ because $(12)(12) = 144$.

d. $\sqrt{225} = 15$ because $(15)(15) = 225$.

11. a. $\sqrt{18} \approx 4.243$

b. $\sqrt{140} \approx 11.832$

c. $\sqrt{39} \approx 6.245$

12. $\begin{aligned} \text{hypotenuse} &= \sqrt{7^2 + 11^2} \\ &= \sqrt{49 + 121} \\ &= \sqrt{170} \approx 13.038 \text{ in.} \end{aligned}$

13. $\text{leg} = \sqrt{20^2 - 12^2} = \sqrt{400 - 144} = \sqrt{256} = 16 \text{ ft}$

14. $\begin{aligned} \text{hypotenuse} &= \sqrt{8^2 + 3^2} \\ &= \sqrt{64 + 9} \\ &= \sqrt{73} \approx 8.5 \end{aligned}$

The ship is about 8.5 miles from the starting point.

15. $y = \dfrac{1}{2}(15 \text{ m}) = 7.5 \text{ m}$

16. $z = \sqrt{2}(10 \text{ in.}) = (1.414)(10 \text{ in.}) = 14.14 \text{ in.}$

17. a. $r = \dfrac{d}{2} = \dfrac{28 \text{ ft}}{2} = 14 \text{ ft}$

    b. $d = 2r = 2(5.5 \text{ in.}) = 11 \text{ in.}$

18. $C = \pi d = (3.14)(6 \text{ m}) = 18.84 \text{ m} \approx 18.8 \text{ m}$

19. $A = \pi r^2$
$$= (3.14)(9 \text{ ft})^2$$
$$= (3.14)(81 \text{ ft}^2)$$
$$= 254.34 \text{ ft}^2 \approx 254.3 \text{ ft}^2$$

20. $V = (4 \text{ ft})(7 \text{ ft})(10 \text{ ft}) = 280 \text{ ft}^3$

21. $V = \pi r^2 h$
$$= (3.14)(3 \text{ in.})^2 (12 \text{ in.})$$
$$= (3.14)(9)(12) \text{ in.}^3$$
$$= 339.12 \text{ in.}^3 \approx 339.1 \text{ in.}^3$$

22. $V = \dfrac{4\pi r^3}{3}$
$$= \dfrac{(4)(3.14)(5 \text{ ft})^3}{3}$$
$$= \dfrac{(4)(3.14)(125)}{3} \text{ ft}^3 \approx 523.3 \text{ ft}^3$$

23. $V = \dfrac{\pi r^2 h}{3}$
$$= \dfrac{(3.14)(2 \text{ m})^2 (14 \text{ m})}{3}$$
$$= \dfrac{(3.14)(4)(14)}{3} \text{ m}^3 \approx 58.6 \text{ m}^3$$

24. $B = (4.5 \text{ ft})(8 \text{ ft}) = 36 \text{ ft}^2$

$V = \dfrac{Bh}{3} = \dfrac{(36 \text{ ft}^2)(10 \text{ ft})}{3} = 120 \text{ ft}^3$

25. $\dfrac{n}{8} = \dfrac{6}{4}$
$$4n = (8)(6)$$
$$\dfrac{4n}{4} = \dfrac{48}{4}$$
$$n = 12 \text{ ft}$$

26. $2 \text{ m} + 3 \text{ m} + 2 \text{ m} + 7 \text{ m} + 7 \text{ m} = 21 \text{ m}$
$$\dfrac{3}{9} = \dfrac{21}{p}$$
$$3p = (9)(21)$$
$$\dfrac{3p}{3} = \dfrac{189}{3}$$
$$p = 63$$
The perimeter of the larger figure is 63 meters.

Chapter 7 Review Problems

1. $90° - 76° = 14°$

2. $180° - 76° = 104°$

3. $\angle b = 146°$
$\angle a = \angle c = 180° - 146° = 34°$

4. $\angle t = \angle x = \angle y = 65°$
$\angle s = \angle u = \angle w = \angle z = 180° - 65° = 115°$

5. $P = 2(9.5 \text{ m}) + 2(2.3 \text{ m})$
$$= 19 \text{ m} + 4.6 \text{ m}$$
$$= 23.6 \text{ m}$$

6. $P = 4s = 4(12.7 \text{ yd}) = 50.8 \text{ yd}$

7. $A = (5.9 \text{ cm})(2.8 \text{ cm}) = 16.52 \text{ cm}^2 \approx 16.5 \text{ cm}^2$

8. $A = s^2 = (7.2 \text{ in.})^2 = 51.84 \text{ in.}^2 \approx 51.8 \text{ in.}^2$

9. $P = 3(8 \text{ ft}) + 2(2 \text{ ft}) + 4 \text{ ft} + 2(3 \text{ ft})$
$$= 24 \text{ ft} + 4 \text{ ft} + 4 \text{ ft} + 6 \text{ ft}$$
$$= 38 \text{ ft}$$

10. $P = 3(11 \text{ ft}) + 2(7 \text{ ft}) + 2(3.5 \text{ ft}) + 4 \text{ ft}$
$$= 33 \text{ ft} + 14 \text{ ft} + 7 \text{ ft} + 4 \text{ ft}$$
$$= 58 \text{ ft}$$

11. $A = (14 \text{ m})(5 \text{ m}) - (1 \text{ m})^2$
$$= 70 \text{ m}^2 - 2 \text{ m}^2$$
$$= 68 \text{ m}^2$$

12. $A = (9 \text{ m})^2 - (2.7 \text{ m})(6.5 \text{ m})$
    $= 81 \text{ m}^2 - 17.55 \text{ m}^2$
    $= 63.45 \text{ m}^2$
    $\approx 63.5 \text{ m}^2$

13. $P = 2(38.5 \text{ m}) + 2(14 \text{ m})$
    $= 77 \text{ m} + 28 \text{ m}$
    $= 105 \text{ m}$

14. $P = 5 \text{ mi} + 22 \text{ mi} + 5 \text{ mi} + 30 \text{ mi} = 62 \text{ mi}$

15. $A = (70 \text{ ft})(50 \text{ ft}) = 3500 \text{ ft}^2$

16. $A = \dfrac{18 \text{ yd}(21 \text{ yd} + 19 \text{ yd})}{2}$
    $= \dfrac{18 \text{ yd}(40 \text{ yd})}{2}$
    $= 360 \text{ yd}^2$

17. $A = \dfrac{(8 \text{ cm})(13 \text{ cm} + 20 \text{ cm})}{2}$
    $\qquad + \dfrac{(20 \text{ cm})(9 \text{ cm} + 20 \text{ cm})}{2}$
    $= \dfrac{(8 \text{ cm})(33 \text{ cm})}{2} + \dfrac{(20 \text{ cm})(29 \text{ cm})}{2}$
    $= 132 \text{ cm}^2 + 290 \text{ cm}^2$
    $= 422 \text{ cm}^2$

18. $A = (15 \text{ m})(17 \text{ m}) + (17 \text{ m})(6 \text{ m})$
    $= 255 \text{ m}^2 + 102 \text{ m}^2$
    $= 357 \text{ m}^2$

19. $P = 18 + 21 + 21 = 60 \text{ ft}$

20. $P = 15.5 + 15.5 + 15.5 = 46.5 \text{ ft}$

21. $180° - (28° + 45°) = 180° - 73° = 107°$

22. $180° - (90° + 35°) = 180° - 125° = 55°$

23. $A = \dfrac{(8.5 \text{ m})(12.3 \text{ m})}{2}$
    $= \dfrac{104.55 \text{ m}^2}{2}$
    $= 52.275 \text{ m}^2$
    $\approx 52.3 \text{ m}^2$

24. $A = \dfrac{(12.5 \text{ m})(9.5 \text{ m})}{2}$
    $= \dfrac{118.75 \text{ m}^2}{2}$
    $= 59.375 \text{ m}^2$
    $\approx 59.4 \text{ m}^2$

25. $A = (18 \text{ m})(22 \text{ m}) + \dfrac{(18 \text{ m})(6 \text{ m})}{2}$
    $= 396 \text{ m}^2 + 54 \text{ m}^2$
    $= 450 \text{ m}^2$

26. $A = (12 \text{ m})(6 \text{ m}) + \dfrac{(6 \text{ m})(3 \text{ m})}{2} + \dfrac{(6 \text{ m})(2 \text{ m})}{2}$
    $= 72 \text{ m}^2 + 9 \text{ m}^2 + 6 \text{ m}^2$
    $= 87 \text{ m}^2$

27. $\sqrt{81} = 9$

28. $\sqrt{64} = 8$

29. $\sqrt{121} = 11$

30. $\sqrt{144} + \sqrt{16} = 12 + 4 = 16$

31. $\sqrt{100} - \sqrt{36} + \sqrt{196} = 10 - 6 + 14 = 18$

32. $\sqrt{62} \approx 7.874$

33. $\sqrt{165} \approx 12.845$

34. $\sqrt{180} \approx 13.416$

35. hypotenuse $= \sqrt{3^2 + 4^2} = \sqrt{9 + 16} = \sqrt{25} = 5 \text{ km}$

36. hypotenuse $= \sqrt{13^2 - 12^2}$
    $= \sqrt{169 - 144}$
    $= \sqrt{25}$
    $= 5 \text{ yd}$

37. leg $= \sqrt{20^2 - 18^2}$
    $= \sqrt{400 - 324}$
    $= \sqrt{76}$
    $\approx 8.72 \text{ cm}$

38. $h = \sqrt{6^2 + 7^2} = \sqrt{36 + 49} = \sqrt{85} \approx 9.22 \text{ m}$

39. hypotenuse $= \sqrt{5^2 + 4^2}$

$\qquad\qquad = \sqrt{25 + 16}$

$\qquad\qquad = \sqrt{41}$

$\qquad\qquad \approx 6.4 \text{ cm}$

The distance is about 6.4 cm.

40. hypotenuse $= \sqrt{18^2 + 1.5^2}$

$\qquad\qquad = \sqrt{324 + 2.25}$

$\qquad\qquad = \sqrt{326.25}$

$\qquad\qquad \approx 18.1 \text{ ft}$

The ramp is 18.1 feet.

41. leg $= \sqrt{11^2 - 9^2} = \sqrt{121 - 81} = \sqrt{40} \approx 6.3 \text{ ft}$

The distance is 6.3 feet.

42. leg $= \sqrt{7^2 - 6^2} = \sqrt{49 - 36} = \sqrt{13} \approx 3.6 \text{ ft}$

The width is 3.6 feet.

43. $d = 2r = 2(53 \text{ cm}) = 106 \text{ cm}$

44. $r = \dfrac{d}{2} = \dfrac{126 \text{ cm}}{2} = 63 \text{ cm}$

45. $C = \pi d = 3.14(20 \text{ m}) = 62.8 \text{ m}$

46. $C = 2\pi r = 2(3.14)(9 \text{ in.}) = 56.52 \text{ in.} \approx 56.5 \text{ in.}$

47. $A = \pi r^2$

$\qquad = 3.14(9 \text{ m})^2$

$\qquad = 3.14(81 \text{ m})^2$

$\qquad = 254.34 \text{ m}^2$

$\qquad \approx 254.3 \text{ m}^2$

48. $r = \dfrac{d}{2} = \dfrac{8.6 \text{ ft}}{2} = 4.3 \text{ ft}$

$A = \pi r^2$

$\qquad = 3.14(4.3 \text{ ft})^2$

$\qquad = 3.14(18.49 \text{ ft})^2$

$\qquad \approx 58.06 \text{ ft}^2$

49. $A = \pi r^2 - \pi r^2$

$\qquad = 3.14(10 \text{ m})^2 - 3.14(6 \text{ m})^2$

$\qquad = 3.14(100 \text{ m}^2) - 3.14(36 \text{ m}^2)$

$\qquad = 314 \text{ m}^2 - 113.04 \text{ m}^2$

$\qquad = 200.96 \text{ m}^2$

$\qquad \approx 201.0 \text{ m}^2$

50. $A = lw + \pi r^2$

$\qquad = (24 \text{ ft})(10 \text{ ft}) + 3.14(5 \text{ ft})^2$

$\qquad = 240 \text{ ft}^2 + 78.5 \text{ ft}^2$

$\qquad = 318.5 \text{ ft}^2$

51. $A = bh - \pi r^2$

$\qquad = (12 \text{ ft})(10 \text{ ft}) - 3.14(2 \text{ ft})^2$

$\qquad = 120 \text{ ft}^2 - 12.56 \text{ ft}^2$

$\qquad = 107.44 \text{ ft}^2$

$\qquad \approx 107.4 \text{ ft}^2$

52. $A = \dfrac{h(b + B)}{2} + \dfrac{1}{2} \times \pi r^2$

$\qquad = \dfrac{5 \text{ m}(8 \text{ m} + 14 \text{ m})}{2} + \dfrac{1}{2} \times (3.14)(4 \text{ m})^2$

$\qquad = \dfrac{5 \text{ m}(22 \text{ m})}{2} + \dfrac{1}{2} \times (3.14)(16 \text{ m}^2)$

$\qquad = 55 \text{ m} + 25.12 \text{ m}^2$

$\qquad = 80.12 \text{ m}^2$

$\qquad \approx 80.1 \text{ m}^2$

53. $V = lwh = (20.8)(7.5)(8.1) = 1263.6 \text{ ft}^3$

The storage area has a volume of 1263.6 ft^3.

54. $V = \dfrac{4\pi r^3}{3}$

$\qquad = \dfrac{4(3.14)(4.5)^3}{3}$

$\qquad = \dfrac{4(3.14)(91.125)}{3}$

$\qquad \approx 381.5 \text{ in.}^3$

The volume of the ball is 381.5 in.^3.

55. $V = \pi r^2 h$

$\qquad = 3.14(1.5 \text{ ft})^2 (3 \text{ ft})$

$\qquad = 3.14(2.25 \text{ ft}^2)(3 \text{ ft})$

$\qquad \approx 21.2 \text{ ft}^3$

The volume of the can is 21.2 ft^3.

56. $B = (7 \text{ m})(7 \text{ m}) = 49 \text{ m}^2$

$V = \dfrac{Bh}{3} = \dfrac{(49 \text{ m}^2)(15 \text{ m})}{3} = 245 \text{ m}^3$

The volume of the sculpture is 245 m^3.

57. $V = \dfrac{\pi r^2 h}{3}$

$\quad = \dfrac{3.14(17 \text{ yd})^2 (30 \text{ yd})}{3}$

$\quad = \dfrac{3.14(289 \text{ yd}^2)(30 \text{ yd})}{3}$

$\quad = \dfrac{27{,}223.8 \text{ yd}^3}{3}$

$\quad = 9074.6 \text{ yd}^3$

The volume of the polluted ground was 9074.6 yd^3.

58. $\dfrac{n}{2} = \dfrac{45}{3}$

$\quad 3n = (2)(45)$

$\quad 3n = 90$

$\quad \dfrac{3n}{3} = \dfrac{90}{3}$

$\quad n = 30$

30 m

59. $\dfrac{n}{20} = \dfrac{6}{36}$

$\quad 36n = (20)(6)$

$\quad 36n = 120$

$\quad \dfrac{36n}{36} = \dfrac{120}{36}$

$\quad n \approx 3.3 \text{ m}$

60. Small figure: $P = 7 + 18 + 7 + 26 = 58 \text{ cm}$

$\quad \dfrac{n}{58} = \dfrac{108}{18}$

$\quad 18n = 108(58)$

$\quad \dfrac{18n}{18} = \dfrac{6264}{18}$

$\quad n = 348$

348 cm

61. Small figure: $P = 13 + 19 + 12 + 26 = 70 \text{ ft}$

$\quad \dfrac{n}{70} = \dfrac{32.5}{13}$

$\quad 13n = 70(32.5)$

$\quad \dfrac{13n}{13} = \dfrac{2275}{13}$

$\quad n = 175$

175 ft

62. $3\dfrac{1}{2} \times 3\dfrac{1}{2} = \dfrac{7}{2} \times \dfrac{7}{2} = \dfrac{49}{4}$

$\quad \dfrac{n}{\frac{49}{4}} = \dfrac{12}{1}$

$\quad n = \left(\dfrac{49}{4}\right)(12) = 147$

The finished banner needs 147 yd^2 of fabric.

63. $V = \dfrac{\pi r^2 h}{3}$

$\quad = \dfrac{3.14(9 \text{ in.})^2 (24 \text{ in.})}{3}$

$\quad = \dfrac{3.14(81 \text{ in.}^2)(24 \text{ in.})}{3}$

$\quad \approx 2034.7 \text{ in.}^3$

$W = 2034.7 \text{ in.}^3 \times \dfrac{16 \text{ g}}{1 \text{ in.}^3} = 32{,}555.2 \text{ g}$

The tank holds 2034.7 in.^3 and the weight of the acid is $32{,}555 \text{ g}$.

64. $A = lw - lw$

$\quad = (14 \text{ yd})(8 \text{ yd}) - (4 \text{ yd})(5 \text{ yd})$

$\quad = 112 \text{ yd}^2 - 20 \text{ yd}^2$

$\quad = 92 \text{ yd}^2$

$\text{Cost} = 92 \text{ yd}^2 \times \dfrac{\$8}{\text{yd}^2} = \$736$

The carpeting will cost \$736.

65. $r = \dfrac{d}{2} = \dfrac{90 \text{ m}}{2} = 45 \text{ m}$

$\quad V = \dfrac{4\pi r^2}{3}$

$\quad = \dfrac{4(3.14)(45 \text{ m})^3}{2}$

$\quad = \dfrac{4(3.14)(91{,}125 \text{ m}^3)}{3}$

$\quad = 381{,}510 \text{ m}^3$

The volume is $381{,}510$ cubic meters.

66. $18 \text{ in.} \times \dfrac{1 \text{ ft}}{12 \text{ in.}} = \dfrac{3}{2}$

$\quad r = \dfrac{d}{2} = \dfrac{\frac{3}{2}}{2} = \dfrac{3}{4} \text{ ft} = 0.75 \text{ ft}$

$V = \pi r^2 h$

$= 3.14(0.75 \text{ ft})^2 (5 \text{ ft})$

$= 3.14(0.5625 \text{ ft}^2)(5 \text{ ft})$

$= 8.83125 \text{ ft}^3$

$\approx 8.8 \text{ ft}^3$

The tank holds approximately 8.8 cubic feet.

67. $V \approx 8.8 \text{ ft}^3$

gallons $= 8.8 \text{ ft}^3 \times \dfrac{7.5 \text{ gal}}{1 \text{ ft}^3} = 66 \text{ gal}$

The tank holds approximately 66 gallons.

68. $A = \dfrac{(35 \text{ ft})(45 \text{ ft} + 50 \text{ ft})}{2}$

$= \dfrac{(35 \text{ ft})(95 \text{ ft})}{2}$

$= \dfrac{3325 \text{ ft}^2}{2}$

$= 1662.5 \text{ ft}^2$

The area of the front lawn is 1662.5 square feet.

69. a. Trip $= 32 \text{ km} + 18 \text{ km} = 50 \text{ km}$

Speed $= \dfrac{50 \text{ km}}{0.5 \text{ hr}} = 100 \text{ km/hr}$

b. Trip $= 26 \text{ km} + 14 \text{ km} + 16 \text{ km} = 56 \text{ km}$

Speed $= \dfrac{56 \text{ km}}{0.8 \text{ hr}} = 70 \text{ km/hr}$

c. The more rapid rate is through Ipswich.

70. a. $V = \pi r^2 h + \dfrac{1}{2} \times \dfrac{4\pi r^3}{3}$

$= 3.14(9 \text{ ft})^2(80 \text{ ft}) + \dfrac{1}{2} \times \dfrac{4(3.14)(9 \text{ ft})^3}{3}$

$= 3.14(81 \text{ ft}^2)(80 \text{ ft}) + \dfrac{1}{2} \times \dfrac{4(3.14)(729 \text{ ft}^3)}{3}$

$= 20,347.2 \text{ ft}^3 + \dfrac{1}{2} \times 3052.08 \text{ ft}^3$

$= 21,873.24 \text{ ft}^3$

The volume is $\approx 21,873.2 \text{ ft}^3$.

b. $B = 21,873.2 \text{ ft}^3 \times \dfrac{0.8 \text{ bushel}}{1 \text{ ft}^3}$

It will hold $\approx 17,498.6$ bushels.

71. $2.757 \text{ billion} \times 1.244 \text{ ft}^3 = 3.429708 \text{ ft}^3$

$= 3,429,708,000 \text{ ft}^3$

$3,429,708,000 \text{ ft}^3$ of storage was needed.

72. $h = \dfrac{V}{lw}$

$= \dfrac{3,429,108,000 \text{ ft}^3}{(10,000 \text{ ft})(20,000 \text{ ft})}$

$= \dfrac{3,429,108,000 \text{ ft}^3}{200,000,000 \text{ ft}^2}$

$\approx 17.1 \text{ ft}$

The bin would need to be 17.1 feet high.

73. $V = lwh = (2.25)(4)(2) = 18 \text{ ft}^3$

$18 \text{ ft}^3 \times \dfrac{62 \text{ lb}}{\text{ft}^3} = 1116 \text{ lb}$

$1116 \text{ lb} \times \dfrac{1 \text{ gal}}{8.6 \text{ lb}} \approx 130 \text{ gal}$

The aquarium holds 1116 lb of water or 130 gallons.

74. $2 \text{ ft} \times \dfrac{12 \text{ in.}}{1 \text{ ft}} = 24 \text{ in.}$

$4 \text{ ft} \times \dfrac{12 \text{ in.}}{1 \text{ ft}} = 48 \text{ in.}$

$V = lwh = 24 \text{ in.}(48 \text{ in.})(1.5 \text{ in.}) = 1728 \text{ in.}^3$

1728 in.^3 of gravel are needed.

75. $C = 2\pi r = 2(3.14)(30 \text{ ft}) = 188.4 \text{ ft}$

$5(188.4) = 942 \text{ ft}$

The pony walks 942 ft for each ride.

76. $r = \dfrac{d}{2} = \dfrac{18 \text{ yd}}{2} = 9 \text{ yd}$

$P = 2l + 2\pi r$

$= 2(25 \text{ yd}) + 2(3.14)(9 \text{ yd})$

$\approx 106.5 \text{ yd}$

The perimeter is 106.5 yards.

77. Cost $= 106.5 \text{ yd} \times \dfrac{3 \text{ ft}}{1 \text{ yd}} \cdot \dfrac{1 \text{ spool}}{150 \text{ ft}}$

$= 2.13$ or order 3 spools

They need to order 3 spools.

78. $\text{leg} = \sqrt{33^2 - 30^2}$
$= \sqrt{1089 - 900}$
$= \sqrt{189}$
$\approx 13.7 \text{ ft}$
The person is about 13.7 feet from the edge of the pond.

How Am I Doing? Chapter 7 Test

1. $\angle b = \angle a = 52°$
$\angle c = 180° - 52° = 128°$
$\angle e = \angle c = 128°$

2. $P = 2(9 \text{ yd}) + 2(11 \text{ yd}) = 18 \text{ yd} + 22 \text{ yd} = 40 \text{ yd}$

3. $P = 4(6.3 \text{ ft}) = 25.2 \text{ ft}$

4. $P = 2(6.5 \text{ m}) + 2(3.5 \text{ m})$
$= 13 \text{ m} + 7 \text{ m}$
$= 20 \text{ m}$

5. $P = 2(13 \text{ m}) + 22 \text{ m} + 32 \text{ m}$
$= 26 \text{ m} + 22 \text{ m} + 32 \text{ m}$
$= 80 \text{ m}$

6. $P = 58.6 \text{ m} + 32.9 \text{ m} + 45.5 \text{ m} = 137 \text{ m}$

7. $A = (10 \text{ yd})(18 \text{ yd}) = 180 \text{ yd}^2$

8. $A = (10.2 \text{ m})^2 = 104.04 \text{ m}^2 \approx 104.0 \text{ m}^2$

9. $A = (13 \text{ m})(6 \text{ m}) = 78 \text{ m}^2$

10. $A = \dfrac{(9 \text{ m})(7 \text{ m} + 25 \text{ m})}{2}$
$= \dfrac{(9 \text{ m})(32 \text{ m})}{2}$
$= \dfrac{288 \text{ m}^2}{2}$
$= 144 \text{ m}^2$

11. $A = \dfrac{(4 \text{ cm})(6 \text{ cm})}{2} = \dfrac{24 \text{ cm}^2}{2} = 12 \text{ cm}^2$

12. $\sqrt{144} = 12$

13. $\sqrt{169} = 13$

14. $90° - 63° = 27°$

15. $180° - 107° = 73°$

16. $180° - (12.5° + 83.5°) = 180° - 96° = 84°$

17. $\sqrt{54} \approx 7.348$

18. $\sqrt{135} \approx 11.619$

19. $\text{hypotenuse} = \sqrt{7^2 + 5^2}$
$= \sqrt{49 + 25}$
$= \sqrt{74}$
$= 8.602$

20. $\text{leg} = \sqrt{26^2 - 24^2} = \sqrt{676 - 576} = \sqrt{100} = 10$

21. $\text{hypotenuse} = \sqrt{5^2 + 3^2}$
$= \sqrt{25 + 9}$
$= \sqrt{34}$
$\approx 5.83 \text{ cm}$
The distance between the holes in 5.83 cm.

22. $\text{hypotenuse} = \sqrt{15^2 - 12^2}$
$= \sqrt{225 - 144}$
$= \sqrt{81}$
$= 9 \text{ ft}$
The ladder is 9 ft from the house.

23. $r = \dfrac{d}{2} = \dfrac{18}{2} = 9 \text{ ft}$
$C = 2\pi r = 2(3.14)(9) \approx 56.52 \text{ ft}$

24. $r = \dfrac{d}{2} = \dfrac{12}{2} = 6 \text{ ft}$
$A = \pi r^2 = 3.14(6)^2 = 3.14(36) = 113.04 \text{ ft}^2$

25. $A = bh - \pi r^2$
$= (15 \text{ in.})(8 \text{ in.}) - (3.14)(2 \text{ in.})^2$
$= 120 \text{ in.}^2 - 12.56 \text{ in.}^2$
$= 107.44 \text{ in.}^2$
$\approx 107.4 \text{ in.}^2$

26. $A = \dfrac{h(b+B)}{2} + \dfrac{1}{2} \times \pi r^2$

$= \dfrac{(7 \text{ in.})(10 \text{ in.} + 20 \text{ in.})}{2} + \dfrac{1}{2} \times (3.14)(5 \text{ in.})^2$

$= \dfrac{(7 \text{ in.})(30 \text{ in.})}{2} + \dfrac{1}{2} \times (3.14)(25 \text{ in.}^2)$

$= 105 \text{ in.}^2 + 39.25 \text{ in.}^2$

$= 144.25 \text{ in.}^2$

$\approx 144.3 \text{ in.}^2$

27. $V = lwh = 3.5(20)(10) = 700 \text{ m}^3$

28. $V = \dfrac{\pi r^2 h}{3}$

$= \dfrac{3.14(8 \text{ m})^2 (12 \text{ m})}{3}$

$= \dfrac{3.14(64 \text{ m}^2)(12 \text{ m})}{3}$

$= 803.84 \text{ m}^3$

$\approx 803.8 \text{ m}^3$

29. $V = \dfrac{4\pi r^3}{3}$

$= \dfrac{4(3.14)(3 \text{ m})^3}{3}$

$= \dfrac{4(3.14)(27 \text{ m}^3)}{3}$

$= 113.04 \text{ m}^3$

$\approx 113.0 \text{ m}^3$

30. $V = \pi r^2 h$

$= 3.14(9 \text{ ft})^2 (2 \text{ ft})$

$= 3.14(81 \text{ ft}^2)(2 \text{ ft})$

$= 508.68 \text{ ft}^3$

$\approx 508.7 \text{ ft}^3$

31. $B = (4 \text{ m})(3 \text{ m}) = 12 \text{ m}^2$

$V = \dfrac{Bh}{3} = \dfrac{(12 \text{ m}^2)(14 \text{ m})}{3} = 56 \text{ m}^3$

32. $\dfrac{n}{18} = \dfrac{13}{5}$

$5n = 18(13)$

$\dfrac{5n}{5} = \dfrac{234}{5}$

$n = 46.8 \text{ m}$

33. $\dfrac{n}{7} = \dfrac{60}{10}$

$10n = 7(60)$

$\dfrac{10n}{10} = \dfrac{420}{10}$

$n = 42 \text{ ft}$

34. $r = \dfrac{d}{2} = 20 \text{ yd}$

$A = lw - \pi r^2$

$= (130 \text{ yd})(40 \text{ yd}) + 3.14(20 \text{ yd})^2$

$= 5200 \text{ yd}^2 + 1256 \text{ yd}^2$

$= 6456 \text{ yd}^2$

The area of the field is 6456 yd^2.

35. $A = 6456 \text{ yd}^2$

$\text{Cost} = 6456 \text{ yd}^2 \times \dfrac{\$0.40}{1 \text{ yd}^2} = \$2582.40$

It will cost $2582.40 to fertilize the field.

Chapter 8

8.1 Exercises

1. Multiply $25\% \times 4000$, which is
$0.25 \times 4000 = 1000$ students.

3. Divide the circle into quarters by drawing two perpendicular lines. Shade in one-quarter of the circle. Label this with the title "within five miles = 1000."

5. Choose the largest sector; rent

7. Utilities is labeled $200.

9. $\$650 + \$150 = \$800$
$800 is allotted for transportation or charitable contributions.

11. $\dfrac{\$650}{\$200} = \dfrac{650 \div 50}{200 \div 50} = \dfrac{13}{4}$

13. $\dfrac{\$1000}{\$2700} = \dfrac{1000 \div 100}{2700 \div 100} = \dfrac{10}{27}$

15. Choose the smallest sector; 85 years or older

17. 65 years old or older but younger than 85 is labeled 47 million or 47,000,000 people.

19. $(89 + 100 + 84)$ million = 282 million or 282,000,000 people

21. $\dfrac{(89+109)\text{ million}}{(84+47+7)\text{ million}} = \dfrac{198\text{ million}}{138\text{ million}} = \dfrac{198}{138} = \dfrac{33}{23}$

23. $\dfrac{89\text{ million}}{336\text{ million}} = \dfrac{89}{336}$

25. $11\% + 8\% = 19\%$

27. Reasonable prices and great food, $22\% + 56\% = 78\%$, make up approximately three-fourths of the circle graph.

29. Difference $= 56\% - 22\% = 34\%$
$n = 34\% \times 1010 = 0.43 \times 1010 \approx 343$ people
About 343 people felt that great food was more important than reasonable prices.

31. China is labeled 32.8%.
Find 32.8% of 42,000,000.
$0.328 \times 42,000,000 = 13,776,000$
13,776,000 vehicles were produced in China.

33. China (32.8%), Japan (18.9%), South Korea (8.4%)
$32.8\% + 18.9\% + 8.4\% = 60.1\%$
60.1% were produced in Asia.

35. China (32.8%), Japan (18.9%)
$32.8\% + 18.9\% = 51.7\%$
$100\% - 51.7\% = 48.3\%$
48.3% were *not* manufactured in China or Japan.

37. China (32.8%), Japan (18.9%), United States (13.6%)
$18.9\% + 13.6\% = 32.5\%$
$32.8\% - 32.5\% = 0.3\%$
0.3% of $42,000,000 = 0.003 \times 42,000,000$
$\qquad\qquad = 126,000$
126,000 more vehicles were produced in China than in Japan and the United States combined.

Cumulative Review

39. $A = \dfrac{bh}{2} = \dfrac{12 \times 20}{2} = 120 \text{ ft}^2$

40. $A = bh = (17 \text{ in.})(12 \text{ in.}) = 204 \text{ in.}^2$

41. $A = 2lw + 2lw$
$\quad = 2(7 \text{ yd})(12 \text{ yd}) + 2(7 \text{ yd})(20 \text{ yd})$
$\quad = 2(84 \text{ yd}^2) + 2(140 \text{ yd}^2)$
$\quad = 168 \text{ yd}^2 + 280 \text{ yd}^2$
$\quad = 448 \text{ yd}^2$
$448 \text{ yd}^2 \times \dfrac{1 \text{ gal}}{28 \text{ yd}^2} = 16 \text{ gal}$
It will take 16 gallons of paint to cover the barn.

42. $A = \pi r^2$
$\quad = 3.14(8 \text{ cm})^2$
$\quad = 3.14(64 \text{ cm}^2)$
$\quad = 200.96 \text{ cm}^2$
$200.96 \text{ cm}^2 \times \dfrac{1 \text{ g}}{64 \text{ cm}^2} = 3.14 \text{ g} \approx 3 \text{ g}$
It will take about 3 grams of silver.

Quick Quiz 8.1

1. $9\% + 42\% = 51\%$
   51% of the vehicles sold were two-door coupes or four-door sedans.

2. $(27\% + 18\%)$ of $850,000 = 45\%$ of $850,000$
   $\phantom{(27\% + 18\%) \text{ of } 850,000} = 0.45 \times 850,000$
   $\phantom{(27\% + 18\%) \text{ of } 850,000} = 382,500$ vehicles
   382,500 vehicles sold were SUVs or minivans.

3. $(100\% - 4\%)$ of $850,000 = 96\%$ of $850,000$
   $\phantom{(100\% - 4\%) \text{ of } 850,000} = 0.96 \times 850,000$
   $\phantom{(100\% - 4\%) \text{ of } 850,000} = 816,000$ vehicles
   816,000 vehicles sold were not station wagons.

4. Answers may vary. Possible solution: Add the percents for the three categories
   $(4\% + 42\% + 18\% = 64\%)$.
   Then find 64% of 850,000.
   $0.64 \times 850,000 = 544,000$ of the vehicles sold were station wagons, four-door sedans, or minivans.

8.2 Exercises

1. The bar rises to 16, which represents an approximate population of 16 million or 16,000,000 people.

3. The bar rises to 24, which represents an approximate population of 24 million or 24,000,000 people.

5. 1960 to 1970: $20 - 16 = 4$ million
   1970 to 1980: $24 - 20 = 4$ million
   1980 to 1990: $30 - 24 = 6$ million
   1990 to 2000: $34 - 30 = 4$ million
   2000 to 2010: $40 - 34 = 6$ million
   2010 to 2020: $45 - 40 = 5$ million
   The population increased by the largest amount from 1980 to 1990 and from 2000 to 2010.

7. The bar rises to 60, which represents an average cost of $6000.

9. $134 - 70 = 64$
   The cost was $6400 higher.

11. 2003–04: $107 - 60 = 47$
    2004–05: $114 - 63 = 51$
    2005–06: $121 - 65 = 56$
    2006–07: $128 - 68 = 60$
    2007–08: $134 - 70 = 64$
    The smallest difference was in the academic year 2003–04.

13. $70 - 68 = 2$
    The cost increased by $200.

15. $60 + 63 + 121 + 128 = 372$
    The total cost is $37,200.

17. $\dfrac{70 - 60}{60} \approx 0.17$ or 17%
    The percent increase in cost is about 17%.

19. The dot for 1997 is at 1.3, which represents $1.3 million or $1,300,000.

21. The line from 2003 to 2005 goes upward at the most shallow angle. This represents the smallest increase.

23. 1999: 1.6
    2001: 2.1
    $2.1 - 1.6 = 0.5$
    The increase from 1999 to 2001 was $0.5 million or $500,000.

25. The dot for September 2010 is about halfway between 2 and 3. This represents 2.5 inches.

27. The line that represents 2010 is below the line that represents 2009 for the months October, November, and December.

29. $4.0 - 2.5 = 1.5$ in.
    1.5 more inches of rain fell in November 2009 than in October 2009.

Cumulative Review

31. $(5+6)^2 - 18 \div 9 \times 3 = 11^2 - 18 \div 9 \times 3$
    $ = 121 - 18 \div 9 \times 3$
    $ = 121 - 2 \times 3$
    $ = 121 - 6$
    $ = 115$

32. $\dfrac{1}{5} \times \left(\dfrac{1}{5} - \dfrac{1}{6}\right) \times \dfrac{2}{3} = \dfrac{1}{5} + \left(\dfrac{6}{30} - \dfrac{5}{30}\right) \times \dfrac{2}{3}$
    $\phantom{\dfrac{1}{5} \times \left(\dfrac{1}{5} - \dfrac{1}{6}\right) \times \dfrac{2}{3}} = \dfrac{1}{5} + \dfrac{1}{30} \times \dfrac{2}{3}$
    $\phantom{\dfrac{1}{5} \times \left(\dfrac{1}{5} - \dfrac{1}{6}\right) \times \dfrac{2}{3}} = \dfrac{1}{5} + \dfrac{1}{45}$
    $\phantom{\dfrac{1}{5} \times \left(\dfrac{1}{5} - \dfrac{1}{6}\right) \times \dfrac{2}{3}} = \dfrac{9}{45} + \dfrac{1}{45}$
    $\phantom{\dfrac{1}{5} \times \left(\dfrac{1}{5} - \dfrac{1}{6}\right) \times \dfrac{2}{3}} = \dfrac{10}{45}$
    $\phantom{\dfrac{1}{5} \times \left(\dfrac{1}{5} - \dfrac{1}{6}\right) \times \dfrac{2}{3}} = \dfrac{2}{9}$

33. 59.2% of $1,821,000 = 0.592 \times 1,821,000$
$$= 1,078,032$$
1,078,032 bachelor's degrees will be awarded to women.

34. 31.25% of $n = 2175$
$$0.3125n = 2175$$
$$n = \frac{2175}{0.3125}$$
$$n = 6960$$
There are 6960 miles of national scenic trails in the United States.

Quick Quiz 8.2

1. The dot for condominiums in 2005 is at 800.

2. Subtract the number of condominiums from the number of homes in 1990.
$$900 - 100 = 800$$
There were 800 more homes than condominiums built in 1990.

3. Find the year closest to where the lines cross; 2000.

4. Answers may vary. Possible solution: Find the increase from 2000 to 2010 ($1000 - 600 = 400$). Since this is a 10-year period, the same increase would be expected for the next 10-year period. $1000 + 400 = 1400$ expected condominiums in 2020.

How Am I Doing? Sections 8.1–8.2

1. The sector labeled Yosemite is labeled 14%.

2. Great Smoky Mountain National Park (the largest sector) is the park where the greatest number of visitors go.

3. $12\% + 12\% = 24\%$
24% of the visitors went to Olympic or Yellowstone National Park.

4. $n = 16\%$ of $27,000,000$
$$n = 0.16 (27,000,000)$$
$$n = 4,320,000$$
About 4,320,000 visitors went to Grand Canyon National Park.

5. $14\% + 12\% = 26\%$
$n = 26\%$ of $27,000,000$
$$= 0.26(27,000,000)$$
$$= 7,020,000$$
About 7,020,000 visitors went to Yosemite or Yellowstone National Park.

6. The bar rises to 450. Therefore, there were 450 housing starts in Springfield in the first quarter of 2009.

7. The bar rises to 550. Therefore, there were 550 housing starts in the second quarter of 2010.

8. The shortest bar represents the 4th quarter of 2009. Therefore, the smallest number of housing starts were during the fourth quarter of 2009.

9. The tallest bar represents the 3rd quarter of 2010. Therefore, the greatest number of housing starts were during the third quarter of 2010.

10. $600 - 350 = 250$
There were 250 more housing starts in the third quarter of 2010 than in the third quarter of 2009.

11. $450 - 300 = 150$
There were 150 fewer housing starts in the first quarter of 2010 than in the first quarter of 2009.

12. Look for the lowest dots on the line representing production of television sets. During August and December the production of television sets was the lowest.

13. Look for the highest dot on the line representing the sales of television sets. During December the sales of television sets was the highest.

14. The first month where the line for sales crosses and is higher than the line for production occurs in November. Therefore, November is the first month in which the production was lower than sales.

15. a. The dot for sales in August is at 20 which represents 20,000 television sets sold in August.

b. The dot for sales in November is halfway between 30 and 40 which is 35 or 35,000 television sets sold in November.

8.3 Exercises

1. The horizontal label for each item in a bar graph is usually a single number or a word title. For a histogram it is a class interval. The vertical bars have a space between them in a bar graph. For a histogram the vertical bars join each other.

3. A class frequency is the number of times a data value occurs in a particular class interval.

5. The bar for 100,000–149,999 rises to 120, so the number of U.S. cities that have a population of 100,000–149,999 is 120 cities.

7. The bar for 1 million or more rises to 10, so the number of U.S. cities that have a population of 1 million or more is 10 cities.

9. Add the heights of the two bars representing 500,000–999,999 (30) and 1 million or more (10).
30 + 10 = 40 cities

11. Add the heights of the two bars representing 100,000–149,999 (120) and 150,000–199,999 (50).
120 + 50 = 170 cities.

13. The bar for $3.00–$4.99 rises to 8000, so 8000 books are priced at $3.00–$4.99.

15. The tallest bar represents $5.00 to $7.99, so the bookstore sold most of the books costing $5.00–$7.99.

17. Add the heights of the bars representing less than $3.00 (3000), $3.00 to $4.99 (8000) and $5.00 to $7.99 (17,000).
3000 + 8000 + 17,000 = 28,000 books

19. Add the heights of the bars representing $5.00 to $7.99 (17,000), $8.00 to $9.99 (10,000), $10.00 to $14.99 (12,000) and $15.00 to $24.99 (13,000).
17,000 + 10,000 + 12,000 + 13,000 = 52,000 books

21.
$$13,000$$
$$+\ 7,000$$
$$\overline{20,000 \text{ books}}$$

$$\text{percent} = \frac{20,000}{70,000} \approx 0.286 = 28.6\%$$

28.6% of the 70,000 books sold were over $14.99.

23. Tally: |||
Frequency: 3

25. Tally: |||| |
Frequency: 6

27. Tally: |||
Frequency: 3

29. Tally: ||
Frequency: 2

31.

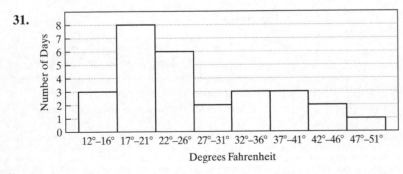

33. Add the frequencies for 12°–16° (3), 17°–21° (8), and 22°–26° (6).
$3 + 8 + 6 = 17$ days

Cumulative Review

34.
$$\frac{182}{m} = \frac{25}{19}$$
$$182 \times 19 = m \times 25$$
$$3458 = 25m$$
$$\frac{3458}{25} = \frac{25m}{25}$$
$$138.32 = m$$

35.
$$\frac{n}{18} = \frac{3.5}{9}$$
$$9n = 18 \times 3.5$$
$$9n = 63$$
$$\frac{9n}{9} = \frac{63}{9}$$
$$n = 7$$

36.
$$\frac{375 \text{ mi}}{7.5 \text{ gal}} = \frac{n}{12.3 \text{ gal}}$$
$$375 \times 12.3 = 7.5 \times n$$
$$4612.5 = 7.5n$$
$$\frac{4612.5}{7.5} = \frac{7.5n}{7.5}$$
$$615 \text{ miles} = n$$
He can drive 615 miles on a full tank.

37.
$$\frac{\text{snow}}{\text{water}}: \frac{23 \text{ in.}}{2 \text{ in.}} = \frac{150 \text{ in.}}{n \text{ in.}}$$
$$23 \times n = 2 \times 150$$
$$23 \times n = 300$$
$$\frac{23 \times n}{23} = \frac{300}{23}$$
$$n \approx 13.0$$
This corresponds to 13.0 in. of water.

Quick Quiz 8.3

1. The bar for 9–12 rises to 700, which represents 700 people.

2. Add the heights of the bars for 5–8, 9–12, 13–16, 17–20, and 21–24.
$400 + 700 + 600 + 200 + 100 = 2000$ people

3. $700 - 200 = 500$ people

4. Answers may vary. Possible solution:
Multiply the height of the bar for 1–4 by 3 and the height of the bar for 5–8 by 2 and add the results.

8.4 Exercises

1. The median of a set of numbers when they are arranged in order from smallest to largest is that value that has the same number of values above it as below it.
The mean of a set of values is the sum of the values divided by the number of values. The mean is most likely to be not typical of the value you would expect if there are many extremely low values or many extremely high values. The median is more likely to be typical of the value you would expect.

3. Mean $= \dfrac{30 + 29 + 28 + 35 + 34 + 37 + 31}{7}$
$$= \frac{224}{7}$$
$$= 32$$
The mean is 32 customers.

5. Mean $= \dfrac{5.3 + 4.7 + 5.8 + 5.6 + 5.1 + 3.3}{6}$
$$= \frac{29.8}{6}$$
$$\approx 5.0$$
The mean amount of rainfall is 5.0 inches.

7. Batting Average $= \dfrac{\text{number of hits}}{\text{times at bat}}$
$$= \frac{0 + 2 + 3 + 2 + 2}{5 + 4 + 6 + 5 + 4}$$
$$= \frac{9}{24}$$
$$= 0.375$$
His batting average is 0.375.

9. $\text{Mean} = \dfrac{107,000+134,000+152,000+182,000+204,000}{5}$

$\phantom{\text{Mean}} = \dfrac{779,000}{5}$

$\phantom{\text{Mean}} = 155,800$

The mean population is 155,800.

11. $\text{Avg miles/gallon} = \dfrac{\text{miles driven}}{\text{gallons used}}$

$\phantom{\text{Avg miles/gallon}} = \dfrac{276+350+391+336}{12+14+17+14}$

$\phantom{\text{Avg miles/gallon}} = \dfrac{1353}{57}$

$\phantom{\text{Avg miles/gallon}} \approx 23.7$

The average is 23.7 mi/gal.

13. 126, 180, 195, 229, 232
Median = 195

15. 11.6, 11.9, 11.9, 12.1, 12.4, 12.5

$\text{Median} = \dfrac{11.9+12.1}{2} = 12.0$

17. $28,500, $32,700, $35,800, $38,250, $40,750, $42,000

$\text{Median} = \dfrac{\$35,800+\$38,250}{2} = \$37,025$

The median salary is $37,025.

19. 10, 12, 18, 21, 25, 28, 31
Median = 21 tables
The median number of tables is 21.

21. $97, $109, $185, $207, $218, $330, $420
Median = $207
The median phone bill is $207.

23. 1.8, 1.9, 2.0, 2.0, 2.4, 3.1, 3.1, 3.7

$\text{Median} = \dfrac{2.0+2.4}{2} = 2.2$

The median GPA was 2.2.

25. $189,481, $192,604, $315,654, $339,938

$\text{Median} = \dfrac{\$192,604+\$315,654}{2}$

$\phantom{\text{Median}} = \$254,129 \text{ million}$

$\phantom{\text{Median}=} \text{or } \$254,129,000,000$

27. $30,000
      74,500
      47,890
      89,000
      57,645
      78,090
     110,370
    + 65,800
    ———————
     $553,295

Mean $= \dfrac{\$553,295}{8} \approx \$69,161.88$

The mean salary is \$69,161.88.

29. 1987, 2576, 3700, 4700, 5000, 7200, 8764, 9365

Median $= \dfrac{4700 + 5000}{2} = 4850$

31.   120.50
       66.74
       80.95
      210.52
    + 45.00
    ———————
      523.71

Mean $= \dfrac{523.71}{5} \approx \$104.74$

45, 66.74, 80.95, 120.5, 210.52
Median = \$80.95
The mean amount is \$104.74 and the median amount is \$80.95.

33. The mode is 60 which occurs twice.

35. The modes are 121 and 150 which both occur twice.

37. 249, 649, 269, 259, 269, 249, 269
Mode = \$269 which occurs three times.

39. Mean $= \dfrac{82.6 + 82.2 + 81.8 + 81.7 + 81.2 + 80.9 + 80.9}{7}$

$= \dfrac{571.3}{7}$

≈ 81.6 years

80.9, 80.9, 81.2, 81.7, 81.8, 82.2, 82.6
median = 81.7 years
mode = 80.9 years, since it occurs twice.

41. a. Mean $= \dfrac{1500 + 1700 + 1650 + 1300 + 1440 + 1580 + 1820 + 1380 + 2900 + 6300}{10}$

$= \dfrac{21,570}{10}$

$= \$2157$

The mean monthly salary is \$2157.

b. 1300, 1380, 1440, 1500, 1580, 1650, 1700, 1820, 2900, 6300

$$\text{Median} = \frac{1580 + 1650}{2} = \$1615$$

The median monthly salary is \$1615.

c. There is no mode.

d. The median, because the mean is affected by the high amount \$6300.

43. a. $\text{Mean} = \dfrac{23 + 3 + 2 + 3 + 7 + 10 + 11}{7}$

≈ 8.4

The mean number of calls is 8.4 phone calls.

b. 2, 3, 3, 7, 10, 11, 23
Median = 7
The median number of calls is 7 phone calls.

c. Mode = 3 phone calls since it occurs 2 times.

d. The median is the most representative. On three nights she gets more calls than 7. On three nights she gets fewer calls than 7. On one night she got 7 calls. The mean is distorted a little because of the very large number of calls on Sunday night. The mode is artificially low because she gets so few calls on Monday and Wednesday and it just happened to be the same number, 3.

Cumulative Review

45. $A = \dfrac{bh}{2}$

$= \dfrac{(7 \text{ in.})(5.5 \text{ in.})}{2}$

$= \dfrac{38.5 \text{ in.}^2}{2}$

$= 19.25 \text{ in.}^2$

$\approx 19.3 \text{ in.}^2$

The area is 19.3 in.^{2}.

46. $A = \pi r^2$

$= 3.14(40 \text{ ft}^2)$

$= 3.14(1600 \text{ ft})^2$

$= 5024 \text{ ft}^2$

$\text{water} = 2 \times 5024 \text{ ft}^2 \times \dfrac{2 \text{ gal}}{1 \text{ ft}^2}$

$= 20{,}096 \text{ gallons per hour}$

20,096 gallons per hour are needed to water the fields.

47. $A = bh = (5 \text{ ft})(4 \text{ ft}) = 20 \text{ ft}^2$

$\text{Cost} = 20 \text{ ft}^2 \times \dfrac{\$16.50}{1 \text{ ft}^2} = \$330$

The cost to make the sign is \$330.

48. $V = \pi r^2 h$

$= 3.14(1.5)^2(7)$

$= 3.14(2.25)(7)$

$= 49.455$

$\text{Cost} = \dfrac{\$1.98}{49.455 \text{ in.}^2} \approx \$0.04/\text{in.}^3$

The cost is about \$0.04 per in.^{3}.

Quick Quiz 8.4

1. 1, 3, 5, 8, 8, 15, 16, 17, 24, 35

$\text{median} = \dfrac{8 + 15}{2} = \dfrac{23}{2} = 11.5$

The median is 11.5 times.

2. $\text{mean} = \dfrac{1 + 3 + 8 + 35 + 16 + 8 + 5 + 17 + 24 + 15}{10}$

$= \dfrac{132}{10}$

$= 13.2$

The mean is 13.2 times.

3. mode = 8 times because it occurs twice.

4. Answers may vary. Possible solution:
The median may be the best measure because it is not affected by very large or very small numbers (whereas the median is). Also, the mode may be misrepresentative because it simply is the value that occurs most often and may not accurately represent the average.

Use Math to Save Money

1. December 2009: \$2.95 × 100 = \$295
November 2009: \$3.55 × 100 = \$355

2. \$3.55 × 100 × 5 = \$1775
He can expect to pay \$1775.

3. 4° lower = 8% savings
8% of $1775 = 0.08 × $1775 = $142
He will save $142.

4. One month's heating cost is $355.
$355 is what percent of $1775?
$$355 = n \times 1775$$
$$\frac{355}{1775} = \frac{n \times 1775}{1775}$$
$$0.2 = n$$
$355 is 20% of $1775.
20% savings = 10° lower
72° − 10° = 62°
He should set his thermostat at 62°.

5. Answers will vary.

6. Answers will vary.

You Try It

1. a. The sector for "over age 50" is labeled 12%. 12% of the police force is over age 50.

   b. 48% of 200 = 0.48 × 200 = 96
   96 people in the police force are between 33 and 50 years old.

2. a. The bar representing sales in the Midwest in 2010 extends to 5. 5000 LCD HDTV sets were sold in the Midwest in 2010.

   b. 7000 sets were sold in the Midwest in 2009. 2000 sets were sold on the West Coast in 2009.
   7000 − 2000 = 5000
   5000 more sets were sold in the Midwest than on the West Coast.

3. a. The point above August on the graph for 2010 is at 6. There were 6000 visitors in August 2010.

   b. The graph for 2010 is above the graph for 2009 for July and August. There were more visitors in 2010 than in 2009 for the months of July and August.

   c. The steepest line on the graph is from July to August 2009. The sharpest increase took place between July and August 2009.

4. a. The bar for scores between 12 and 15 extends to 16. 16 students had scores between 12 and 15.

b. 12 + 20 + 16 = 48 students
48 students had scores of more than 3.

5. $$\frac{32+28+23+30+35+29}{6} = \frac{177}{6} = 29.5$$
The mean is 29.5.

6. a. 14, 15, 18, 19, 20, 21, 25
The middle number is 19. The median is 19.

   b. 30, 45, 54, 56, 68, 79
   The two middle numbers are 54 and 56.
   $$\frac{54+56}{2} = 55$$
   The median is 55.

7. a. The mode is 8, since it occurs twice.

   b. The modes are 9 and 12, since each occurs twice.

Chapter 8 Review Problems

1. 13 computers were manufactured by Lenovo.

2. 43 + 25 = 68 computers were manufactured by Dell or Compaq.

3. $\frac{13}{21}$ is the ratio of Lenovo to Hewlett-Packard.

4. $\frac{43}{32}$ is the ratio of Dell to Apple.

5. $\frac{25}{140} \approx 0.179 = 17.9\%$
Approximately 17.9% of the computers were manufactured by Compaq.

6. $\frac{32}{140} \approx \frac{32}{140} \approx 0.229 = 22.9\%$
Approximately 22.9% of the computers were manufactured by Apple.

7. 23% + 25% = 48%
48% of the students are majoring in business or social sciences.

8. 100% − 23% = 77%
77% of the students are majoring in an area other than business.

9. Art is the smallest sector.

10. $8\% + 1\% = 20\% = \dfrac{1}{5}$

Art and education together make up one-fifth of the graph.

11. $n = 15\%(8000) = 0.15(8000) = 1200$ students
1200 students are majoring in language arts.

12. $23\% - 12\% = 11\%$
$n = 11\%(8000) = 0.11(8000) = 880$ students
880 more students are majoring in business than education.

13. The height of the bar for prescription drugs in 2000 is 121, which represents \$121 billion or \$121,000,000,000.

14. The height of the bar for physician/clinical services in 2010 is 552, which represents \$552 billion or \$552,000,000,000.

15. $552 - 430 = 122$
\$122 billion or \$122,000,000,000 was the increase.

16. $204 - 121 = 83$
\$83 billion or \$83,000,000,000 was the increase.

17. The greatest difference in bar heights for any 5-year period for prescription drugs is from 2000 to 2005.

18. $\dfrac{256}{552} = \dfrac{32}{69}$ is the ratio.

19. The height of the bar is 12.1, which represents 12.1 billion or 12,100,000,000 bushels.

20. The height of the bar is 3.2, which represents 3.2 billion or 3,200,000,000 bushels.

21. The greatest difference in bar heights for soybeans for any two consecutive years is between 2008 and 2010.

22. $13.2 + 3.5 = 16.7$
16.7 billion or 16,700,000,000 bushels of corn and soybeans were produced in 2010.

23. The greatest difference in bar heights for any one year is 2010.

24. The smallest difference in bar heights for any one year is 2002.

25. $\text{mean} = \dfrac{9.0 + 11.8 + 10.5 + 12.1 + 13.2}{5}$
$= \dfrac{56.6}{5}$
$= 11.32$
The average corn production is 11.32 billion or 11,320,000,000 bushels.

26. $13.2 - 10.5 = 2.7$
$13.2 + 2.7 = 15.9$
The estimated production of corn in 2014 is 15.9 billion or 15,900,000,000 bushels.

27. The dot for 2009 is about halfway between 3 and 4, which represents 350 students.

28. $7 - 2 = 5$ or 500 more students graduated in 2010 than in 2005.

29. $5.5 - 3.5 = 2.0$ or 200 fewer students graduated in 2009 than in 2008.

30. The line between any two consecutive points that is the most gradual is between 2005 and 2006.

31. $\text{mean} = \dfrac{2 + 2.5 + 4 + 5.5 + 3.5 + 7}{6}$
$= \dfrac{24.5}{6}$
≈ 4.08
The average is about 408 students.

32. $\dfrac{5.5 - 4}{4} = \dfrac{1.5}{4} = 0.375$ or 37.5%
The percentage of increase is 37.5%.

33. The dot on the graph is at 45, which represents 45,000 cones.

34. The dot on the graph is at 30, which represents 30,000 cones.

35. $20,000 - 10,000 = 10,000$ more cones were purchased in May 2009 than May 2010.

36. $60,000 - 30,000 = 30,000$ more cones were purchased in August 2010 than August 2009.

37. $20 + 30 + 55 + 30 = 135$ or 135,000 cones were purchased between May and August of 2009.

38. Since August was cold and rainy, significantly fewer people wanted ice cream during August.

39. The dot on the graph is at 35, which represents 3500 degrees.

40. 52 − 44 = 8 or 800 more Mathematics and Statistics degrees were awarded.

41. The line representing physical science is above the line representing mathematics and statistics for the years 2000 and 2002.

42. The line representing mathematics and statistics is above the line representing physical science for the years 2004, 2006, and 2008.

43. The line increases the most between years 2002 to 2004.

44. The line increases the most between years 2002 to 2004.

45. Between 2002 and 2004, the number of master's degrees in mathematics and statistics increased by 900.

46. 52 − 43 = 9
52 + 9 = 61 or 6100 degrees is the expected number of degrees in 2012.

47. The bar rises to 65, which represents 65 pairs.

48. The bar rises to 10, which represents 10 pairs.

49. 55 + 65 + 25 = 145 pairs sold between size 7 and size 9.5.

50. Sold = 5 + 20 + 55 + 65 + 25 + 10 = 180 pairs
$p = \dfrac{180}{200} = 0.9 = 90\%$ of the shoes sold during the grand opening.

51. Difference = 65 − 20 = 45 pairs
45 more pairs of size 8–8.5 sold than of size 6–6.5.

52. $\dfrac{25}{180} = \dfrac{5}{36}$ is the ratio of size 9–9.5 to the total pairs sold.

53. Tally: 卌 卌
Frequency: 10

54. Tally: 卌 |||
Frequency: 8

55. Tally: |||
Frequency: 3

56. Tally: 卌
Frequency: 5

57. Tally: ||
Frequency: 2

58.

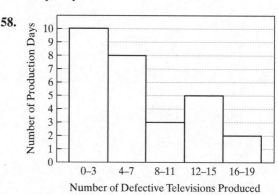

59. 10 + 8 = 18 times there were between 0 and 7 defective television sets.

60. $\text{Mean} = \dfrac{86 + 83 + 88 + 95 + 97 + 100 + 81}{7}$
$= \dfrac{630}{7}$
$= 90$
$= 90$
The mean temperature is 90°F.

61. $\text{Mean} = \dfrac{145 + 162 + 95 + 67 + 43 + 26}{6} = 89.67$
The mean gas bill is $89.67.

62. Mean
$= \dfrac{12,000 + 17,000 + 24,000 + 29,000 + 19,000}{5}$
$= 20,200$
The mean number of people is 20,200.

63. $\text{Mean} = \dfrac{882 + 913 + 1017 + 1592 + 1778 + 1936}{6}$
$= \dfrac{8118}{6}$
$= 1353$
The mean number of employees is 1353.

64. $21,690, $28,500, $29,300, $35,000, $37,000, $38,600, $43,600, $45,300

$$\text{Median} = \frac{\$35,000 + \$37,000}{2} = \$36,000$$

The median cost is $36,000.

65. $98,000, $120,000, $126,000, $135,000, $139,000, $144,000, $150,000, $154,000, $156,000, $170,000

$$\text{Median} = \frac{\$139,000 + \$144,000}{2} = \$141,500$$

The median cost is $141,500.

66. 4, 7, 8, 10, 15, 28, 28, 30, 31, 34, 35, 38, 43, 54, 77, 79

$$\text{Median} = \frac{30 + 31}{2} = 30.5 \text{ years}$$

Mode = 28 years

67. 3, 9, 13, 14, 15, 15, 16, 18, 19, 21, 24, 25, 26, 28, 31, 36

$$\text{Median} = \frac{18 + 19}{2} = 18.5 \text{ deliveries}$$

Mode = 15 deliveries

68. The median is better because the mean is skewed by one low score, 31

69. The median is better because the mean is skewed by one high data item, 39.

70. a. $\text{Mean} = \dfrac{2 + 3 + 2 + 4 + 7 + 12 + 5}{7} = 5 \text{ hours}$

The mean is 5 hours.

b. 2, 2, 3, 4, 5, 7, 12
Median = 4 hours
The median is 4 hours.

c. Mode = 2 hours since it occurs twice.
The mode is 2 hours.

d. The median is the most representative. On three days she uses the computer more than 4 hours and on three days she uses the computer less than 4 hours. One day she used it exactly 7 hours. The mean is distorted a little because of the very large number of hours on Friday. The mode is artificially low because she happened to use the computers only two hours on Sunday and Tuesday. All other days it was more than this.

How Am I Doing? Chapter 8 Test

1. 37% is the sector marked Passed inspections.

2. 21% is the sector marked 2 violations.

3. Add the percents from the three sectors, 3 violations, 4 violations, and more than 4 violations.
6% + 2% + 4% = 12%

4. $30\% \times 300,000 = 0.30 \times 300,000$
$\qquad\qquad\qquad\quad = 90,000$ automobiles
had one safety violation.

5. 21% + 6% = 27%
$27\% \times 300,000 = 0.27 \times 300,000$
81,000 automobiles had 2 or 3 violations.

6. The bar representing private in 1991–1992 rises to 12, which is $12,000.

7. The bar representing public in 1996–1997 rises to 3, which is $3000.

8. 22,000 – 12,000 = 10,000
The increase was $10,000.

9. 6000 – 3000 = 3000
The increase was $3000.

10. 14,000 – 3000 = 11,000
The difference in cost was $11,000.

11. 22,000 – 6000 = 16,000
The difference in cost was $16,000.

12. The dot on the graph is at 20, or 20 years.

13. The dot on the graph is at 26, or 26 years.

14. 48 – 36 = 12 years
He is expected to live 12 years longer.

15. The difference between the two graphs is the greatest at age 35.

16. The difference between the two graphs is the least at age 65.

17. The bar rises to 60, which represents 60,000 televisions.

18. The bar rises to 25, which represents 25,000 televisions.

19. Add the heights of the bars for 12–14 and 15–17.
15,000 + 5000 = 20,000 televisions last more than 11 years.

20. Add the heights of the bars for 9–11 and 12–14.
45,000 + 15,000 = 60,000 televisions last 9–14 years.

21. $10 + 16 + 15 + 12 + 18 + 17 + 14 + 10 + 13 + 20 = 145$

$$\text{Mean} = \frac{145}{10} = 14.5$$

The mean quiz score is 14.5.

22. 10, 10, 12, 13, 14, 15, 16, 17, 18, 20

$$\text{Median} = \frac{14 + 15}{2} = 14.5$$

The median quiz score is 14.5.

23. Mode = 10 since it occurs twice.

24. Mean or median because they are the same and representative of the typical score.

Chapter 9

9.1 Exercises

1. Find the absolute value of each number. Then add those two absolute values. Use the common sign in the answer.

3. Since -9 lies to the left of 2, $-9 < 2$.

5. Since -3 lies to the right of -5, $-3 > -5$.

7. Since 5 lies to the right of -2, $5 > -2$.

9. Since -12 lies to the left of -10, $-12 < -10$.

11. $|7| = 7$

13. $|-16| = 16$

15. $-6 + (-11) = -17$

17. $-4.9 + (-2.1) = -7$

19. $8.9 + 7.6 = 16.5$

21. $\dfrac{1}{5} + \dfrac{2}{7} = \dfrac{7}{35} + \dfrac{10}{35} = \dfrac{17}{35}$

23. $-2\dfrac{1}{2} + \left(-\dfrac{1}{2}\right) = -2 + \left(-\dfrac{2}{2}\right) = -3$

25. $14 + (-5) = 9$

27. $-17 + 12 = -5$

29. $-36 + 58 = 22$

31. $-9.3 + 6.05 = -3.25$

33. $\dfrac{1}{12} + \left(-\dfrac{3}{4}\right) = \dfrac{1}{12} + \left(-\dfrac{9}{12}\right) = -\dfrac{8}{12} = -\dfrac{2}{3}$

35. $\dfrac{7}{9} + \left(-\dfrac{2}{9}\right) = \dfrac{5}{9}$

37. $-18 + (-4) = -22$

39. $1.48 + (-2.2) = -0.72$

41. $-125 + (-238) = -363$

43. $13 + (-9) = 4$

45. $-3\dfrac{3}{4} + \left(-1\dfrac{7}{10}\right) = -4 + \left(-\dfrac{15}{20} - \dfrac{14}{20}\right)$
$= -4\dfrac{29}{20}$
$= -5\dfrac{9}{20}$

47. $-7.56 + 13.8 = 6.24$

49. $-5 + \left(-\dfrac{1}{2}\right) = -\dfrac{10}{2} + \left(-\dfrac{1}{2}\right) = -\dfrac{11}{2}$ or $-5\dfrac{1}{2}$

51. $-20.5 + 18.1 + (-12.3) = -32.8 + 18.1 = -14.7$

53. $11 + (-9) + (-10) + 8 = 2 + (-10) + 8$
$= -8 + 8$
$= 0$

55. $-7 + 6 + (-2) + 5 + (-3) + (-5)$
$= [-7 + (-2) + (-3) + (-5)] + (6 + 5)$
$= -17 + 11$
$= -6$

57. $\left(-\dfrac{1}{5}\right) + \left(-\dfrac{2}{3}\right) + \dfrac{4}{25} = \left(-\dfrac{15}{75}\right) + \left(-\dfrac{50}{75}\right) + \dfrac{12}{75}$
$= -\dfrac{65}{75} + \dfrac{12}{75}$
$= -\dfrac{53}{75}$

59. $-\$43{,}000 + (-\$51{,}000) = -\$94{,}000$
The report shows $-\$94{,}000$ (a loss of \$94,000).

61. $\$9500 + (-\$17{,}000) = -\$7500$
The report shows $-\$7500$ (a loss of \$7500).

63. $-\$18{,}500 + \$12{,}300 + \$15{,}000$
$= -\$6200 + \$15{,}000$
$= \$8800$
The report shows \$8800 (a profit of \$8800).

65. $-5° + (-13°) = -18°F$
The morning temperature is $-18°F$.

67. $-5° + 4° = -1°F$
The new temperature is $-1°F$.

69. $0.12 + (-0.23) + (-0.38) + (-0.38) + 0.05$
$= 0.17 + (-0.99)$
$= -0.82$
The net loss was -0.82.

71. $-8 + 13 + (-6) = 5 + (-6)$
$\qquad\qquad\qquad = -1$ yard
The total was a loss of 1 yard.

73. $\$89.50 + (-\$50.00) + (-\$2.50)$
$= \$39.50 + (-\$2.50)$
$= \$37.00$
The actual balance was $\$37.00$.

Cumulative Review

75. $V = \dfrac{4\pi r^3}{3} = \dfrac{4(3.14)(6 \text{ ft})^3}{3} \approx 904.3 \text{ ft}^3$
The volume is 904.3 ft^3.

76. $B = 9 \times 7 = 63$
$V = \dfrac{Bh}{3} = \dfrac{(63)(10)}{3} = 210$ cubic meters
The volume is 210 m^3.

77. 57, 59, 60, 60, 61, 62
$\text{Median} = \dfrac{60 + 60}{2} = 60$

78. $\text{Mean} = \dfrac{36 + 42 + 39 + 39 + 41 + 43}{6} = \dfrac{240}{6} = \$40$

Quick Quiz 9.1

1. $16 + (-3) + 5 + (-12) = 13 + 5 + (-12)$
$\qquad\qquad\qquad\qquad = 18 + (-12)$
$\qquad\qquad\qquad\qquad = 6$

2. $5.9 + (-7.4) = -1.5$

3. $-4\dfrac{2}{3} + 1\dfrac{1}{3} = -3\dfrac{1}{3}$

4. Answers may vary. Possible solution:
The solution may have fewer steps of the second method is used.

9.2 Exercises

1. $-9 - (-3) = -9 + 3 = -6$

3. $-12 - (-7) = -12 + 7 = -5$

5. $3 - 9 = 3 + (-9) = -6$

7. $-14 - 3 = -14 + (-3) = -17$

9. $-16 - (-25) = -16 + 25 = 9$

11. $46 - (-39) = 46 + 39 = 85$

13. $12 - 30 = 12 + (-30) = -18$

15. $-12 - (-15) = -12 + 15 = 3$

17. $150 - 210 = 150 + (-210) = -60$

19. $300 - (-256) = 300 + 256 = 556$

21. $-2.5 - 4.2 = -2.5 + (-4.2) = -6.7$

23. $6.2 - 14.9 = 6.2 + (-14.9) = -8.7$

25. $-10.9 - (-2.3) = -10.9 + 2.3 = -8.6$

27. $20.23 - (-12.71) = 20.23 + 12.71 = 32.94$

29. $\dfrac{1}{4} - \left(-\dfrac{3}{4}\right) = \dfrac{1}{4} + \dfrac{3}{4} = \dfrac{4}{4} = 1$

31. $-\dfrac{5}{6} - \dfrac{1}{3} = -\dfrac{5}{6} + \left(-\dfrac{1}{3}\right)$
$\qquad\qquad = -\dfrac{5}{6} + \left(-\dfrac{2}{6}\right)$
$\qquad\qquad = -\dfrac{7}{6}$ or $-1\dfrac{1}{6}$

33. $-2\dfrac{3}{10} - \left(-3\dfrac{5}{6}\right) = -2\dfrac{3}{10} + 3\dfrac{5}{6}$
$\qquad\qquad\qquad = -2\dfrac{9}{30} + 3\dfrac{25}{30}$
$\qquad\qquad\qquad = 1\dfrac{16}{30}$
$\qquad\qquad\qquad = 1\dfrac{8}{15}$

35. $\dfrac{2}{9} - \dfrac{5}{7} = \dfrac{2}{9} + \left(-\dfrac{5}{7}\right) = \dfrac{14}{63} + \left(-\dfrac{45}{63}\right) = -\dfrac{31}{63}$

37. $2 - (-8) + 5 = 2 + 8 + 5 = 10 + 5 = 15$

39. $-5 - 6 - (-11) = -5 + (-6) + 11 = -11 + 11 = 0$

41. $21 - (-15) - (-10) = 21 + 15 + 10 = 46$

43. $-16-(-6)-12 = -16+6+(-12)$
$$= -10+(-12)$$
$$= -22$$

45. $9-3-2-6 = 9+(-3)+(-2)+(-6)$
$$= 6+(-2)+(-6)$$
$$= 4+(-6)$$
$$= -2$$

47. $-2.4-7.1+1.3-(-2.8)$
$$= -2.4+(-7.1)+1.3+2.8$$
$$= -9.5+4.1$$
$$= -5.4$$

49. $14,494-(-282) = 14,494+282 = 14,776$ ft
Mt. Whitney is 14,776 feet above Death Valley.

51. $23°-(-19°) = 23°+19° = 42°F$
The difference in temperature is 42°F.

53. $-29°+16° = -13°F$
The new temperature is $-13°F$.

55. $18,700+(-34,700) = -\$16,000$
The change in status is $-\$16,000$.

57. $-6,300+43,600+(-12,400)$
$$= 37,300+(-12,400)$$
$$= \$24,900$$
The change in status is $24,900.

59. $\$15\frac{1}{2}-\$1\frac{1}{2}+\$2\frac{3}{4}-\$3\frac{1}{4}$
$$= \$15\frac{1}{2}+\left(-\$1\frac{1}{2}\right)+\$2\frac{3}{4}+\left(-\$3\frac{1}{4}\right)$$
$$= \$14+\$2\frac{3}{4}+\left(-\$3\frac{1}{4}\right)$$
$$= \$16\frac{3}{4}+\left(-\$3\frac{1}{4}\right)$$
$$= \$13\frac{2}{4}$$
$$= \$13\frac{1}{2}$$

The value of one share is $\$13\frac{1}{2}$ or $13.50.

Cumulative Review

61. $20 \times 2 \div 10+4-3 = 40 \div 10+4-3$
$$= 4+4-3$$
$$= 8-3$$
$$= 8+(-3)$$
$$= 5$$

62. $2+3 \times (5+7) \div 9 = 2+3 \times 12 \div 9$
$$= 2+36 \div 9$$
$$= 2+4$$
$$= 6$$

Quick Quiz 9.2

1. $\frac{3}{7}-\left(-\frac{9}{14}\right) = \frac{3}{7}+\frac{9}{14} = \frac{6}{14}+\frac{9}{14} = \frac{15}{14}$ or $1\frac{1}{14}$

2. $-8.7-(-3.2) = -8.7+3.2 = -5.5$

3. $-34-48 = -34+(-48) = -82$

4. Answers may vary. Possible solution:
Change $8-(-13)$ to $8+13$ by adding the opposite. Then evaluate from left to right.
$8-(-13)+(-5) = 8+13+(-5) = 21+(-5) = 16$

9.3 Exercises

1. To multiply two numbers with the same sign, multiply the absolute values. The sign of the result is positive.

3. $(12)(3) = 36$

5. $(-20)(-3) = 60$

7. $(-20)(8) = -160$

9. $3(-22) = -66$

11. $(2.5)(-0.6) = -1.5$

13. $(-12.5)(-2.25) = 28.125$

15. $\left(-\frac{2}{5}\right)\left(\frac{3}{7}\right) = -\frac{6}{35}$

17. $\left(-\frac{6}{5}\right)\left(-\frac{5}{2}\right) = \frac{30}{10} = 3$

19. $-64 \div 8 = -8$

21. $\dfrac{48}{-6} = -8$

23. $\dfrac{-150}{-25} = 6$

25. $-25 \div (-5) = 5$

27. $-\dfrac{4}{9} \div \left(-\dfrac{16}{27}\right) = -\dfrac{4}{9} \times \left(-\dfrac{27}{16}\right) = \dfrac{3}{4}$

29. $\dfrac{-\frac{4}{5}}{-\frac{7}{10}} = -\dfrac{4}{5} \div \left(-\dfrac{7}{10}\right) = -\dfrac{4}{5} \times \left(-\dfrac{10}{7}\right) = \dfrac{8}{7} \text{ or } 1\dfrac{1}{7}$

31. $50.28 \div (-6) = -8.38$

33. $\dfrac{45.6}{-8} = -5.7$

35. $\dfrac{-21,000}{-700} = 30$

37. $5(-9) = -45$

39. $(-12)(-4) = 48$

41. $\dfrac{15}{-3} = -5$

43. $-30 \div (-3) = 10$

45. $(-1.4)(2) = -2.8$

47. $0.028 \div (-1.4) = -0.02$

49. $\left(-\dfrac{3}{5}\right)\left(-\dfrac{5}{7}\right) = \dfrac{3}{7}$

51. $\dfrac{12}{5} \div \left(-\dfrac{3}{10}\right) = \dfrac{12}{5} \cdot \left(-\dfrac{10}{3}\right) = -8$

53. $10(-5)(-3) = -50(-3) = 150$

55. $(-6)(7)(-2) = -42(-2) = 84$

57. $2(-8)(3)\left(-\dfrac{1}{3}\right) = -16(3)\left(-\dfrac{1}{3}\right) = -48\left(-\dfrac{1}{3}\right) = 16$

59. $(-20)(6)(-30)(-5) = (-120)(-30)(-5)$
$$= (3600)(-5)$$
$$= -18,000$$

61. $8(-3)(-5)(0)(-2) = -24(-5)(0)(-2)$
$$= 120(0)(-2)$$
$$= 0(-2)$$
$$= 0$$

63. $\left(-\dfrac{2}{3}\right)\left(-\dfrac{3}{4}\right)\left(-\dfrac{5}{6}\right) = \left(\dfrac{1}{2}\right)\left(-\dfrac{5}{6}\right) = -\dfrac{5}{12}$

65. $70(2.60) + 120(-0.90) = 182 + (-108) = 74$
He gained \$74.

67. $\dfrac{-12° + (-14°) + (-3°) + (-1°) + (-10°) + (-23°) + 5° + 8°}{8} = -\dfrac{50°}{8}$
$$= -\dfrac{25°}{4}$$
$$= -6.25°$$

The average temperature was −6.25°F.

69. $7(-10) = -70$
He dropped 70 feet.

71. $11(+3) = +33$

73. $6(+2) + 4(-1) = +12 + (-4) = +8$

75. $-8(-1) = 8$ or $+8$

77. $2(-1) + (-2) + 2(2) = -2 + (-2) + 4 = -4 + 4 = 0$
0; he is at par.

Cumulative Review

79. $A = bh = (15\ \text{in.})(6\ \text{in.}) = 90$ square inches
The area is $90\ \text{in.}^2$.

80. $A = \dfrac{h(b + B)}{2}$
$$= \dfrac{(12\ \text{m})(18\ \text{m} + 26\ \text{m})}{2}$$
$$= \dfrac{(12\ \text{m})(44\ \text{m})}{2}$$
$$= \dfrac{528\ \text{m}^2}{2}$$
$$= 264\ \text{square meters}$$
The area is $264\ \text{m}^2$.

Quick Quiz 9.3

1. $(-5)(-9) = 45$

2. $(-2)(3)(-4)(-3) = -6(-4)(-3) = 24(-3) = -72$

3. $-156 \div (-4) = 39$

4. Answers may vary. Possible solution: Addition of two negative numbers is always negative. Multiplication of two negative numbers is always positive.

How Am I Doing? Sections 9.1–9.3

1. $-7 + (-12) = -19$

2. $-23 + 19 = -4$

3. $7.6 + (-3.1) = 4.5$

4. $8 + (-5) + 6 + (-9) = 14 + (-14) = 0$

5. $\dfrac{8}{9} + \left(-\dfrac{2}{3}\right) = \dfrac{8}{9} + \left(-\dfrac{6}{9}\right) = \dfrac{2}{9}$

6. $-\dfrac{5}{6} + \left(-\dfrac{1}{3}\right) = -\dfrac{5}{6} + \left(-\dfrac{2}{6}\right) = -\dfrac{7}{6}$ or $-1\dfrac{1}{6}$

7. $-2.8 + (-4.2) = -7$

8. $-3.7 + 5.4 = 1.7$

9. $13 - 21 = 13 + (-21) = -8$

10. $-26 - 15 = -26 + (-15) = -41$

11. $\dfrac{5}{17} - \left(-\dfrac{9}{17}\right) = \dfrac{5}{17} + \dfrac{9}{17} = \dfrac{14}{17}$

12. $-19 - (-7) = -19 + 7 = -12$

13. $-12.5 - 3.8 = -12.5 + (-3.8) = -16.3$

14. $3.5 - 9.1 = 3.5 + (-9.1) = -5.6$

15. $21 - (-21) = 21 + 21 = 42$

16. $\dfrac{2}{3} - \left(-\dfrac{3}{5}\right) = \dfrac{2}{3} + \dfrac{3}{5} = \dfrac{10}{15} + \dfrac{9}{15} = \dfrac{19}{15}$ or $1\dfrac{4}{15}$

17. $(-3)(-8) = 24$

18. $-48 \div (-12) = 4$

19. $-72 \div 9 = -8$

20. $(5)(-4)(2)(-1)\left(-\dfrac{1}{4}\right) = -20(2)(-1)\left(-\dfrac{1}{4}\right)$
$$= -40(-1)\left(-\dfrac{1}{4}\right)$$
$$= 40\left(-\dfrac{1}{4}\right)$$
$$= -10$$

21. $\dfrac{72}{-3} = -24$

22. $\dfrac{-\frac{2}{3}}{-\frac{11}{12}} = -\dfrac{2}{3} \div \left(-\dfrac{11}{12}\right) = -\dfrac{2}{3}\left(-\dfrac{12}{11}\right) = \dfrac{8}{11}$

23. $(-8)(-2)(-4) = 16(-4) = -64$

24. $120 \div (-12) = -10$

25. $18 - (-6) = 18 + 6 = 24$

26. $-7(-3) = 21$

27. $-15 \div 10 = -1.5$

28. $1.6 + (-1.8) + (-3.4) = -0.2 + (-3.4) = -3.6$

29. $2.9 - 3.5 = 2.9 + (-3.5) = -0.6$

30. $\left(-\dfrac{1}{3}\right) + \left(-\dfrac{2}{5}\right) = \left(-\dfrac{5}{15}\right) + \left(-\dfrac{6}{15}\right) = -\dfrac{11}{15}$

31. $\left(-\dfrac{7}{10}\right)\left(-\dfrac{2}{7}\right) = \dfrac{1}{5}$

32. $5\dfrac{1}{2} \div \left(-\dfrac{1}{2}\right) = \dfrac{11}{2} \cdot \left(-\dfrac{2}{1}\right) = -11$

33. Average $= \dfrac{2 + (-6) + (-10) + (-8) + (-3) + 4}{6}$
$$= \dfrac{-21}{6}$$
$$= -3.5°F$$
The average temperature was $-3.5°F$.

9.4 Exercises

1. $-8 \div (-4)(3) = 2(3) = 6$

3. $50 \div (-25)(4) = (-2)(4) = -8$

5. $16 + 32 \div (-4) = 16 + (-8) = 8$

7. $24 \div (-3) + 16 \div (-4) = -8 + (-4) = -12$

9. $3(-4) + 5(-2) - (-3) = -12 + (-10) + 3$
$$= -22 + 3$$
$$= -19$$

11. $-4(1.5 - 2.3) = -4(-0.8) = 3.2$

13. $5 - 30 \div 3 = 5 - 10 = 5 + (-10) = -5$

15. $36 \div 12(-2) = 3(-2) = -6$

17. $3(-4) + 6(-2) - 3 = -12 + (-12) + (-3)$
$$= -24 + (-3)$$
$$= -27$$

19. $11(-6) - 3(12) = -66 - 36 = -66 + (-36) = -102$

21. $16 - 4(8) + 18 \div (-9) = 16 - 32 + (-2)$
$$= 16 + (-32) + (-2)$$
$$= -16 + (-2)$$
$$= -18$$

23. $\dfrac{8 + 6 - 12}{3 - 6 + 5} = \dfrac{2}{2} = 1$

25. $\dfrac{6(-2) + 4}{6 - 3 - 5} = \dfrac{-12 + 4}{6 - 3 - 5} = \dfrac{-8}{-2} = 4$

27. $\dfrac{2(8) \div 4 - 5}{-35 \div (-7)} = \dfrac{16 \div 4 - 5}{5} = \dfrac{4 - 5}{5} = -\dfrac{1}{5}$

29. $\dfrac{24 \div (-3) - (6 - 2)}{-5(4) + 8} = \dfrac{24 \div (-3) - 4}{-20 + 8}$
$$= \dfrac{-8 - 4}{-12}$$
$$= \dfrac{-12}{-12}$$
$$= 1$$

31. $\dfrac{12 \div 3 + (-2)(2)}{9 - 9 \div (-3)} = \dfrac{4 + (-4)}{9 - (-3)} = \dfrac{4 + (-4)}{9 + 3} = \dfrac{0}{12} = 0$

33. $3(2 - 6) + 4^2 = 3(-4) + 4^2$
$$= 3(-4) + 16$$
$$= -12 + 16$$
$$= 4$$

35. $12 \div (-6) + (7 - 2)^3 = 12 \div (-6) + 5^3$
$$= 12 \div (-6) + 125$$
$$= -2 + 125$$
$$= 123$$

37. $\left(-1\dfrac{1}{2}\right)(-4) - 3\dfrac{1}{4} \div \dfrac{1}{4} = \left(-\dfrac{3}{2}\right)(-4) - \dfrac{13}{4} \div \dfrac{1}{4}$
$$= \left(-\dfrac{3}{2}\right)(-4) - \dfrac{13}{4}\left(\dfrac{4}{1}\right)$$
$$= 6 - 13$$
$$= -7$$

39. $\left(\dfrac{3}{5} - \dfrac{2}{5}\right)^2 + \dfrac{3}{2}\left(-\dfrac{1}{5}\right) = \left(\dfrac{1}{5}\right)^2 + \left(-\dfrac{3}{10}\right)$
$$= \dfrac{1}{25} + \left(-\dfrac{3}{10}\right)$$
$$= \dfrac{2}{50} + \left(-\dfrac{15}{50}\right)$$
$$= -\dfrac{13}{50}$$

41. $(1.2)^2 - 3.6(-1.5) = 1.44 - 3.6(-1.5)$
$$= 1.44 + 5.4$$
$$= 6.84$$

43. $\dfrac{-18 + (-27) + (-21)}{3} = \dfrac{-66}{3} = -22°F$

The average low temperature is $-22°F$.

45. $[-27 + (-21) + (-14) + (-3) + 17 + 38] \div 6$
$$= -10 \div 6$$
$$\approx -1.7°F$$
The average low temperature is $-1.7°F$.

47. $[-13 + (-11) + 0 + 13 + 29 + 50 + 64 + 59 + 44$
$+ 21 + (-1) + (-9)] \div 12$
$$= \dfrac{246}{12}$$
$$= 20.5°F$$
The average yearly high temperature is $20.5°F$.

Cumulative Review

49. $3840 \text{ m} = \dfrac{3840}{1000} \text{ km} = 3.84 \text{ km}$

The telephone wire is 3.84 km long.

50. $36.8 \text{ grams} = 36.8 \times 1000 \text{ milligrams}$
$$= 36,800 \text{ milligrams}$$
It contains $36,800$ mg of protein.

Quick Quiz 9.4

1. $5(-4)-6(2-5)^2 = 5(-4)-6(-3)^2$
   $$= 5(-4)-6(9)$$
   $$= -20-54$$
   $$= -20+(-54)$$
   $$= -74$$

2. $-3.2-8.5-(-3.0)+2(0.4)$
   $$= -3.2+(-8.5)+3.0+0.8$$
   $$= -11.7+3.0+0.8$$
   $$= -8.7+0.8$$
   $$= -7.9$$

3. $\dfrac{5+26\div(-2)}{-2(3)+4(-3)} = \dfrac{5+(-13)}{-6+(-12)} = \dfrac{-8}{-18} = \dfrac{4}{9}$

4. Answers may vary. Possible solution:
   Perform operations inside parentheses. Then simplify expression with exponent. Then multiply and finally add.
   $4(3-9)+5^2 = 4(-6)+5^2$
   $$= 4(-6)+25$$
   $$= -24+25$$
   $$= 1$$

9.5 Exercises

1. Our number system is structured according to base 10. By making scientific notation also in base 10, the calculations are easier to perform.

3. The first part is a number greater than or equal to 1 but smaller than 10. It has at least one non-zero digit. The second part is 10 raised to some integer power.

5. $120 = 1.2\times10^2$

7. $1900 = 1.9\times10^3$

9. $26,300 = 2.63\times10^4$

11. $288,000 = 2.88\times10^5$

13. $10,000 = 1\times10^4$

15. $12,000,000 = 1.2\times10^7$

17. $0.0931 = 9.31\times10^{-2}$

19. $0.00279 = 2.79\times10^{-3}$

21. $0.82 = 8.2\times10^{-1}$

23. $0.00054 = 5.4\times10^{-4}$

25. $0.00000531 = 5.31\times10^{-6}$

27. $0.000008 = 8\times10^{-6}$

29. $5.36\times10^4 = 53,600$

31. $5.334\times10^3 = 5334$

33. $4.6\times10^{12} = 4,600,000,000,000$

35. $6.2\times10^{-2} = 0.062$

37. $8.99\times10^{-3} = 0.00899$

39. $9\times10^{11} = 900,000,000,000$

41. $3.862\times10^{-8} = 0.00000003862$

43. $35,689 = 3.5689\times10^4$

45. $3.3\times10^{-4} = 0.00033$

47. $0.00278 = 2.78\times10^{-3}$

49. $1.88\times10^6 = 1,880,000$

51. $5,878,000,000,000 = 5.878\times10^{12}$
    In 1 year, light will travel 5.878×10^{12} miles.

53. $0.000000000000092 = 9.2\times10^{-14}$
    The average volume is 9.2×10^{-14} L.

55. $10^{-9} = 0.000000001$
    It can travel one foot in 0.000000001 sec.

57. $7.5\times10^{-5} = 0.000075$
    The diameter is 0.000075 cm.

59. $1.4\times10^{10} = 14,000,000,000$
    There were 14,000,000,000 tons of pumice carried into the air.

61.

$$3.38 \times 10^7$$
$$+ \, 5.63 \times 10^7$$
$$9.01 \times 10^7 \text{ dollars}$$

63.

$$5.87 \times 10^{21}$$
$$+ \, 4.81 \times 10^{21}$$
$$10.68 \times 10^{21} = 1.068 \times 10^{22} \text{ tons}$$

65.

$$4.00 \times 10^8 \qquad 40.00 \times 10^7$$
$$- \, 3.76 \times 10^7 \qquad - \, 3.76 \times 10^7$$
$$ 36.24 \times 10^7 = 3.624 \times 10^8 \text{ feet}$$

67.

$$1.76 \times 10^7$$
$$- \, 1.16 \times 10^7$$
$$0.60 \times 10^7 = 6.0 \times 10^6$$

Asia is 6.0×10^6 square miles larger than Africa.

69.
$$20(5.88 \text{ trillion}) = 117.6 \text{ trillion}$$
$$= 117.6 \times 10^{12}$$
$$= 1.176 \times 10^{14}$$

It is 1.176×10^{14} miles from Earth.

Cumulative Review

71.
$$
\begin{array}{r}
12.5 \\
\times \; 0.21 \\
\hline
125 \\
2\,50 \\
\hline
2.625
\end{array}
$$

72.
$$
\begin{array}{r}
0.258 \\
0.53_\wedge \overline{)0.13_\wedge 674} \\
\underline{10\;6} \\
3\;07 \\
\underline{2\;65} \\
424 \\
\underline{424} \\
0
\end{array}
$$

73. $\text{Cost} = 4(2 \times 9 + 2 \times 13)$
$$= 4(18 + 26)$$
$$= 4(44)$$
$$= \$176$$
It will cost \$176.

74.
$$\frac{70}{95} = \frac{x}{800}$$
$$70 \times 800 = 95x$$
$$\frac{56,000}{95} = \frac{95x}{95}$$
$$589 \approx x$$
The cliff is about 589 feet tall.

Quick Quiz 9.5

1. $0.000345 = 3.45 \times 10^{-4}$

2. $568,300 = 5.683 \times 10^5$

3. $8.34 \times 10^{-6} = 0.00000834$

4. Answers may vary. Possible solution:
Move the decimal point 8 places to the right.
You will need to add 5 zeros.
$5.398 \times 10^8 = 539,800,000$

Use Math to Save Money

1. $4 \times \dfrac{2}{3} = \dfrac{4}{1} \times \dfrac{2}{3} = \dfrac{8}{3}$ or $2\dfrac{2}{3}$

Lucy's family eats $\dfrac{8}{3}$ or $2\dfrac{2}{3}$ cups of rice each week.

2. $\dfrac{21 \text{ ounces}}{1 \text{ cup}} \times \dfrac{8}{3} \text{ cups} = 56 \text{ ounces}$

Lucy's family eats 56 ounces of rice each week.

3. $\dfrac{1 \text{ lb}}{16 \text{ oz}} \times 56 \text{ oz} = \dfrac{7}{2}$ or $3\dfrac{1}{2}$ lb

Lucy's family eats $3\dfrac{1}{2}$ pounds of rice each week.

4. $3.5 \text{ lb} \times \dfrac{\$2.66}{1 \text{ lb}} = \$9.31$

Lucy spends \$9.31 per week on the name brand rice.

5. $3.5 \text{ lb} \times \dfrac{\$0.88}{1 \text{ lb}} = \$3.08$

$\$9.31 - \$3.08 = \$6.23$
Lucy could save \$6.23 per week by buying the store brand.

6. $\dfrac{6.23}{9.31} \approx 0.67 = 67\%$

Lucy saves approximately 67%.

7. $\dfrac{\$162}{1\ \text{week}} \times 52\ \text{weeks} = \$8424$

67% of $8424 = 0.67(\$8424) = \$5644.08$
Lucy could save $5644.08 per year.

8. $\$162 - \$130 = \$32$

$\dfrac{\$32}{1\ \text{week}} \times 52\ \text{weeks} = \$1664$
Lucy will save $1664.

You Try It

1. a. $|-10| = 10$

 b. $|18| = 18$

 c. $|-2.5| = 2.5$

2. a. $13 + 4 = 17$

 b. $-12 + (-5) = -17$

 c. $-3.2 + (-3.5) = -6.7$

 d. $-\dfrac{3}{8} + \left(-\dfrac{2}{8}\right) = -\dfrac{5}{8}$

3. a. $9 + (-8) = 1$

 b. $-22 + 5 = -17$

 c. $-9.5 + 4.7 = -4.8$

 d. $\dfrac{1}{12} + \left(-\dfrac{5}{12}\right) = -\dfrac{4}{12} = -\dfrac{1}{3}$

4. a. $8.1 - (-1.3) = 8.1 + 1.3 = 9.4$

 b. $-10 - 3 = -10 + (-3) = -13$

 c. $-34 - (-16) = -34 + 16 = -18$

 d. $1.2 - 1.9 = 1.2 + (-1.9) = -0.7$

 e. $-\dfrac{2}{5} - \left(-\dfrac{3}{5}\right) = -\dfrac{2}{5} + \dfrac{3}{5} = \dfrac{1}{5}$

5. a. $(-5)(9) = -45$

 b. $(-0.5)(24) = -12$

 c. $63 \div (-3) = -21$

 d. $\dfrac{-17}{3.4} = -5$

6. a. $\left(\dfrac{1}{3}\right)(9) = 3$

 b. $(-7)(-8) = 56$

 c. $\dfrac{-30}{-15} = 2$

 d. $\dfrac{-\frac{1}{2}}{-\frac{1}{4}} = \left(-\dfrac{1}{2}\right)\left(-\dfrac{4}{1}\right) = 2$

7. $(-2)^2 + 3(-5) + 5(9 - 12) = (-2)^2 + 3(-5) + 5(-3)$
$= 4 + 3(-5) + 5(-3)$
$= 4 + (-15) + (-15)$
$= -26$

8. $\dfrac{-2(5-2)+8}{3(-4)}$

The numerator is
$-2(5 - 2) + 8 = -2(3) + 8 = -6 + 8 = 2.$
The denominator is $3(-4) = -12$. The fraction

becomes $\dfrac{2}{-12} = -\dfrac{1}{6}.$

9. a. $3124 = 3.124 \times 10^3$

 b. $0.000588 = 5.88 \times 10^{-4}$

 c. $180,000,000 = 1.8 \times 10^8$

 d. $0.000009 = 9 \times 10^{-6}$

10. a. $4 \times 10^5 = 400,000$

 b. $4 \times 10^{-3} = 0.004$

 c. $2.526 \times 10^6 = 2,526,000$

 d. $8.13 \times 10^{-2} = 0.0813$

Chapter 9 Review Problems

1. $-20 + 5 = -15$

2. $-18 + 4 = -14$

3. $-3.6 + (-5.2) = -8.8$

4. $10.4 + (-7.8) = 2.6$

5. $-\dfrac{1}{5} + \left(-\dfrac{1}{3}\right) = -\dfrac{3}{15} + \left(-\dfrac{5}{15}\right) = -\dfrac{8}{15}$

6. $\dfrac{9}{10} + \left(-\dfrac{5}{2}\right) = \dfrac{9}{10} + \left(-\dfrac{25}{10}\right) = -\dfrac{16}{10} = -\dfrac{8}{5}$

7. $20 + (-14) = 6$

8. $12 + (-7) + (-8) + 3 = 15 + (-15) = 0$

9. $25 - 36 = 25 + (-36) = -11$

10. $12 - 40 = 12 + (-40) = -28$

11. $14.5 - (-6) = 14.5 + 6 = 20.5$

12. $-11.4 - 5.8 = -11.4 + (-5.8) = -17.2$

13. $-5.2 - 7.1 = -5.2 + (-7.1) = -12.3$

14. $-\dfrac{2}{5} - \left(-\dfrac{1}{3}\right) = -\dfrac{2}{5} + \dfrac{1}{3} = -\dfrac{6}{15} + \dfrac{5}{15} = -\dfrac{1}{15}$

15. $5 - (-2) - (-6) = 5 + 2 + 6 = 7 + 6 = 13$

16. $-15 - (-3) + 9 = -15 + 3 + 9 = -12 + 9 = -3$

17. $\begin{aligned} 9 - 8 - 6 - 4 &= 9 + (-8) + (-6) + (-4) \\ &= 1 + (-6) + (-4) \\ &= -5 + (-4) \\ &= -9 \end{aligned}$

18. $\left(-\dfrac{2}{7}\right)\left(-\dfrac{1}{5}\right) = \dfrac{2}{35}$

19. $(5.2)(-1.5) = -7.8$

20. $-60 \div (-20) = 3$

21. $-18 \div (-3) = 6$

22. $\dfrac{70}{-14} = -5$

23. $\dfrac{-13.2}{-2.2} = 6$

24. $\dfrac{-\frac{3}{4}}{\frac{1}{6}} = -\dfrac{3}{4} \div \dfrac{1}{6} = -\dfrac{3}{4} \times \dfrac{6}{1} = -\dfrac{9}{2}$ or $-4\dfrac{1}{2}$

25. $\dfrac{-\frac{1}{3}}{-\frac{7}{9}} = -\dfrac{1}{3} \div \left(-\dfrac{7}{9}\right) = -\dfrac{1}{3} \times \left(-\dfrac{9}{7}\right) = \dfrac{3}{7}$

26. $3(-5)(-2) = -15(-2) = 30$

27. $(-2)(3)(-6)(-1) = -6(-6)(-1) = 36(-1) = -36$

28. $10 + 40 \div (-4) = 10 + (-10) = 0$

29. $\begin{aligned} 2(-6) + 3(-4) - (-13) &= -12 + (-12) + 13 \\ &= -24 + 13 \\ &= -11 \end{aligned}$

30. $36 \div (-12) + 50 \div (-25) = -3 + (-2) = -5$

31. $50 \div 25(-4) = 2(-4) = -8$

32. $-3.5 \div (-5) - 1.2 = 0.7 - 1.2 = -0.5$

33. $2.5(-2) + 3.8 = -5 + 3.8 = -1.2$

34. $\dfrac{8 - 17 + 1}{6 - 10} = \dfrac{-8}{-4} = 2$

35. $\dfrac{9 - 3 + 4(-3)}{2 - (-6)} = \dfrac{9 + (-3) + (-12)}{2 + 6} = \dfrac{-6}{8} = -\dfrac{3}{4}$

36. $\dfrac{20 \div (-5) - (-6)}{(2)(-2)(-5)} = \dfrac{-4 + 6}{20} = \dfrac{2}{20} = \dfrac{1}{10}$

37. $\begin{aligned} 2(7 - 11)^2 - 4^3 &= 2(-4)^2 - 4^3 \\ &= 2(16) - 64 \\ &= 32 + (-64) \\ &= -32 \end{aligned}$

38. $\begin{aligned} -50 \div (-10) + (5 - 3)^4 &= -50 \div (-10) + 2^4 \\ &= -50 \div (-10) + 16 \\ &= 5 + 16 \\ &= 21 \end{aligned}$

39. $\left(\dfrac{2}{3}\right)^2 - \dfrac{3}{8}\left(\dfrac{8}{5}\right) = \dfrac{4}{9} - \dfrac{3}{8}\left(\dfrac{8}{5}\right)$

$\qquad\qquad\qquad\quad = \dfrac{4}{9} - \dfrac{3}{5}$

$\qquad\qquad\qquad\quad = \dfrac{20}{45} - \dfrac{27}{45}$

$\qquad\qquad\qquad\quad = -\dfrac{7}{45}$

40. $(1.2)^2 + (2.8)(-0.5) = 1.44 + (2.8)(-0.5)$

$\qquad\qquad\qquad\qquad = 1.44 + (-1.4)$

$\qquad\qquad\qquad\qquad = 0.04$

41. $1.4(4.7 - 4.9) - 12.8 \div (-0.2)$

$\quad = 1.4(-0.2) - 12.8 \div (-0.2)$

$\quad = -0.28 + 64$

$\quad = 63.72$

42. $4160 = 4.16 \times 10^3$

43. $3,700,000 = 3.7 \times 10^6$

44. $200,000 = 2 \times 10^5$

45. $0.007 = 7.0 \times 10^{-3}$

46. $0.0000218 = 2.18 \times 10^{-5}$

47. $0.00000763 = 7.63 \times 10^{-6}$

48. $1.89 \times 10^4 = 18,900$

49. $3.76 \times 10^3 = 3760$

50. $3.14 \times 10^5 = 314,000$

51. $7.52 \times 10^{-2} = 0.0752$

52. $6.61 \times 10^{-3} = 0.00661$

53. $9 \times 10^{-7} = 0.0000009$

54. $\begin{array}{r} 2.42 \times 10^7 \\ + 5.76 \times 10^7 \\ \hline 8.18 \times 10^7 \end{array}$

55. $\begin{array}{r} 6.11 \times 10^{10} \\ + 3.87 \times 10^{10} \\ \hline 9.98 \times 10^{10} \end{array}$

56. $\begin{array}{r} 3.42 \times 10^{14} \\ - 1.98 \times 10^{14} \\ \hline 1.44 \times 10^{14} \end{array}$

57. $\begin{array}{r} 1.76 \times 10^{26} \\ - 1.08 \times 10^{26} \\ \hline 0.68 \times 10^{26} = 6.8 \times 10^{25} \end{array}$

58. $93,000,000 = 9.3 \times 10^7$

$\quad 9.3 \times 10^7 \text{ miles} \times \dfrac{5.28 \times 10^3 \text{ feet}}{1 \text{ mi}}$

$\quad = 49.104 \times 10^{10} \text{ ft}$

$\quad = 4.9104 \times 10^{11}$

The distance is 4.9104×10^{11} feet.

59. $280,000 \times 93,000,000 = (2.8 \times 10^5) \times (9.3 \times 10^7)$

$\qquad\qquad\qquad\qquad\quad = 2.604 \times 10^{13}$

The distance is 2.604×10^{13} miles.

60. $1.67 \text{ yg} = 1.67 \times 10^{-24} \text{ grams}$

$\quad 0.00091 \text{ yg} = 9.1 \times 10^{-28} \text{ grams}$

61. $2.5 \times 10^8 = 250,000,000$

The diameter is 250,000,000 m.

62. $384.4 \times 10^6 = 384,400,000$

The distance is 384,400,000 m.

63. $-5 + 6 + (-7) = 1 + (-7) = -6$

The total loss was 6 yards.

64. Top of Fred's head: $-282 \text{ ft} + 6 \text{ ft} = -276 \text{ ft}$

Distance to plane:

$2400 - (-276) = 2400 + 276 = 2676$

The distance is 2676 feet.

65. $-\$18 + (-\$20) + \$40 = -\$38 + \$40 = \$2$

The balance is $2.

66. $\dfrac{-16° + (-18°) + (-5°) + 3° + (-12°)}{5} = \dfrac{-48°}{5}$
$$= -9.6°$$
The average temperature was –9.6°F.

67. $2(-1) + (-2) + 4(1) + 2 = -2 + (-2) + 4 + 2$
$$= -4 + 4 + 2$$
$$= 0 + 2$$
$$= 2$$
Frank was 2 points above par.

How Am I Doing? Chapter 9 Test

1. $-26 + 15 = -11$

2. $-31 + (-12) = -43$

3. $12.8 + (-8.9) = 3.9$

4. $-3 + (-6) + 7 + (-4) = -9 + 7 + (-4)$
$$= -2 + (-4)$$
$$= -6$$

5. $-5\dfrac{3}{4} + 2\dfrac{1}{4} = -3\dfrac{2}{4} = -3\dfrac{1}{2}$

6. $-\dfrac{1}{4} + \left(-\dfrac{5}{8}\right) = -\dfrac{2}{8} + \left(-\dfrac{5}{8}\right) = -\dfrac{7}{8}$

7. $-32 - 6 = -32 + (-6) = -38$

8. $23 - 18 = 23 + (-18) = 5$

9. $\dfrac{4}{5} - \left(-\dfrac{1}{3}\right) = \dfrac{4}{5} + \dfrac{1}{3} = \dfrac{12}{15} + \dfrac{5}{15} = \dfrac{17}{15}$ or $1\dfrac{2}{15}$

10. $-50 - (-7) = -50 + 7 = -43$

11. $-2.5 - (-6.5) = -2.5 + 6.5 = 4$

12. $-8.5 - 2.8 = -8.5 + (-2.8) = -11.3$

13. $\dfrac{1}{12} - \left(-\dfrac{5}{6}\right) = \dfrac{1}{12} + \dfrac{5}{6} = \dfrac{1}{12} + \dfrac{10}{12} = \dfrac{11}{12}$

14. $-15 - (-15) = -15 + 15 = 0$

15. $(-20)(-6) = 120$

16. $27 \div \left(-\dfrac{3}{4}\right) = 27 \times \left(-\dfrac{4}{3}\right) = -36$

17. $-40 \div (-4) = 10$

18. $(-9)(-1)(-2)(4)\left(\dfrac{1}{4}\right) = 9(-2)(4)\left(\dfrac{1}{4}\right)$
$$= -18(4)\left(\dfrac{1}{4}\right)$$
$$= -72\left(\dfrac{1}{4}\right)$$
$$= -18$$

19. $\dfrac{-39}{-13} = 3$

20. $\dfrac{-\dfrac{3}{5}}{\dfrac{6}{7}} = -\dfrac{3}{5} \div \dfrac{6}{7} = -\dfrac{3}{5} \times \dfrac{7}{6} = -\dfrac{7}{10}$

21. $(-12)(0.5)(-3) = (-6)(-3) = 18$

22. $96 \div (-3) = -32$

23. $7 - 2(-5) = 7 + 10 = 17$

24. $-2.5 - 1.2 \div (-0.4) = -2.5 - (-3) = -2.5 + 3 = 0.5$

25. $18 \div (-3) + 24 \div (-12) = -6 + (-2) = -8$

26. $-6(-3) - 4(3 - 7)^2 = -6(-3) - 4(-4)^2$
$$= -6(-3) - 4(16)$$
$$= 18 - 64$$
$$= -46$$

27. $1.3 - 9.5 - (-2.5) + 3(-0.5)$
$$= 1.3 + (-9.5) + 2.5 + (-1.5)$$
$$= -8.2 + 2.5 + (-1.5)$$
$$= -5.7 + (-1.5)$$
$$= -7.2$$

28. $-48 \div (-6) - 7(-2)^2 = -48 \div (-6) - 7(4)$
$$= 8 + (-28)$$
$$= -20$$

29. $\dfrac{3 + 8 - 5}{(-4)(6) + (-6)(3)} = \dfrac{3 + 8 - 5}{-24 + (-18)} = \dfrac{6}{-42} = -\dfrac{1}{7}$

30. $\dfrac{5 + 28 \div (-4)}{7 - (-5)} = \dfrac{5 + (-7)}{7 + 5} = \dfrac{-2}{12} = -\dfrac{1}{6}$

31. $80{,}540 = 8.054 \times 10^4$

32. $0.000007 = 7 \times 10^{-6}$

33. $9.36 \times 10^{-5} = 0.0000936$

34. $7.2 \times 10^{4} = 72,000$

35. $\dfrac{-14° + (-8°) + (-5°) + 7° + (-11°)}{5} = -\dfrac{-31°}{5}$
$$= -6.2°\text{F}$$
The average temperature is −6.2°F.

36.
$$
\begin{array}{r}
2 \times 5.8 \times 10^{-5} \\
+\; 2 \times 7.8 \times 10^{-5} \\
\hline
\end{array}
$$

$$
\begin{array}{r}
11.6 \times 10^{-5} \\
+\; 15.6 \times 10^{-5} \\
\hline
27.2 \times 10^{-5} \text{ or } 2.72 \times 10^{-4} \text{ meter}
\end{array}
$$

The perimeter is 2.72×10^{-4} m.

37. $58.3 - (-128.6) = 58.3 + 128.6 = 186.9°\text{F}$
The difference in temperatures is 186.9°F.

Chapter 10

10.1 Exercises

1. A variable is a symbol, usually a letter of the alphabet, that stands for a number.

3. All the exponents for like terms must be the same. The exponent for x must be the same. The exponent for y must be the same. In this case x is raised to the second power in the first term but y is raised to the second power in the second term.

5. $G = 5xy$: variables are G, x, y.

7. $p = \dfrac{4ab}{3}$: variables are p, a, b.

9. $r = 3 \times m + 5 \times n$
   $r = 3m + 5n$

11. $H = 2 \times a - 3 \times b$
    $H = 2a - 3b$

13. $-16x + 26x = 10x$

15. $2x - 8x + 5x = -6x + 5x = -1x = -x$

17. $-\dfrac{1}{2}x + \dfrac{3}{4}x + \dfrac{1}{12}x = -\dfrac{6}{12}x + \dfrac{9}{12}x + \dfrac{1}{12}x$
    $= \dfrac{4}{12}x$
    $= \dfrac{1}{3}x$

19. $8x - x + 10 - 6 = 7x + 4$

21. $1.3x + 10 - 2.4x - 3.6 = 1.3x - 2.4x + 10 - 3.6$
    $= -1.1x + 6.4$

23. $16x + 9y - 11 + 21x = 16x + 21x + 9y - 11$
    $= (16 + 21)x + 9y - 11$
    $= 37x + 9y - 11$

25. $\left(3\dfrac{1}{2}\right)x - 32 - \left(1\dfrac{1}{6}\right)x - 18$
    $= \left(3\dfrac{1}{2}\right)x - \left(1\dfrac{1}{6}\right)x - 32 - 18$
    $= \left(3\dfrac{3}{6} - 1\dfrac{1}{6}\right)x - 50$
    $= \left(2\dfrac{1}{3}\right)x - 50 \text{ or } \dfrac{7}{3}x - 50$

27. $7a - c + 6b - 3c - 10a$
    $= 7a - 10a + 6b - c - 3c$
    $= (7 - 10)a + 6b + (-1 - 3)c$
    $= -3a + 6b - 4c$

29. $\dfrac{1}{2}x + \dfrac{1}{7}y - \dfrac{3}{4}x + \dfrac{5}{21}y = \dfrac{1}{2}x - \dfrac{3}{4}x + \dfrac{1}{7}y + \dfrac{5}{21}y$
    $= \left(\dfrac{1}{2} - \dfrac{3}{4}\right)x + \left(\dfrac{1}{7} + \dfrac{5}{21}\right)y$
    $= \left(\dfrac{2}{4} - \dfrac{3}{4}\right)x + \left(\dfrac{3}{21} + \dfrac{5}{21}\right)y$
    $= -\dfrac{1}{4}x + \dfrac{8}{21}y$

31. $7.3x + 1.7x + 4 - 6.4x - 5.6x - 10$
    $= 7.3x + 1.7x - 6.4x - 5.6x + 4 - 10$
    $= -3x - 6$

33. $-7.6n + 1.2 + 11.2m - 3.5n - 8.1m$
    $= 11.2m - 8.1m - 7.6n - 3.5n + 1.2$
    $= 3.1m - 11.1n + 1.2$

35. a.   Perimeter $= 4x - 2 + 3x + 6 + 5x - 3$
         $= 4x + 3x + 5x - 2 + 6 - 3$
         $= 12x + 1$

    b.   It is doubled.
         $2(12x + 1) = 24x + 2$

Cumulative Review

37. $3 \times n = 36$
    $\dfrac{3 \times n}{3} = \dfrac{36}{3}$
    $n = 12$

38. $8 \times n = 64$
    $\dfrac{8 \times n}{8} = \dfrac{64}{8}$
    $n = 8$

39. $\dfrac{n}{6} = \dfrac{12}{15}$
    $n \times 15 = 6 \times 12$
    $\dfrac{n \times 15}{15} = \dfrac{72}{15}$
    $n = 4.8$

40.

$$\frac{n}{9} = \frac{36}{40}$$

$$n \times 40 = 9 \times 36$$

$$n \times 40 = 324$$

$$\frac{n \times 40}{40} = \frac{324}{40}$$

$$n = 8.1$$

41. 10% of $80 = 0.10 \times 80 = 8$

42. 50% of $80 = 0.50 \times 80 = 40$

Quick Quiz 10.1

1. $6a - 5b - 3a - 9b = 6a - 3a - 5b - 9b = 3a - 14b$

2.
$$\frac{1}{3}x - \frac{4}{5}y - \frac{3}{4}x + \frac{3}{25}y$$
$$= \frac{1}{3}x - \frac{3}{4}x - \frac{4}{5}y + \frac{3}{25}y$$
$$= \left(\frac{4}{12} - \frac{9}{12}\right)x + \left(-\frac{20}{25} + \frac{3}{25}\right)y$$
$$= -\frac{5}{12}x - \frac{17}{25}y$$

3. $-12x + 22y - 34 - 6x - 7y - 13$
$$= -12x - 6x + 22y - 7y - 34 - 13$$
$$= -18x + 15y - 47$$

4. Answers may vary. Possible solution:
Change all subtraction to addition. Use the commutative property to rearrange terms. Then simplify.
$-8.2x - 3.4y + 6.7z - 3.1x + 5.6y - 9.8z$
$= -8.2x + (-3.4y) + 6.7z + (-3.1x) + 5.6y$
$\qquad + (-9.8z)$
$= -8.2x + (-3.1x) + (-3.4y) + 5.6y + 6.7z$
$\qquad + (-9.8z)$
$= -11.3x + 2.2y - 3.1z$

10.2 Exercises

1. A <u>variable</u> is a symbol, usually a letter of the alphabet, that stands for a number.

3. $3x$ and x, $2y$ and $-3y$

5. $9(3x - 2) = 9(3x) - 9(2) = 27x - 18$

7. $-2(x + y) = -2(x) - 2(y) = -2x - 2y$

9. $-6(-2.4x + 5y) = -6(-2.4x) - 6(5y)$
$\qquad = 14.4x - 30y$

11. $(-3x + 7y)(-10) = (-3x)(-10) + (7y)(-10)$
$\qquad = 30x - 70y$

13. $(6a - 5b)(8) = (6a)8 + (-5b)8 = 48a - 40b$

15. $(-8y - 7z)(-3) = (-8y)(-3) + (-7z)(-3)$
$\qquad = 24y + 21z$

17. $4(p + 9q - 10) = 4(p) + 4(9q) - 4(10)$
$\qquad = 4p + 36q - 40$

19. $3\left(\frac{1}{5}x + \frac{2}{3}y - \frac{1}{4}\right) = 3\left(\frac{1}{5}x\right) + 3\left(\frac{2}{3}y\right) - 3\left(\frac{1}{4}\right)$
$$= \frac{3}{5}x + 2y - \frac{3}{4}$$

21. $-15(-2a - 3.2b + 4.5)$
$= -15(-2a) + (-15)(-3.2b) + (-15)(4.5)$
$= 30a + 48b - 67.5$

23. $(8a + 12b - 9c - 5)(4)$
$= (8a)(4) + (12b)(4) + (-9c)(4) + (-5)(4)$
$= 32a + 48b - 36c - 20$

25. $(-2)(1.3x - 8.5y - 5z + 12)$
$= (-2)(1.3x) + (-2)(-8.5y) + (-2)(-5z)$
$\qquad + (-2)(12)$
$= -2.6x + 17y + 10z - 24$

27. $\frac{1}{2}\left(2x - 3y + 4z - \frac{1}{2}\right)$
$= \frac{1}{2}(2x) + \frac{1}{2}(-3y) + \frac{1}{2}(4z) + \frac{1}{2}\left(-\frac{1}{2}\right)$
$= x - \frac{3}{2}y + 2z - \frac{1}{4}$

29. $-\frac{1}{3}(9s - 30t - 63) = -\frac{1}{3}(9s) - \frac{1}{3}(-30t) - \frac{1}{3}(-63)$
$\qquad = -3s + 10t + 21$

31. $P = 2(l + w) = 2l + 2w$

33. $A = \frac{h(B + b)}{2}$
$A = \frac{hB + hb}{2}$

35. $4(5x - 1) + 7(x - 5) = 20x - 4 + 7x - 35$
$\qquad = 27x - 39$

37. $10(4a+5b)-8(6a+2)=40a+50b-48a-16$
$$=-8a+5b-16$$

39. $1.5(x+2.2y)+3(2.2x+1.6y)$
$$=1.5x+3.3y+6.6x+4.8y$$
$$=8.1x+8.1y$$

41. $2(3b+c-2a)-5(a-2c+5b)$
$$=6b+2c-4a-5a+10c-25b$$
$$=-9a-19b+12c$$

43. $A=ab+ac$
$A=a(b+c)$
Hence, $a(b+c)=ab+ac$.

Cumulative Review

45. $P=2l+2w$
$$=2(7.5\text{ ft})+2(4\text{ ft})$$
$$=15\text{ ft}+8\text{ ft}$$
$$=23\text{ ft}$$
The perimeter is 23 feet.

46. $A=\dfrac{1}{2}bh=\dfrac{1}{2}(8.5\text{ in.})(15\text{ in.})=63.75\text{ in.}^2$
The area is 63.75 square inches.

Quick Quiz 10.2

1. $3\left(\dfrac{5}{6}x-\dfrac{7}{12}y\right)=3\left(\dfrac{5}{6}x\right)+3\left(-\dfrac{7}{12}y\right)$
$$=\dfrac{15}{6}x-\dfrac{21}{12}y$$
$$=\dfrac{5}{2}x-\dfrac{7}{4}y$$

2. $-3.5(2x-3y+z-4)$
$$=-3.5(2x)-3.5(-3y)-3.5(z)-3.5(-4)$$
$$=-7x+10.5y-3.5z+14$$

3. $2(-3x+7y)-5(2x-9y)$
$$=2(-3x)+2(7y)-5(2x)-5(-9y)$$
$$=-6x+14y-10x+45y$$
$$=-16x+59y$$

4. Answers may vary. Possible solution:
Use the distributive property to remove parentheses. Then combine like terms to simplify.
$-3(2x+5y)+4(5x-1)$
$$=-3(2x)-3(5y)+4(5x)-4(1)$$
$$=-6x-15y+20x-4$$
$$=14x-15y-4$$

10.3 Exercises

1. An <u>equation</u> is a mathematical statement that says that two expressions are equal.

3. To use the addition property, we add to both sides of the equation the <u>opposite</u> of the number we want to remove from one side of the equation.

5. $y-12=20$
$$y-12+12=20+12$$
$$y=32$$

7. $x+6=15$
$$x+6+(-6)=15+(-6)$$
$$x=9$$

9. $x+16=-2$
$$x+16+(-16)=-2+(-16)$$
$$x=-18$$

11. $14+x=-11$
$$14+(-14)+x=-11+(-14)$$
$$x=-25$$

13. $-12+x=7$
$$-12+12+x=7+12$$
$$x=19$$

15. $5.2=x-4.6$
$$5.2+4.6=x-4.6+4.6$$
$$9.8=x$$

17. $y+8.2=-3.4$
$$y+8.2+(-8.2)=-3.4+(-8.2)$$
$$y=-11.6$$

19. $x-25.2=-12$
$$x-25.2+25.2=-12+25.2$$
$$x=13.2$$

21.
$$\frac{4}{5} = x + \frac{2}{5}$$
$$\frac{4}{5} + \left(-\frac{2}{5}\right) = x + \frac{2}{5} + \left(-\frac{2}{5}\right)$$
$$\frac{2}{5} = x$$

23.
$$x - \frac{3}{5} = \frac{2}{5}$$
$$x - \frac{3}{5} + \frac{3}{5} = \frac{2}{5} + \frac{3}{5}$$
$$x = \frac{5}{5} = 1$$

25.
$$x + \frac{2}{3} = -\frac{5}{6}$$
$$x + \frac{2}{3} + \left(-\frac{2}{3}\right) = -\frac{5}{6} + \left(-\frac{2}{3}\right)$$
$$x = -\frac{5}{6} + \left(-\frac{4}{6}\right)$$
$$= -\frac{9}{6}$$
$$= -\frac{3}{2} \text{ or } -1\frac{1}{2}$$

27.
$$\frac{1}{4} + y = 2\frac{3}{8}$$
$$-\frac{1}{4} + \frac{1}{4} + y = -\frac{1}{4} + 2\frac{3}{8}$$
$$y = -\frac{2}{8} + 2\frac{3}{8}$$
$$y = 2\frac{1}{8} \text{ or } \frac{17}{8}$$

29.
$$3x - 5 = 2x + 9$$
$$3x + (-2x) - 5 = 2x + (-2x) + 9$$
$$x - 5 = 9$$
$$x - 5 + 5 = 9 + 5$$
$$x = 14$$

31.
$$5x + 12 = 4x - 1$$
$$5x + (-4x) + 12 = 4x + (-4x) - 1$$
$$x + 12 = -1$$
$$x + 12 + (-12) = -1 + (12)$$
$$x = -13$$

33.
$$7x - 9 = 6x - 7$$
$$7x + (-6x) - 9 = 6x + (-6x) - 7$$
$$x - 9 = -7$$
$$x - 9 + 9 = -7 + 9$$
$$x = 2$$

35.
$$18x + 28 = 17x + 19$$
$$18x + (-17x) + 28 = 17x + (-17x) + 19$$
$$x + 28 = 19$$
$$x + 28 + (-28) = 19 + (-28)$$
$$x = -9$$

37.
$$y - \frac{1}{2} = 6$$
$$y - \frac{1}{2} + \frac{1}{2} = 6 + \frac{1}{2}$$
$$y = 6\frac{1}{2} \text{ or } \frac{13}{2}$$

39.
$$5 = z + 13$$
$$5 + (-13) = z + 13 + (-13)$$
$$-8 = z$$

41.
$$-5.9 + y = -4.7$$
$$-5.9 + 5.9 + y = -4.7 + 5.9$$
$$y = 1.2$$

43.
$$2x - 1 = x + 5$$
$$2x + (-x) - 1 = x + (-x) + 5$$
$$x - 1 = 5$$
$$x - 1 + 1 = 5 + 1$$
$$x = 6$$

45.
$$3.6x - 8 = 2.6 + 4$$
$$3.6x + (-2.6x) - 8 = 2.6x + (-2.6x) + 4$$
$$x - 8 = 4$$
$$x - 8 + 8 = 4 + 8$$
$$x = 12$$

47.
$$6x - 12 = 7x - 5$$
$$-6x + 6x - 12 = 7x + (-6x) - 5$$
$$-12 = x - 5$$
$$-12 + 5 = x - 5 + 5$$
$$-7 = x$$

49. To solve the equation $3x = 12$, divide both sides of the equation by 3 so that x stands alone on one side of the equation.

Cumulative Review

51. $5x - y + 3 - 2x + 4y = 5x - 2x - y + 4y + 3$
$$= 3x + 3y + 3$$

52. $7(2x + 3y) - 3(5x - 1)$
$$= 7(2x) + 7(3y) + (-3)(5x) + (-3)(-1)$$
$$= 14x + 21y - 15x + 3$$
$$= 14x - 15x + 21y + 3$$
$$= -x + 21y + 3$$

53. $\text{mean} = \dfrac{85 + 78 + 92 + 83 + 72}{5} = \dfrac{410}{5} = 82$

The mean is $82.

54. List in order: 80, 84, 85, 86, 90, 93, 98
median = 86

Quick Quiz 10.3

1. $x - 8.4 = -10.6$
$$x - 8.4 + 8.4 = -10.6 + 8.4$$
$$x = -2.2$$

2. $8x - 15 = 7x + 20$
$$8x + (-7x) - 15 = 7x + (-7x) + 20$$
$$x - 15 = 20$$
$$x - 15 + 15 = 20 + 15$$
$$x = 35$$

3. $5 + 6x = 5x - 5$
$$5 + 6x + (-5x) = 5x + (-5x) - 5$$
$$5 + x = -5$$
$$-5 + 5 + x = -5 - 5$$
$$x = -10$$

4. Answers may vary. Possible solution:
Add $8x$ to both sides. Then add -5 to both sides.
$$-7x + 5 = -8x - 13$$
$$8x - 7x + 5 = 8x - 8x - 13$$
$$x + 5 = -13$$
$$x + 5 + (-5) = -13 + (-5)$$
$$x = -18$$

10.4 Exercises

1. A sample answer is: To maintain the balance, whatever you do to one side of the scale, you need to do the exact same thing to the other side of the scale.

3. To change $\dfrac{3}{4}x = 5$ to a simpler equation, multiply both sides of the equation by $\dfrac{4}{3}$.

5. $4x = 36$
$$\dfrac{4x}{4} = \dfrac{36}{4}$$
$$x = 9$$

7. $7y = -28$
$$\dfrac{7y}{7} = \dfrac{-28}{7}$$
$$y = -4$$

9. $-9y = 16$
$$\dfrac{-9y}{-9} = \dfrac{16}{-9}$$
$$y = -\dfrac{16}{9} \text{ or } -1\dfrac{7}{9}$$

11. $-12x = -144$
$$\dfrac{-12x}{-12} = \dfrac{-144}{-12}$$
$$x = 12$$

13. $-64 = -4m$
$$\dfrac{-64}{-4} = \dfrac{-4m}{-4}$$
$$16 = m$$

15. $0.6x = 6$
$$\dfrac{0.6x}{0.6} = \dfrac{6}{0.6}$$
$$x = 10$$

17. $17.5 = 2.5t$
$$\dfrac{17.5}{2.5} = \dfrac{2.5t}{2.5}$$
$$7 = t$$

19. $-0.5x = 6.75$
$$\dfrac{-0.5x}{-0.5} = \dfrac{6.75}{-0.5}$$
$$x = -13.5$$

21. $\dfrac{5}{8}x = 5$
$$\dfrac{8}{5} \cdot \dfrac{5}{8}x = \dfrac{8}{5} \cdot 5$$
$$x = 8$$

23.
$$\frac{2}{5}y = 4$$
$$\frac{5}{2} \cdot \frac{2}{5}y = 4 \cdot \frac{5}{2}$$
$$y = 10$$

25.
$$\frac{3}{5}n = \frac{3}{4}$$
$$\frac{5}{3} \cdot \frac{3}{5}n = \frac{3}{4} \cdot \frac{5}{3}$$
$$n = \frac{5}{4} \text{ or } 1\frac{1}{4}$$

27.
$$-\frac{2}{9}x = \frac{4}{5}$$
$$-\frac{9}{2} \cdot \left(-\frac{2}{9}\right)x = -\frac{9}{2} \cdot \frac{4}{5}$$
$$x = -\frac{18}{5} \text{ or } -3\frac{3}{5}$$

29.
$$\frac{1}{2}x = -2\frac{1}{4}$$
$$\frac{2}{1} \cdot \frac{1}{2}x = -2\frac{1}{4} \cdot 2$$
$$x = -\frac{9}{4} \cdot 2 = -\frac{9}{2} \text{ or } -4\frac{1}{2}$$

31.
$$\left(-3\frac{1}{3}\right)z = -20$$
$$-\frac{10}{3}z = -20$$
$$-\frac{3}{10} \cdot \left(-\frac{10}{3}\right)z = -\frac{3}{10} \cdot (-20)$$
$$z = 6$$

33.
$$-60 = -10x$$
$$\frac{-60}{-10} = \frac{-10x}{-10}$$
$$6 = x$$

35.
$$\frac{2}{3}x = -6$$
$$\left(\frac{3}{2}\right)\left(\frac{2}{3}x\right) = \left(\frac{3}{2}\right)(-6)$$
$$x = -9$$

37.
$$1.5x = 0.045$$
$$\frac{1.5x}{1.5} = \frac{0.045}{1.5}$$
$$x = 0.03$$

39.
$$12 = -\frac{3}{5}x$$
$$\left(-\frac{5}{3}\right)(12) = \left(-\frac{5}{3}\right)\left(-\frac{3}{5}x\right)$$
$$-20 = x$$

Cumulative Review

41. $6 - 3x + 5y + 7x - 12y$
$$= -3x + 7x + 5y - 12y + 6$$
$$= 4x - 7y + 6$$

42. $-2(3a - 5b + c) + 5(-a + 2b - 5c)$
$$= (-2)(3a) + (-2)(-5b) + (-2)(c) + 5(-a) + 5(2b)$$
$$\quad + 5(-5c)$$
$$= -6a + 10b - 2c - 5a + 10b - 25c$$
$$= -6a - 5a + 10b + 10b - 2c - 25c$$
$$= -11a + 20b - 27c$$

43. decrease $= \$10.10 - \$9.50 = \$0.60$
$$\text{percent} = \frac{0.60}{10.10} \approx 0.059 = 5.9\%$$
The percentage of decrease was 5.9%.

44. percent covered: $\dfrac{1,755,637}{2,166,086} \approx 0.811$ or 81.1%

percent not covered: $100\% - 81.1\% = 18.9\%$

Quick Quiz 10.4

1.
$$-4x = 15$$
$$\frac{-4x}{-4} = \frac{15}{-4}$$
$$x = -\frac{15}{4} \text{ or } -3\frac{3}{4} \text{ or } -3.75$$

2.
$$-3.5 = -0.5x$$
$$\frac{-3.5}{-0.5} = \frac{-0.5x}{-0.5}$$
$$7 = x$$

3.
$$-\frac{3}{4}x = \frac{9}{2}$$
$$-\frac{4}{3} \cdot \left(-\frac{3}{4}x\right) = -\frac{4}{3} \cdot \frac{9}{2}$$
$$x = -6$$

4. Answers may vary. Possible solution:
Simply the left side. Then divide both sides by
-14.

$$12 - 5 = -14x$$
$$7 = -14x$$
$$\frac{7}{-14} = \frac{-14x}{-14}$$
$$-\frac{1}{2} = x$$

10.5 Exercises

1. You want to obtain the x-terms all by itself on
one side of the equation. So you want to remove
the -6 from the left-hand side of the equation.
Therefore you would add the opposite of -6.
This means you would add 6 to each side.

3. $2x + 5 = 7 - 4x$
$2(3) + 5 \stackrel{?}{=} 7 - 4(3)$
$6 + 5 \stackrel{?}{=} 7 - 12$
$\quad 11 \neq -5$
No

5. $8x - 2 = 10 - 16x$
$8\left(\frac{1}{2}\right) - 2 \stackrel{?}{=} 10 - 16\left(\frac{1}{2}\right)$
$4 - 2 \stackrel{?}{=} 10 - 8$
$2 = 2$ Yes

7. $15x - 10 = 35$
$15x - 10 + 10 = 35 + 10$
$15x = 45$
$\frac{15x}{15} = \frac{45}{15}$
$x = 3$

9. $6x - 9 = -12$
$6x - 9 + 9 = -12 + 9$
$6x = -3$
$\frac{6x}{6} = \frac{-3}{6}$
$x = -\frac{1}{2}$

11. $-9x = 3x - 10$
$-9x + (-3x) = 3x + (-3x) - 10$
$-12x = -10$
$\frac{-12x}{-12} = \frac{-10}{-12}$
$x = -\frac{10}{12} = \frac{5}{6}$

13. $14x - 10 = 18$
$14x - 10 + 10 = 18 + 10$
$14x = 28$
$\frac{14x}{14} = \frac{28}{14}$
$x = 2$

15. $0.26 = 2x - 0.34$
$0.26 + 0.34 = 2x - 0.34 + 0.34$
$0.6 = 2x$
$\frac{0.6}{2} = \frac{2x}{2}$
$0.3 = x$

17. $\frac{2}{3}x - 5 = 17$
$\frac{2}{3}x - 5 + 5 = 17 + 5$
$\frac{2}{3}x = 22$
$\frac{3}{2} \cdot \frac{2}{3}x = 22 \cdot \frac{3}{2}$
$x = 33$

19. $18 - 2x = 4x + 6$
$18 - 2x + 2x = 4x + 2x + 6$
$18 = 6x + 6$
$18 + (-6) = 6x + 6 + (-6)$
$12 = 6x$
$\frac{12}{6} = \frac{6x}{6}$
$2 = x$

21. $9 - 8x = 3 - 2x$
$9 - 8x + 8x = 3 - 2x + 8x$
$9 = 3 + 6x$
$-3 + 9 = -3 + 3 + 6x$
$6 = 6x$
$\frac{6}{6} = \frac{6x}{6}$
$1 = x$

23. $5z + 6 = 3z - 2$
$5z + (-3z) + 6 = 3z + (-3z) - 2$
$2z + 6 = -2$
$2z + 6 + (-6) = -2 + (-6)$
$2z = -8$
$\frac{2z}{2} = \frac{-8}{2}$
$z = -4$

25.
$$1.2 + 0.3x = 0.6x - 2.1$$
$$1.2 + 0.3x + (-0.3x) = 0.6x + (-0.3x) - 2.1$$
$$1.2 = 0.3x - 2.1$$
$$1.2 + 2.1 = 0.3x - 2.1 + 2.1$$
$$3.3 = 0.3x$$
$$\frac{3.3}{0.3} = \frac{0.3x}{0.3}$$
$$11 = x$$

27.
$$0.2x + 0.6 = -0.8 - 1.2x$$
$$0.2x + 1.2x + 0.6 = -0.8 - 1.2x + 1.2x$$
$$1.4x + 0.6 = -0.8$$
$$1.4x + 0.6 + (-0.6) = -0.8 + (-0.6)$$
$$1.4x = -1.4$$
$$\frac{1.4x}{1.4} = \frac{-1.4}{1.4}$$
$$x = -1$$

29.
$$-10 + 6y + 2 = 3y - 26$$
$$-8 + 6y = 3y - 26$$
$$-8 + 6y + (-3y) = 3y + (-3y) - 26$$
$$-8 + 3y = -26$$
$$-8 + 8 + 3y = -26 + 8$$
$$3y = -18$$
$$\frac{3y}{3} = \frac{-18}{3}$$
$$y = -6$$

31.
$$-y + 7 = 14 + 2y - 6$$
$$-y + 7 = 8 + 2y$$
$$-y + (-2y) + 7 = 8 + 2y + (-2y)$$
$$-3y + 7 = 8$$
$$-3y + 7 + (-7) = 8 + (-7)$$
$$-3y = 1$$
$$\frac{-3y}{-3} = \frac{1}{-3}$$
$$y = -\frac{1}{3}$$

33.
$$-30 - 12y + 18 = -24y + 13 + 7y$$
$$-12 - 12y = -17y + 13$$
$$-12 - 12y + 17y = -17y + 17y + 13$$
$$-12 + 5y = 13$$
$$-12 + 12 + 5y = 13 + 12$$
$$5y = 25$$
$$\frac{5y}{5} = \frac{25}{5}$$
$$y = 5$$

35.
$$3(2x - 5) - 5x = 1$$
$$6x - 15 - 5x = 1$$
$$x - 15 = 1$$
$$x - 15 + 15 = 1 + 15$$
$$x = 16$$

37.
$$5(y - 2) = 2(2y + 3) - 16$$
$$5y - 10 = 4y + 6 - 16$$
$$5y - 10 = 4y - 10$$
$$5y + (-4y) - 10 = 4y + (-4y) - 10$$
$$y - 10 = -10$$
$$y - 10 + 10 = -10 + 10$$
$$y = 0$$

39.
$$8x + 4(4 - x) = 2x - 18$$
$$8x + 16 - 4x = 2x - 18$$
$$4x + 16 = 2x - 18$$
$$4x + (-2x) + 16 = 2x + (-2x) - 18$$
$$2x + 16 = -18$$
$$2x + 16 - 16 = -18 - 16$$
$$2x = -34$$
$$\frac{2x}{2} = \frac{-34}{2}$$
$$x = -17$$

41.
$$5x + 9 = \frac{1}{3}(3x - 6)$$
$$5x + 9 = x - 2$$
$$5x + (-x) + 9 = x + (-x) - 2$$
$$4x + 9 = -2$$
$$4x + 9 + (-9) = -2 + (-9)$$
$$4x = -11$$
$$\frac{4x}{4} = -\frac{11}{4}$$
$$x = -\frac{11}{4} \text{ or } -2\frac{3}{4}$$

43.
$$-2x - 5(x + 1) = -3(2x + 5)$$
$$-2x - 5x - 5 = -6x - 15$$
$$-7x - 5 = -6x - 15$$
$$-7x + 7x - 5 = -6x + 7x - 15$$
$$-5 = x - 15$$
$$-5 + 15 = x - 15 + 15$$
$$10 = x$$

Cumulative Review

45. $V = \dfrac{4\pi r^3}{3}$

$= \dfrac{4(3.14)(46 \text{ cm})^3}{3}$

$= \dfrac{4(3.14)(97{,}336 \text{ cu cm})}{3}$

$\approx 407{,}513.4 \text{ cu cm}$

The volume is $407{,}513.4 \text{ cm}^3$.

46. Area = Area of Square − Area of Circle

$= 6^2 - \pi(2)^2$

$= 36 - 3.14(4)$

≈ 23.4

The area is 23.4 square inches.

Quick Quiz 10.5

1.
$$0.7 - 0.3x = 0.9x - 4.1$$
$$0.7 - 0.3x + (-0.9x) = 0.9x + (-0.9x) - 4.1$$
$$0.7 - 1.2x = -4.1$$
$$0.7 + (-0.7) - 1.2x = -4.1 + (-0.7)$$
$$-1.2x = -4.8$$
$$\frac{-1.2x}{-1.2} = \frac{-4.8}{-1.2}$$
$$x = 4$$

2.
$$-6 - 2x + 9 = 12 + 3x + 11$$
$$-2x + 3 = 3x + 23$$
$$-2x + (-3x) + 3 = 3x + (-3x) + 23$$
$$-5x + 3 = 23$$
$$-5x + 3 + (-3) = 23 + (-3)$$
$$-5x = 20$$
$$\frac{-5x}{-5} = \frac{20}{-5}$$
$$x = -4$$

3.
$$-4(x + 6) + 9 = 12 + 3(x + 5)$$
$$-4x - 24 + 9 = 12 + 3x + 15$$
$$-4x - 15 = 27 + 3x$$
$$-4x + (-3x) - 15 = 27 + 3x + (-3x)$$
$$-7x - 15 = 27$$
$$-7x - 15 + 15 = 27 + 15$$
$$-7x = 42$$
$$\frac{-7x}{-7} = \frac{42}{-7}$$
$$x = -6$$

4. Answers may vary. Possible solution: Remove parentheses using the distributive property. Combine like terms on each side. Gather the variable terms on one side and the numbers on the other side. Divide both sides of the equation by the numerical coefficient of the variable term.

$$7x - 3(x - 6) = 2(x - 3) + 8$$
$$7x - 3x + 18 = 2x - 6 + 8$$
$$4x + 18 = 2x + 2$$
$$2x = -16$$
$$x = -8$$

How Am I Doing? Sections 10.1–10.5

1. $23x - 40x = -17x$

2. $-8y + 12y - 3y = y$

3. $6a - 5b - 9a + 7b = 6a - 9a - 5b + 7b$
         $= -3a + 2b$

4. $8y - 9 + x + 3y - 7x - 1$
 $= x - 7x + 8y + 3y - 9 - 1$
 $= -6x + 11y - 10$

5. $7x - 14 + 5y + 8 - 7y + 9x$
 $= 7x + 9x + 5y - 7y - 14 + 8$
 $= 16x - 2y - 6$

6. $4a - 7b + 3c - 5b = 4a - 7b - 5b + 3c$
         $= 4a - 12b + 3c$

7. $6(7x - 3y) = 6(7x) - 6(3y) = 42x - 18y$

8. $-4\left(\dfrac{1}{2}a - \dfrac{1}{4}b + 3\right) = -4\left(\dfrac{1}{2}a\right) - 4\left(-\dfrac{1}{4}b\right) - 4(3)$
         $= -2a + b - 12$

9. $-2(1.5a + 3b - 6c - 5)$
 $= (-2)(1.5a) + (-2)(3b) + (-2)(-6c) + (-2)(-5)$
 $= -3a - 6b + 12c + 10$

10. $4(x + 3y) - 2(5x - y)$
 $= 4(x) + 4(3y) + (-2)(5x) + (-2)(-y)$
 $= 4x + 12y - 10x + 2y$
 $= 4x - 10x + 12y + 2y$
 $= -6x + 14y$

11. $(9x + 4y)(-2) = 9x(-2) + 4y(-2) = -18x - 8y$

12. $(7x - 3y)(-3) = 7x(-3) + (-3y)(-3) = -21x + 9y$

13.
$$5 + x = 42$$
$$5 + (-5) + x = 42 + (-5)$$
$$x = 37$$

14.
$$x + 2.5 = 6$$
$$x + 2.5 + (-2.5) = 6 + (-2.5)$$
$$x = 3.5$$

15.
$$y + \frac{4}{5} = -\frac{3}{10}$$
$$y + \frac{4}{5} + \left(-\frac{4}{5}\right) = -\frac{3}{10} + \left(-\frac{4}{5}\right)$$
$$y = -\frac{3}{10} + \left(-\frac{8}{10}\right)$$
$$y = -\frac{11}{10} \text{ or } -1\frac{1}{10}$$

16.
$$-12 = -20 + x$$
$$-12 + 20 = -20 + 20 + x$$
$$8 = x$$

17.
$$-9y = -72$$
$$\frac{-9y}{-9} = \frac{-72}{-9}$$
$$y = 8$$

18.
$$2.7y = 27$$
$$\frac{2.7y}{2.7} = \frac{27}{2.7}$$
$$y = 10$$

19.
$$\frac{3}{5}x = \frac{9}{10}$$
$$\frac{5}{3}\left(\frac{3}{5}x\right) = \frac{5}{3}\left(\frac{9}{10}\right)$$
$$x = \frac{3}{2} \text{ or } 1\frac{1}{2}$$

20.
$$84 = -7x$$
$$\frac{84}{-7} = \frac{-7x}{-7}$$
$$-12 = x$$

21.
$$-7 + 6m = 25$$
$$-7 + 7 + 6m = 25 + 7$$
$$6m = 32$$
$$\frac{6m}{6} = \frac{32}{6}$$
$$m = \frac{16}{3} \text{ or } 5\frac{1}{3}$$

22.
$$12 - 5a = -8a + 15$$
$$12 - 5a + 8a = -8a + 8a + 15$$
$$12 + 3a = 15$$
$$-12 + 12 + 3a = -12 + 15$$
$$3a = 3$$
$$\frac{3a}{3} = \frac{3}{3}$$
$$a = 1$$

23.
$$5(x - 1) = 7 - 3(x - 4)$$
$$5x + 5(-1) = 7 + (-3)x + (-3)(-4)$$
$$5x - 5 = 7 - 3x + 12$$
$$5x - 5 = -3x + 19$$
$$5x - 5 + 5 = -3x + 19 + 5$$
$$5x = -3x + 24$$
$$5x + 3x = -3x + 3x + 24$$
$$8x = 24$$
$$\frac{8x}{8} = \frac{24}{8}$$
$$x = 3$$

24.
$$3x + 7 = 5(5 - x)$$
$$3x + 7 = 5(5) + 5(-x)$$
$$3x + 7 = 25 - 5x$$
$$3x + 7 + (-7) = 25 + (-7) - 5x$$
$$3x = 18 - 5x$$
$$3x + 5x = 18 - 5x + 5x$$
$$8x = 18$$
$$\frac{8x}{8} = \frac{18}{8}$$
$$x = \frac{9}{4} \text{ or } 2\frac{1}{4}$$

25.
$$5x - 18 = 2(x + 3)$$
$$5x - 18 = 2x + 2(3)$$
$$5x - 18 = 2x + 6$$
$$5x - 18 + 18 = 2x + 6 + 18$$
$$5x = 2x + 24$$
$$5x - 2x = 2x - 2x + 24$$
$$3x = 24$$
$$\frac{3x}{3} = \frac{24}{3}$$
$$x = 8$$

26.
$$8x - 5(x+2) = -3(x-5)$$
$$8x + (-5)x + (-5)(2) = (-3x) + (-3)(-5)$$
$$8x - 5x - 10 = -3x + 15$$
$$3x - 10 = -3x + 15$$
$$3x - 10 + 10 = -3x + 15 + 10$$
$$3x = -3x + 25$$
$$3x + 3x = -3x + 3x + 25$$
$$6x = 25$$
$$\frac{5x}{6} = \frac{25}{6}$$
$$x = \frac{25}{6} \text{ or } 4\frac{1}{6}$$

27.
$$12 + 4y - 7 = 6y - 9$$
$$4y + 5 = 6y - 9$$
$$4y + 5 + 9 = 6y - 9 + 9$$
$$4y + 14 = 6y$$
$$4y - 4y + 14 = 6y - 4y$$
$$14 = 2y$$
$$\frac{14}{2} = \frac{2y}{2}$$
$$7 = y$$

28.
$$0.3x + 0.4 = 0.7x - 1.2$$
$$0.3x + 0.4 + 1.2 = 0.7x - 1.2 + 1.2$$
$$0.3x + 1.6 = 0.7x$$
$$0.3x - 0.3x + 1.6 = 0.7x - 0.3x$$
$$1.6 = 0.4x$$
$$\frac{1.6}{0.4} = \frac{0.4x}{0.4}$$
$$4 = x$$

10.6 Exercises

1. $h = 34 + r$

3. $b = n - 107$

5. $n = a + 14$

7. $l = 2w + 7$

9. $l = 3w - 2$

11. $m = 3t + 10$

13. $t + l = 32$

15. $ht = 500$

17. p = airfare to Phoenix
$p + 135$ = airfare to San Diego

19. b = number of degrees in angle B
$b - 46$ = number of degrees in angle A

21. w = height of Willis Tower in feet
$w + 1267$ = height of Burj Khalifa in feet

23. a = number of books Aaron read
$2a$ = number of books Nina read
$a + 5$ = number of books Molly read

25. h = height; $h + 5$ = length; $3h$ = width

27. x = first angle; $2x$ = second angle;
$x - 14$ = third angle

Cumulative Review

29. $-6 - (-7)(2) = -6 - (-14) = -6 + 14 = 8$

30. $5 - 5 + 8 - (-4) + 2 - 15$
$= 5 + (-5) + 8 + 4 + 2 + (-15)$
$= 0 + 8 + 4 + 2 + (-15)$
$= 12 + 2 + (-15)$
$= 14 + (-15)$
$= -1$

31.
$$-2(3x+5) + 12 = 8$$
$$-6x - 10 + 12 = 8$$
$$-6x + 2 = 8$$
$$-6x + 2 + (-2) = 8 + (-2)$$
$$-6x = 6$$
$$\frac{-6x}{-6} = \frac{6}{-6}$$
$$x = -1$$

32.
$$3y - 4 - 5y = 10 - y$$
$$-4 - 2y = 10 - y$$
$$-4 + 4 - 2y = 10 + 4 - y$$
$$-2y = 14 - y$$
$$-2y + y = 14 - y + y$$
$$-y = 14$$
$$\frac{-y}{-1} = \frac{14}{-1}$$
$$y = -14$$

33. Korver: $\dfrac{59}{110} \approx 0.536$

Miller: $\dfrac{82}{0.480} \approx 171$

Gibson: $149 \times 0.477 \approx 71$

Quick Quiz 10.6

1. Charlie's truck, t, gets 12 miles per gallon *less* than his car, c.

   $t = c - 12$

2. w = width of the rectangle
   $2w + 3$ = the length of the rectangle

3. c = the number of compact cars in the college parking lot

   $0.5c$ or $\dfrac{1}{2}c$ or $\dfrac{c}{2}$ = the number of SUVs in the college parking lot

   $c + 35$ = the number of trucks in the college parking lot

4. Answers may vary. Possible solution: Both are right. In either case, it shows that 12 more students have part-time jobs than full-time jobs.

10.7 Exercises

1. x = length of shorter piece
   $x + 5.5$ = length of longer piece
   $$x + x + 5.5 = 16$$
   $$2x = 10.5$$
   $$\frac{2x}{2} = \frac{10.5}{2}$$
   $$x = 5.25$$
   shorter piece = 5.25 feet
   longer piece = 5.25 + 5.5 = 10.75 feet

3. x = number of points scored by France
   $x - 22$ = number of points scored by Japan
   $$x + x - 22 = 80$$
   $$2x - 22 = 80$$
   $$2x - 22 + 22 = 80 + 22$$
   $$2x = 102$$
   $$\frac{2x}{2} = \frac{102}{2}$$
   $$x = 51$$
   51 points scored by France
   51 − 22 = 29 points scored by Japan

5. x = number of cars in November
   $x + 84$ = number of cars in May
   $x - 43$ = number of cars in July
   $$x + x + 84 + x - 43 = 398$$
   $$3x + 41 = 398$$
   $$3x = 357$$
   $$x = 119$$

119 cars in November
119 + 84 = 203 cars in May
119 − 43 = 76 cars in July

In problems 7–21, the check is left for the student.

7. x = length of longer piece
   $x - 47$ = length of shorter piece
   $$x + x - 4.7 = 12$$
   $$2x - 4.7 = 12$$
   $$2x = 16.7$$
   $$\frac{2x}{2} = \frac{16.7}{2}$$
   $$x = 8.35$$
   longer piece = 8.35 feet
   shorter piece = 8.35 − 4.7 = 3.65 feet

9. x = width of board
   $2x - 4$ = length of board
   $$2(x) + 2(2x - 4) = 76$$
   $$2x + 4x - 8 = 76$$
   $$6x - 8 = 76$$
   $$6x = 84$$
   $$\frac{6x}{6} = \frac{84}{6}$$
   $$x = 14$$
   width = 14 inches
   length = 2(14) − 4 = 28 − 4 = 24 inches

11. x = length of the first side
    $x + 20$ = length of the second side
    $x - 4$ = length of the third side
    $$x + x + 20 + x - 4 = 199$$
    $$3x + 16 = 199$$
    $$3x = 183$$
    $$x = 61$$
    first side = 61 mm
    second side = 61 + 20 = 81 mm
    third side = 61 − 4 = 57 mm

13. x = length of the first side
    $2x$ = length of the second side
    $x + 12$ = length of the third side
    $$x + 2x + x + 12 = 64$$
    $$4x + 12 = 44$$
    $$4x = 32$$
    $$\frac{4x}{4} = \frac{32}{4}$$
    $$x = 8$$
    first side = 8 cm
    second side = 2(8) = 16 cm
    third side = 8 + 12 = 20 cm

15. x = number of degrees in angle A
$3x$ = number of degrees in angle B
$x + 40$ = number of degrees in angle C
$$x + 3x + x + 40 = 180$$
$$5x + 40 = 180$$
$$5x = 140$$
$$\frac{5x}{5} = \frac{140}{5}$$
$$x = 28$$
$3x = 3(28) = 84$
$x + 40 = 28 + 40 = 68$
Angle A measures = 28°; angle B
measures 84°; angle C measures = 68°

17. x = total sales
$0.05x$ = commission
$$0.05x + 1200 = 5000$$
$$0.05x = 3800$$
$$\frac{0.05x}{0.05} = \frac{3800}{0.05}$$
$$x = 76,000$$
total sales = \$76,000

19. x = yearly rent
$0.12x$ = commission
$$100 + 0.12x = 820$$
$$0.12x = 720$$
$$\frac{0.12x}{0.12} = \frac{720}{0.12}$$
$$x = 6000$$
yearly rent = \$6000

21. x = length of the adult section
$x + 6.2$ = length of the children section
$$x + x + 6.2 = 32$$
$$2x + 6.2 = 32$$
$$2x = 25.8$$
$$\frac{2x}{2} = \frac{25.8}{2}$$
$$x = 12.9$$
Adult section = 12.9 ft
Children section = 12.9 + 6.2 = 19.1 ft

23. x = number of pancreas transplants
$5x + 316$ = number of heart transplants
$17x - 123$ = number of liver transplants
$$x + 5x + 316 + 17x - 123 = 8910$$
$$23x + 193 = 8910$$
$$23x + 193 - 193 = 8910 - 193$$
$$23x = 8717$$
$$\frac{23x}{23} = \frac{8717}{23}$$
$$x = 379$$

$5x + 316 = 5(379) + 316 = 2211$
$17x - 123 = 17(379) - 123 = 6320$
2211 heart transplants
6320 liver transplants
379 pancreas transplants

25. x = value for customer service representatives
x = value for registered nurses
mean = 2323

$$2323 = \frac{4198 + 3396 + 2540 + 2293 + x + x + 929 + 978}{8}$$

$$2323 = \frac{14,334 + 2x}{8}$$

$$8 \cdot 2323 = 8 \cdot \frac{14,334 + 2x}{8}$$

$$18,584 = 14,334 + 2x$$

$$18,584 + (-14,334) = 14,334 + 2x + (-14,334)$$

$$4250 = 2x$$

$$\frac{4250}{2} = \frac{2x}{2}$$

$$2125 = x$$

Each occupation has 2125 thousand or 2,125,000.

Cumulative Review

26. What percent of 20 is 12?

$$\frac{n}{100} = \frac{12}{20}$$

$$n \times 20 = 100 \times 12$$

$$\frac{n \times 20}{20} = \frac{1200}{20}$$

$$n = 60 \text{ or } 60\%$$

27. 38% of what number is 190?

$$\frac{38}{100} = \frac{190}{b}$$

$$38 \times b = 100 \times 190$$

$$\frac{38 \times b}{38} = \frac{19,000}{38}$$

$$b = 500$$

28.

$$\frac{x}{12} = \frac{10}{15}$$

$$15x = 12(10)$$

$$\frac{15x}{15} = \frac{120}{15}$$

$$x = 8$$

29. $5 \text{ pounds} \times \left(\dfrac{16 \text{ ounces}}{1 \text{ pound}} \right) = 80 \text{ ounces}$

Quick Quiz 10.7

1. x = Barbara's earnings
    $x - 125$ = Melinda's earnings
    $$x + x - 125 = 437$$
    $$2x - 125 = 437$$
    $$2x - 125 + 125 = 437 + 125$$
    $$2x = 562$$
    $$\frac{2x}{2} = \frac{562}{2}$$
    $$x = 281$$
    $x - 15 = 156$
    Barbara earns $281 per week.
    Melinda earns $156 per week.

2. x = number of students who work full time
    $2x$ = number of students who work part time
    $2x - 1200$ = number of students who do not
    work
    $$x + 2x + 2x - 1200 = 6000$$
    $$5x - 1200 = 6000$$
    $$5x - 1200 + 1200 = 6000 + 1200$$
    $$5x = 7200$$
    $$\frac{5x}{5} = \frac{7200}{5}$$
    $$x = 1440$$
    $2x = 2880$
    $2x - 1200 = 2(1440) - 1200 = 1680$

3. x = width
    $2x + 7$ = length
    $$P = 2l + 2w$$
    $$176 = 2(2x + 7) + 2x$$
    $$176 = 4x + 14 + 2x$$
    $$176 = 6x + 14$$
    $$176 + (-14) = 6x + 14 + (-14)$$
    $$162 = 62$$
    $$\frac{162}{6} = \frac{6x}{6}$$
    $$27 = x$$
    $2x + 7 = 2(27) + 7 = 61$
    The width is 27 yards. The length is 61 yards.

4. Answers may vary. Possible solution:
    Let x = degrees of 2nd angle. State the degrees
    of the first and third angles in terms of x (degrees
    of 2nd angle). Since the sum of the interior
    angles of a triangle is 180°, write an equation.
    Solve the equation by getting all terms with x on
    one side of the equation and all numbers on the
    other. Then divide by the coefficient of the
    x-term.

Use Math to Save Money

1. Find 5% of $2500 for 1 month.
    Then multiply by 12 for one year's savings.
    $0.05 \times \$2500 = \$125$
    $12 \times \$125 = \$1500$
    Michael will have saved $1500.

2. Find 15% of $2500 for 1 month.
    Then multiply by 12 for one year's food cost.
    $0.15 \times \$2500 = \$375$
    $12 \times \$375 = \$4500$
    Michael will spend $4500 for food in one year.

3. $3000 + $450 = $3450
    The total cost for one semester is $3450.

4. Michael saves $125 each month and two
    semesters cost $2 \times \$3450 = \$6900$.
    $$\$6900 \times \frac{1 \text{ month}}{\$125} \approx 55 \text{ months}$$
    Michael will have to save for about 55 months or
    4 years and 7 months to pay for one year.

5. 10% of $2500 = 0.10 \times \$2500 = \$250$
    $$\$6900 \times \frac{1 \text{ month}}{\$250} \approx 28 \text{ months}$$
    If he increases savings to 10%, it will take about
    28 months or 2 years and 4 months to pay for
    one year.

6. After 5 years, his monthly salary will be
    $$\$2500 + \frac{\$5800}{12} \approx \$2500 + \$483.33 = \$2983.33.$$
    If he saves 5% of this, then he saves
    $0.05 \times \$2983.33 \approx \$149.17$ each month.

7. 20% of $2983.33 = 0.20 \times \$2983.33 \approx \$596.67$
    Michael will have $596.67 for miscellaneous
    spending each month.

8. Answers will vary.

9. Answers will vary.

10. Answers will vary.

You Try It

1. **a.** $10a - a + 3a = 9a + 3a = 12a$

    b. $8a + 10b - 5a - 12b = 8a - 5a + 10b - 12b$
    $$= 3a - 2b$$

c. $18x - 6 - 7y + x = 18x + x - 7y - 6$
$$= 19x - 7y - 6$$

2. a. $3(a - 6b) = 3 \cdot a + 3(-6b) = 3a - 18b$

b. $-4(2x - 3y - 1)$
$$= -4(2x) + (-4)(-3y) + (-4)(-1)$$
$$= -8x + 12y + 4$$

c. $(5a - b)8 = 5a(8) - b(8) = 40a - 8b$

3. $-3(4x - 5y) - (x + 3y)$
$$= -3(4x) + (-3)(-5y) - x - 3y$$
$$= -12x + 15y - x - 3y$$
$$= -12x - x + 15y - 3y$$
$$= -13x + 12y$$

4. $x + 4.2 = 9$
$$x + 4.2 + (-4.2) = 9 + (-4.2)$$
$$x = 4.8$$
Check: $x + 4.2 = 9$
$$4.8 + 4.2 \stackrel{?}{=} 9$$
$$9 = 9 ✓$$

5. $-15x = 90$
$$\frac{-15x}{-15} = \frac{90}{-15}$$
$$x = -6$$
Check: $-15x = 90$
$$-15(-6) \stackrel{?}{=} 90$$
$$90 = 90 ✓$$

6. $-\frac{2}{3}x = \frac{5}{9}$
$$-\frac{3}{2} \cdot \left(-\frac{2}{3}x\right) = -\frac{3}{2} \cdot \frac{5}{9}$$
$$x = -\frac{5}{6}$$
Check: $-\frac{2}{3}x = \frac{5}{9}$
$$-\frac{2}{3}\left(-\frac{5}{6}\right) \stackrel{?}{=} \frac{5}{9}$$
$$\frac{5}{9} = \frac{5}{9} ✓$$

7. $7(2x - 5) - (8x + 1) = -5(x - 2) - 13$
$$14x - 35 - 8x - 1 = -5x + 10 - 13$$
$$6x - 36 = -5x - 3$$
$$6x + 5x - 36 = -5x + 5x - 3$$
$$11x - 36 = -3$$
$$11x - 36 + 36 = -3 + 36$$
$$11x = 33$$
$$\frac{11x}{11} = \frac{33}{11}$$
$$x = 3$$
Check: $7(2x - 5) - (8x + 1) = -5(x - 2) - 13$
$$7(2 \cdot 3 - 5) - (8 \cdot 3 + 1) \stackrel{?}{=} -5(3 - 2) - 13$$
$$7(1) - 25 \stackrel{?}{=} -5(1) - 13$$
$$-18 = -18 ✓$$

8. Ana's age is 32 less than her father's age.
$$a = f - 32$$

9. $x =$ the number of degrees in the second angle
$x + 12 =$ the number of degrees in the first angle
$x - 18 =$ the number of degrees in the third angle

10. $x =$ the number of students on the first floor
$2x =$ the number of students on the second floor
$x + 32 =$ the number of students on the third floor
$$x + 2x + x + 32 = 260$$
$$4x + 32 = 260$$
$$4x + 32 + (-32) = 260 + (-32)$$
$$4x = 228$$
$$\frac{4x}{4} = \frac{228}{4}$$
$$x = 57$$
$2x = 2(57) = 114$
$x + 32 = 57 + 32 = 89$
There are 57 students on the first floor, 114 on the second floor, and 89 on the third floor.

Chapter 10 Review Problems

1. $-8a + 6 - 5a - 3 = -8a - 5a + 6 - 3 = -13a + 3$

2. $\frac{1}{3}x + \frac{1}{3} + \frac{5}{9} + \frac{1}{2}x = \frac{1}{3}x + \frac{1}{2}x + \frac{1}{3} + \frac{5}{9}$
$$= \frac{2}{6}x + \frac{3}{6}x + \frac{3}{9} + \frac{5}{9}$$
$$= \frac{5}{6}x + \frac{8}{9}$$

3. $5x + 2y - 7x - 9y = 5x - 7x + 2y - 9y$
$$= -2x - 7y$$

4. $3x - 7y + 8x + 2y = 3x + 8x - 7y + 2y$
$= 11x - 5y$

5. $5x - 9y - 12 - 6x - 3y + 18$
$= 5x - 6x - 9y - 3y - 12 + 18$
$= -x - 12y + 6$

6. $8a - 11b + 15 - b + 5a - 19$
$= 8a + 5a - 11b - b + 15 - 19$
$= 13a - 12b - 4$

7. $-3(5x + y) = -3(5x) + (-3)(y) = -15x - 3y$

8. $-4(2x + 3y) = -4(2x) + (-4)(3y) = -8x - 12y$

9. $2(x - 3y + 4) = 2(x) + 2(-3y) + 2(4)$
$= 2x - 6y + 8$

10. $5(6a - 8b + 5) = 5(6a) - 5(8b) + 5(5)$
$= 30a - 40b + 25$

11. $-12\left(\dfrac{3}{4}a - \dfrac{1}{6}b - 1\right)$
$= (-12)\left(\dfrac{3}{4}a\right) + (-12)\left(-\dfrac{1}{6}b\right) + (-12)(-1)$
$= -9a + 2b + 12$

12. $5(1.2x + 3y - 5.5) = 5(1.2x) + 5(3y) + 5(-5.5)$
$= 6x + 15y - 27.5$

13. $2(x + 3y) - 4(x - 2y) = 2x + 6y - 4x + 8y$
$= -2x + 14y$

14. $2(5x - y) - 3(x + 2y) = 10x - 2y - 3x - 6y$
$= 7x - 8y$

15. $-2(a + b) - 3(2a + 8) = -2a - 2b - 6a - 24$
$= -8a - 2b - 24$

16. $-4(a - 2b) + 3(5 - a) = -4a + 8b + 15 - 3a$
$= -7a + 8b + 15y$

17. $x - 3 = 9$
$x + (-3) + 3 = 9 + 3$
$x = 12$

18. $x + 8.3 = 20$
$x + 8.3 + (-8.3) = 20 + (-8.3)$
$x = 11.7$

19. $-8 = x - 12$
$-8 + 12 = x - 12 + 12$
$4 = x$

20. $2.4 = x - 5$
$2.4 + 5 = x - 5 + 5$
$7.4 = x$

21. $3.1 + x = -9$
$3.1 + (-3.1) + x = -9 + (-3.1)$
$x = -12.1$

22. $x + \dfrac{1}{2} = 3\dfrac{3}{4}$
$x + \dfrac{1}{2} + \left(-\dfrac{1}{2}\right) = 3\dfrac{3}{4} + \left(-\dfrac{1}{2}\right)$
$x = 3\dfrac{3}{4} + \left(-\dfrac{2}{4}\right)$
$x = 3\dfrac{1}{4}$ or $\dfrac{13}{4}$

23. $y + \dfrac{5}{8} = -\dfrac{1}{8}$
$y + \dfrac{5}{8} + \left(-\dfrac{5}{8}\right) = -\dfrac{1}{8} + \left(-\dfrac{5}{8}\right)$
$y = -\dfrac{6}{8}$
$y = -\dfrac{3}{4}$

24. $2x + 20 = 25 + x$
$2x + (-x) + 20 = 25 + x + (-x)$
$x + 20 = 25$
$x + 20 + (-20) = 25 + (-20)$
$x = 5$

25. $7y + 12 = 8y + 3$
$7y + 12 + (-7y) = 8y + 3 + (-7y)$
$12 = y + 3$
$12 + (-3) = y + 3 + (-3)$
$9 = y$

26. $8x = -20$
$\dfrac{8x}{8} = \dfrac{-20}{8}$
$x = \dfrac{-20}{8}$
$x = -\dfrac{5}{2}$ or $-2\dfrac{1}{2}$

27. $-12y = 60$

$$\frac{-12y}{-12} = \frac{60}{-12}$$

$$y = -5$$

28. $1.5x = 9$

$$\frac{1.5x}{1.5} = \frac{9}{1.5}$$

$$x = 6$$

29. $-1.4y = -12.6$

$$\frac{-1.4y}{-1.4} = \frac{-12.6}{-1.4}$$

$$y = 9$$

30. $\dfrac{3}{4}x = 6$

$$\frac{4}{3} \cdot \frac{3}{4}x = \frac{4}{3} \cdot 6$$

$$x = 8$$

31. $\dfrac{2}{9}x = \dfrac{5}{18}$

$$\frac{9}{2} \cdot \frac{2}{9}x = \frac{9}{2} \cdot \frac{5}{18}$$

$$x = \frac{5}{4} \text{ or } 1\frac{1}{4}$$

32. $5x - 3 = 27$

$$5x - 3 + 3 = 27 + 3$$

$$5x = 30$$

$$\frac{5x}{5} = \frac{30}{5}$$

$$x = 6$$

33. $8x - 5 = 19$

$$8x - 5 + 5 = 19 + 5$$

$$8x = 24$$

$$\frac{8x}{8} = \frac{24}{8}$$

$$x = 3$$

34. $10 - x = -3x - 6$

$$10 + (-10) - x = -3x - 6 + (-10)$$

$$-x = -3x - 16$$

$$-x + 3x = -3x + 3x - 16$$

$$2x = \frac{-16}{2}$$

$$x = -8$$

35. $9x - 3x + 18 = 36$

$$6x + 18 = 36$$

$$6x + 18 + (-18) = 36 + (-18)$$

$$6x = 18$$

$$\frac{6x}{6} = \frac{18}{6}$$

$$x = 3$$

36. $4 + 3x - 8 = 12 + 5x + 4$

$$3x - 4 = 5x + 16$$

$$3x + (-3x) - 4 = 5x + (-3x) + 16$$

$$-4 = 2x + 16$$

$$-4 + (-16) = 2x + 16 + (-16)$$

$$-20 = 2x$$

$$\frac{-20}{2} = \frac{2x}{2}$$

$$-10 = x$$

37. $-2(3x + 5) = 4x + 8 - x$

$$-6x - 10 = 3x + 8$$

$$-6x + (-3x) - 10 = 3x + 8 + (-3x)$$

$$-9x - 10 = 8$$

$$-9x - 10 + 10 = 8 + 10$$

$$-9x = 18$$

$$\frac{-9x}{-9} = \frac{18}{-9}$$

$$x = -2$$

38. $2(3x - 4) = 7 - 2x + 5x$

$$6x - 8 = 7 + 3x$$

$$6x + (-3x) - 8 = 7 + 3x + (-3x)$$

$$3x - 8 = 7$$

$$3x - 8 + 8 = 7 + 8$$

$$3x = 15$$

$$\frac{3x}{3} = \frac{15}{3}$$

$$x = 5$$

39. $5 - (y + 7) = 10 + 3(y - 4)$

$$5 - y - 7 = 10 + 3y - 12$$

$$-y - 2 = -2 + 3y$$

$$-y - 2 + (-3y) = -2 + 3y + (-3y)$$

$$-4y - 2 = -2$$

$$-4y - 2 + 2 = -2 + 2$$

$$-4y = 0$$

$$\frac{-4y}{-4} = \frac{0}{4}$$

$$y = 0$$

40. $w = c + 3000$

41. $e = 12 + a$

42. $A = 3B$

43. $l = 2w - 3$

44. r = Roberto's salary
$r + 2050$ = Michael's salary

45. x = length of the first side
$2x$ = length of the second side

46. d = number of days Dennis worked
$2d + 12$ = number of days Carmen worked

47. n = number of nonfiction books sold
$n + 225$ = number of fiction books sold

48. x = length of shorter piece
$x + 6.5$ = length of longer piece
$$x + x + 6.5 = 60$$
$$2x + 6.5 = 60$$
$$2x + 6.5 + (-6.5) = 60 + (-6.5)$$
$$2x = 53.5$$
$$\frac{2x}{2} = \frac{53.5}{2}$$
$$x = 26.75$$
shorter piece = 26.75 feet
longer piece = 26.75 + 6.5 = 33.25 feet

49. x = experienced employee's salary
$x - 28$ = new employee's salary
$$x + x - 28 = 412$$
$$2x - 28 = 412$$
$$2x - 28 + 28 = 412 + 28$$
$$2x = 440$$
$$\frac{2x}{2} = \frac{440}{2}$$
$$x = 220$$
experienced employee = \$220
new employee = 220 − 28 = \$192

50. x = number of customers in February
$2x$ = number of customers in March
$x + 3000$ = number of customers in April
$$x + 2x + x + 3000 = 45,200$$
$$4x + 3000 = 45,200$$
$$4x + 3000 + (-3000) = 45,200 + (-3000)$$
$$4x = 42,200$$
$$\frac{4x}{4} = \frac{42,200}{4}$$
$$x = 10,550$$
February = 10,550
March = 2(10,550) = 21,100
April = 10,550 + 3000 = 13,550

51. x = miles on Friday
$2x$ = miles on Saturday
$x + 30$ = miles on Sunday
$$x + 2x + x + 30 = 670$$
$$4x + 30 = 670$$
$$4x + 30 + (-30) = 670 + (-30)$$
$$4x = 640$$
$$\frac{4x}{4} = \frac{640}{4}$$
$$x = 160$$
$2x = 2(160) = 320$
$x + 30 = 160 + 30 = 190$
Friday = 160 mi
Saturday = 320 mi
Sunday = 190 mi

52. w = width
$2w - 3$ = length
$$72 = 2w + 2(2w - 3)$$
$$72 = 2w + 4w - 6$$
$$72 = 6w - 6$$
$$72 + 6 = 6w + (-6) + 6$$
$$78 = 6w$$
$$\frac{78}{6} = \frac{6w}{6}$$
$$w = 13$$
width = 13 in.
length = 2(13) − 3 = 23 in.

53. x = measure of angle Z
$2x$ = measure of angle Y
$x - 12$ = measure of angle X
$$x + 2x + x - 12 = 180$$
$$4x - 12 = 180$$
$$4x - 12 + 12 = 180 + 12$$
$$4x = 192$$
$$\frac{4x}{4} = \frac{192}{4}$$
$$x = 48$$
Angle Z measures 48°
Angle Y measures 2(48°) = 96°
Angle X measures 48° − 12° = 36°

54. $p = 2l + 2w$

$x = \text{length}$

$x - 67 = \text{width}$

$$346 = 2(x) + 2(x - 67)$$
$$346 = 2x + 2x + 2(-67)$$
$$346 = 4x - 134$$
$$346 + 134 = 4x - 134 + 134$$
$$480 = 4x$$
$$\frac{480}{4} = \frac{4x}{4}$$
$$120 = x$$

Length = 120 yards

Width = 120 − 67 = 53 yards

55. $x = \text{miles an Saturday}$

$x + 106 = \text{miles on Sunday}$

$$x + x + 106 = 810$$
$$2x + 106 = 810$$
$$2x + 106 + (-106) = 810 + (-106)$$
$$2x = 704$$
$$\frac{2x}{2} = \frac{704}{2}$$
$$x = 352$$

$x + 106 = 352 + 106 = 458$

Saturday = 352 mi

Sunday = 458 mi

56. $x = \text{first week}$

$x + 156 = \text{second week}$

$x - 142 = \text{third week}$

$$x + x + 156 + x - 142 = 800$$
$$3x + 14 = 800$$
$$3x + 14 + (-14) = 800 + (-14)$$
$$3x = 786$$
$$\frac{3x}{3} = \frac{786}{3}$$
$$x = 262$$

first week = 262

second week = 262 + 156 = 418

third week = 262 − 142 = 120

57. $x = \text{cost of the furniture}$

$0.08x = \text{commission}$

$$0.08x + 1500 = 3050$$
$$0.08x + 1500 + (-1500) = 3050 + (-1500)$$
$$0.08x = 1550$$
$$\frac{0.08x}{0.08} = \frac{1550}{0.08}$$
$$x = 19{,}375$$

The cost of the furniture was $19,375.

How Am I Doing? Chapter 10 Test

1. $5a - 11a = -6a$

2. $\dfrac{1}{3}x + \dfrac{5}{8}y - \dfrac{1}{5}x + \dfrac{1}{2}y = \dfrac{1}{3}x - \dfrac{1}{5}x + \dfrac{5}{8}y + \dfrac{1}{2}y$

$\qquad\qquad\qquad\qquad = \dfrac{5}{15}x - \dfrac{3}{15}x + \dfrac{5}{8}y + \dfrac{4}{8}y$

$\qquad\qquad\qquad\qquad = \dfrac{2}{15}x + \dfrac{9}{8}y$

3. $\dfrac{1}{4}a - \dfrac{2}{3}b + \dfrac{3}{8}a = \dfrac{2}{8}a + \dfrac{3}{8}a - \dfrac{2}{3}b = \dfrac{5}{8}a - \dfrac{2}{3}b$

4. $6a - 5b - 5a - 3b = 6a - 5a - 5b - 3b = a - 8b$

5. $7x - 8y + 2z - 9z + 8y = 7x - 8y + 8y + 2z - 9z$

$\qquad\qquad\qquad\qquad\qquad = 7x - 7z$

6. $x + 5y - 6 - 5x - 7y + 11$

$\quad = x - 5x + 5y - 7y - 6 + 11$

$\quad = -4x - 2y + 5$

7. $5(12x - 5y) = 5(12x) + 5(-5y) = 60x - 25y$

8. $4\left(\dfrac{1}{2}x - \dfrac{5}{6}y\right) = 4\left(\dfrac{1}{2}x\right) - 4\left(\dfrac{5}{6}y\right) = 2x - \dfrac{10}{3}y$

9. $-1.5(3a - 2b + c - 8)$

$\quad = (-1.5)(3a) + (-1.5)(-2b) + (-1.5)(c)$

$\qquad\quad + (-1.5)(-8)$

$\quad = -4.5a + 3b - 1.5c + 12$

10. $2(-3a + 2b) - 5(a - 2b) = -6a + 4b - 5a + 10b$

$\qquad\qquad\qquad\qquad\qquad = -11a + 14b$

11. $\qquad -5 - 3x = 19$

$\qquad -5 + 5 - 3x = 19 + 5$

$\qquad\qquad -3x = 24$

$\qquad\qquad \dfrac{-3x}{-3} = \dfrac{24}{-3}$

$\qquad\qquad\qquad x = -8$

12. $\qquad x - 3.45 = -9.8$

$\qquad x - 3.45 + 3.45 = -9.8 + 3.45$

$\qquad\qquad\qquad x = -6.35$

13. $\qquad -5x + 9 = -4x - 6$

$\qquad -5x + 5x + 9 = -4x + 5x - 6$

$\qquad\qquad\qquad 9 = x - 6$

$\qquad\qquad 9 + 6 = x - 6 + 6$

$\qquad\qquad\qquad 15 = x$

14.
$$8x - 2 - x = 3x - 9 - 10x$$
$$7x - 2 = -7x - 9$$
$$7x + 7x - 2 = -7x + 7x - 9$$
$$14x - 2 = -9$$
$$14x - 2 + 2 = -9 + 2$$
$$14x = -7$$
$$\frac{14x}{14} = \frac{-7}{14}$$
$$x = -\frac{1}{2}$$

15.
$$0.5x + 0.6 = 0.2x - 0.9$$
$$0.5x + 0.6 + (-0.6) = 0.2x - 0.9 + (-0.6)$$
$$0.5x = 0.2x - 1.5$$
$$0.5x - 0.2x = 0.2x - 0.2x - 1.5$$
$$0.3x = -1.5$$
$$\frac{0.3x}{0.3} = \frac{-1.5}{0.3}$$
$$x = -5$$

16.
$$-\frac{5}{6}x = \frac{7}{12}$$
$$\left(-\frac{6}{5}\right)\left(-\frac{5}{6}\right)x = \left(-\frac{6}{5}\right)\left(\frac{7}{12}\right)$$
$$x = -\frac{7}{10}$$

17. $s = f + 15$

18. $n = s - 15{,}000$

19. $\frac{1}{2}s$ = measure of the first angle

s = measure of the second angle

$2s$ = measure of the third angle

20. w = width; $2w - 5$ = length

21. x = acres in Prentice farm

$3x$ = acres in Smithfield farm
$$x + 3x = 348$$
$$4x = 348$$
$$\frac{4x}{4} = \frac{348}{4}$$
$$x = 87$$
Prentice farm = 87 acres

Smithfield farm = 3(87) = 261 acres

22. x = Marcia's earnings

$x - 1500$ = Sam's earnings
$$x + x - 1500 = 46{,}500$$
$$2x - 1500 = 46{,}500$$
$$2x - 1500 + 1500 = 46{,}500 + 1500$$
$$2x = 48{,}000$$
$$\frac{2x}{2} = \frac{48{,}000}{2}$$
$$x = 24{,}000$$
Marcia = \$24,000

Sam = 24,000 − 1500 = \$22,500

23. x = number of afternoon students

$x - 24$ = number of morning students

$x + 12$ = number of evening students
$$x + x - 24 + x + 12 = 183$$
$$3x - 12 = 183$$
$$3x - 12 + 12 = 183 + 12$$
$$3x = 195$$
$$\frac{3x}{3} = \frac{195}{3}$$
$$x = 65$$
Number of afternoon students = 65

Number of morning students = 65 − 24 = 41

Number of evening students = 65 + 12 = 77

24. x = length

$\frac{1}{2}x + 8$ = width

$$2(x) + 2\left(\frac{1}{2}x + 8\right) = 118$$
$$2x + x + 16 = 118$$
$$3x + 16 = 118$$
$$3x + 16 + (-16) = 118 + (-16)$$
$$3x = 102$$
$$\frac{3x}{3} = \frac{102}{3}$$
$$x = 34$$
length = 34 feet

width $= \frac{1}{2}(34) + 8 = 25$ feet

Practice Final Examination

1. 82,367 = Eighty-two thousand, three hundred sixty-seven

2.
$$
\begin{array}{r}
13,428 \\
+\,16,905 \\
\hline
30,333
\end{array}
$$

3.
$$
\begin{array}{r}
19 \\
23 \\
16 \\
45 \\
+\,70 \\
\hline
173
\end{array}
$$

4.
$$
\begin{array}{r}
89,071 \\
-\,54,968 \\
\hline
34,103
\end{array}
$$

5.
$$
\begin{array}{r}
78 \\
\times\,54 \\
\hline
312 \\
390 \\
\hline
4212
\end{array}
$$

6.
$$
\begin{array}{r}
2035 \\
\times\,107 \\
\hline
14\,245 \\
203\,50 \\
\hline
217,745
\end{array}
$$

7.
$$
\begin{array}{r}
158 \\
7\overline{)1106} \\
7 \\
\hline
40 \\
35 \\
\hline
56 \\
56 \\
\hline
0
\end{array}
$$

8.
$$
\begin{array}{r}
606 \\
26\overline{)15,756} \\
15\,6 \\
\hline
156 \\
156 \\
\hline
0
\end{array}
$$

9. $3^4 + 20 \div 4 \times 2 + 5^2 = 81 + 10 + 25 = 116$

10. $512 \div 16 = 32$
The car achieved 32 miles/gallon.

11. $\dfrac{14}{30} = \dfrac{14 \div 2}{30 \div 2} = \dfrac{7}{15}$

12. $3\dfrac{9}{11} = \dfrac{3 \times 11 + 9}{11} = \dfrac{42}{11}$

13. $\dfrac{1}{10} + \dfrac{3}{4} + \dfrac{4}{5} = \dfrac{1}{10} \times \dfrac{2}{2} + \dfrac{3}{4} \times \dfrac{5}{5} + \dfrac{4}{5} \times \dfrac{4}{4}$
$$= \dfrac{2}{20} + \dfrac{15}{20} + \dfrac{16}{20}$$
$$= \dfrac{33}{20}$$
$$= 1\dfrac{13}{20}$$

14. $2\dfrac{1}{3} + 3\dfrac{3}{5} = 2\dfrac{5}{15} + 3\dfrac{9}{15} = 5\dfrac{14}{15}$

15.
$$
\begin{array}{r}
4\dfrac{5}{7} \\
-\,2\dfrac{1}{2} \\
\hline
\end{array}
\qquad
\begin{array}{r}
4\dfrac{10}{14} \\
-\,2\dfrac{7}{14} \\
\hline
2\dfrac{3}{14}
\end{array}
$$

16. $1\dfrac{1}{4} \times 3\dfrac{1}{5} = \dfrac{5}{4} \times \dfrac{16}{5} = \dfrac{5 \times 4 \times 4}{4 \times 5} = 4$

17. $\dfrac{7}{9} \div \dfrac{5}{18} = \dfrac{7}{9} \times \dfrac{18}{5} = \dfrac{14}{5}$ or $2\dfrac{4}{5}$

18. $\dfrac{5\frac{1}{2}}{3\frac{1}{4}} = \dfrac{\frac{11}{2}}{\frac{13}{4}} = \dfrac{11}{2} \times \dfrac{4}{13} = \dfrac{22}{13}$ or $1\dfrac{9}{13}$

19. $1\dfrac{1}{2} + 3\dfrac{1}{4} + 2\dfrac{1}{10} = 1\dfrac{10}{20} + 3\dfrac{5}{20} + 2\dfrac{2}{20}$

She jogged $6\dfrac{17}{20}$ miles.

20. $11\dfrac{2}{3} \div 2\dfrac{1}{3} = \dfrac{35}{3} \div \dfrac{7}{3} = \dfrac{35}{3} \times \dfrac{3}{7} = 5$
5 packages can be made.

21. $\dfrac{719}{1000} = 0.719$

22. $0.86 = \dfrac{86}{100} = \dfrac{43}{50}$

23. $0.315 > 0.309$

24. 506.3782 rounds to 506.38.

25.
$$\begin{array}{r} 9.6 \\ 3.82 \\ 1.05 \\ + 7.3 \\ \hline 21.77 \end{array}$$

26.
$$\begin{array}{r} 3.610 \\ - 2.853 \\ \hline 0.757 \end{array}$$

27.
$$\begin{array}{r} 1.23 \\ \times\ 0.4 \\ \hline 0.492 \end{array}$$

28.

$$\begin{array}{r} 3.69 \\ 0.24_\wedge \overline{)0.88_\wedge 56} \\ \underline{72\ \ } \\ 16\ 5 \\ \underline{14\ 4} \\ 2\ 16 \\ \underline{2\ 16} \\ 0 \end{array}$$

29.

$$\begin{array}{r} 0.8125 \\ 16\overline{)13.0000} \\ \underline{12\ 8\ \ } \\ 20 \\ \underline{16} \\ 40 \\ \underline{32} \\ 80 \\ \underline{80} \\ 0 \end{array}$$

$\dfrac{13}{16} = 0.8125$

30. $0.7 + (0.2)^3 - 0.08(0.03) = 0.7 + 0.008 - 0.0024$
$$= 0.708 - 0.0024$$
$$= 0.7056$$

31. $\dfrac{7000}{215} = \dfrac{7000 \div 5}{215 \div 5} = \dfrac{1400 \text{ students}}{43 \text{ faculty}}$

32.
$$\dfrac{12}{15} \overset{?}{=} \dfrac{17}{21}$$
$$12 \times 21 \overset{?}{=} 15 \times 17$$
$$252 \neq 255$$
No, it is not a proportion.

33.
$$\dfrac{5}{9} = \dfrac{n}{17}$$
$$5 \times 17 = 9 \times n$$
$$\dfrac{85}{9} = \dfrac{9 \times n}{9}$$
$$n \approx 9.4$$

34.
$$\dfrac{3}{n} = \dfrac{7}{18}$$
$$3 \times 18 = 7 \times n$$
$$\dfrac{54}{7} = \dfrac{7 \times n}{7}$$
$$7.7 \approx n$$

35.
$$\dfrac{n}{12} = \dfrac{5}{4}$$
$$n \times 4 = 12 \times 5$$
$$\dfrac{n \times 4}{4} = \dfrac{60}{4}$$
$$n = 15$$

36.
$$\dfrac{n}{7} = \dfrac{36}{28}$$
$$n \times 28 = 7 \times 36$$
$$\dfrac{n \times 28}{28} = \dfrac{252}{28}$$
$$n = 9$$

37.
$$\dfrac{2000}{3} = \dfrac{n}{5}$$
$$2000 \times 5 = 3 \times n$$
$$\dfrac{10,000}{3} = \dfrac{3 \times n}{3}$$
$$n \approx 3333.33$$
He would earn \$3333.33.

38.
$$\frac{200}{6} = \frac{325}{n}$$
$$200 \times n = 6 \times 325$$
$$\frac{200 \times n}{200} = \frac{1950}{200}$$
$$n = 9.75$$
They will appear 9.75 inches apart.

39.
$$\frac{115}{12} = \frac{6000}{n}$$
$$115 \times n = 12 \times 6000$$
$$\frac{115 \times n}{115} = \frac{72,000}{115}$$
$$n \approx \$626.09$$
$626.09 will be withheld.

40.
$$\frac{18}{1.2} = \frac{24}{n}$$
$$18 \times n = 1.2 \times 24$$
$$\frac{18 \times n}{18} = \frac{28.8}{18}$$
$$n = 1.6$$
She needs 1.6 lb of butter.

41. $0.0063 = 0.63\%$

42.
$$\frac{17}{80} = \frac{n}{100}$$
$$17 \times 100 = 80 \times n$$
$$\frac{1700}{80} = \frac{80 \times n}{80}$$
$$21.25 = n$$
21.25%

43. $164\% = 1.64$

44.
$$300 \times n = 52$$
$$\frac{300 \times n}{300} = \frac{52}{300}$$
$$n \approx 0.1733 = 17.33\%$$

45. 6.3% of 4800
$$6.3\% \times 4800 = n$$
$$0.063 \times 4800 = n$$
$$n = 302.4$$

46.
$$\frac{58}{100} = \frac{145}{b}$$
$$58 \times b = 100 \times 145$$
$$\frac{58 \times b}{58} = \frac{14,500}{58}$$
$$b = 250$$

47. 126% of 3400
$$126\% \times 3400 = n$$
$$1.26 \times 3400 = n$$
$$n = 4284$$

48. $18,800 - 0.08(18,800) = 18,800 - 1504 = 17,296$
She paid $17,296 for the car.

49.
$$\frac{28}{100} = \frac{1260}{b}$$
$$28 \times b = 100 \times 1260$$
$$\frac{28 \times b}{28} = \frac{126,000}{28}$$
$$b = 4500$$
4500 students are in the student body.

50. Difference $= 11.28 - 8.40 = 2.88$
$$\text{percent} = \frac{2.88}{8.40} \approx 0.343 = 34.3\%$$
The percent increase is 34.3%.

51. $17 \text{ qt} \times \dfrac{1 \text{ gallon}}{4 \text{ quarts}} = 4.25 \text{ gallons}$

52. $3.25 \text{ tons} \times \dfrac{2000 \text{ lb}}{1 \text{ ton}} = 6500 \text{ lb}$

53. $16 \text{ ft} \times \dfrac{12 \text{ in.}}{1 \text{ ft}} = 192 \text{ in.}$

54. $5.6 \text{ km} = 5600 \text{ m}$

55. $6.98 \text{ g} = 0.0698 \text{ kg}$

56. $2.48 \text{ ml} = 0.00248 \text{ L}$

57. $12 \text{ mi} \times \dfrac{1.61 \text{ km}}{1 \text{ mi}} = 19.32 \text{ km}$

58. $9.62 \text{ cm} = 0.0962 \text{ m}$

59. $3 \text{ mi} \times \dfrac{5280 \text{ ft}}{1 \text{ mi}} = 15,840 \text{ ft}$

60. $0.623 \text{ cm} + 0.74 \text{ cm} + 0.0428 \text{ cm}$
$= 0.623 \text{ cm} + 0.74 \text{ cm} + 0.00428 \text{ cm}$
$= 1.36728 \text{ cm}$
The total thickness is 1.36728 cm.

61. $P = 2l + 2w = 2(6) + 2(1.2) = 12 + 2.4 = 14.4 \text{ m}$

62. $P = 82 + 13 + 98 + 13 = 206 \text{ cm}$

63. $A = \dfrac{bh}{2} = \dfrac{6(1.8)}{2} = 5.4 \text{ ft}^2$

64. $A = \dfrac{h(b+B)}{2} = \dfrac{7.5(8+12)}{2} = \dfrac{7.5(20)}{2} = 75 \text{ m}^2$

65. $A = \pi r^2 = 3.14(6)^2 \approx 113.04 \text{ m}^2$

66. $C = \pi d$
$C = 3.14(18) = 56.52 \text{ m}$

67. $V = \dfrac{\pi r^2 h}{3} = \dfrac{\pi(4)^2(10)}{3} \approx 167.46 \text{ cm}^3$

68. $V = \dfrac{12(19)(2.7)}{3} = 205.2$

205.2 ft^3

69. Total area = Area of square + Area of triangle
$$= s^2 + \dfrac{bh}{2}$$
$$= (5)^2 + \dfrac{3(5)}{2}$$
$$= 32.5 \text{ m}^2$$

70. $\dfrac{n}{130} = \dfrac{30}{120}$
$n \times 120 = 30 \times 130$
$\dfrac{n \times 120}{120} = \dfrac{3900}{120}$
$n = 32.5$

71. The profit was 8 million dollars.

72. $9 - 8 = 1$
The profits were greater by one million dollars.

73. Find the height of the line at 1990: 50°F

74. The line is the steepest going down from left to right from 2000 to 2010.

75. Find the height of the bar representing 17–22: 600 students are between 17–22 years old.

76. Add the heights of the bars for 23–28 and 29–34.
$10 + 4 = 14$
1400 students

77. Mean $= \dfrac{8+12+16+17+20+22}{6} = \dfrac{95}{6} \approx 15.83$

Median $= \dfrac{16+17}{2} = 16.5$

78. $\sqrt{49} + \sqrt{81} = 7 + 9 = 16$

79. $\sqrt{123} \approx 11.091$

80. hypotenuse $= \sqrt{9^2 + 12^2}$
$= \sqrt{81 + 144}$
$= \sqrt{225}$
$= 15 \text{ feet}$

81. $-8 + (-2) + (-3) = -10 + (-3) = -13$

82. $-\dfrac{1}{4} + \dfrac{3}{8} = -\dfrac{1}{4} \times \dfrac{2}{2} + \dfrac{3}{8} = -\dfrac{2}{8} + \dfrac{3}{8} = \dfrac{-2+3}{8} = \dfrac{1}{8}$

83. $9 - 12 = 9 + (-12) = -3$

84. $-20 - (-3) = -20 + 3 = -17$

85. $2(-3)(4)(-1) = -6(4)(-1) = -24(-1) = 24$

86. $-\dfrac{2}{3} \div \dfrac{1}{4} = -\dfrac{2}{3} \times \dfrac{4}{1} = -\dfrac{8}{3} \text{ or } -2\dfrac{2}{3}$

87. $(-16) \div (-2) + (-4) = 8 + (-4) = 4$

88. $12 - 3(-5) = 12 + 15 = 27$

89. $7 - (-3) + 12 \div (-6) = 7 + 3 + (-2) = 10 + (-2) = 8$

90. $\dfrac{(-3)(-1) + (-4)(2)}{(0)(6) + (-5)(2)} = \dfrac{3 + (-8)}{0 + (-10)} = \dfrac{-5}{-10} = \dfrac{1}{2} \text{ or } 0.5$

91. $5x - 3y - 8x - 4y = 5x - 8x - 3y - 4y = -3x - 7y$

92. $5 + 2a - 8b - 12 - 6a - 9b$
$= 5 + (-12) + 2a + (-6a) + (-8b) + (-9b)$
$= -7 - 4a - 17b$

93. $-2(x - 3y - 5) = -2(x) + (-2)(-3y) + (-2)(-5)$
$= -2x + 6y + 10$

94. $-2(4x + 2) - 3(x + 3y) = -8x - 4 - 3x - 9y$
$= -11x - 9y - 4$

95.
$$5 - 4x = -3$$
$$5 + (-5) - 4x = -3 + (-5)$$
$$-4x = -8$$
$$\frac{-4x}{-4} = \frac{-8}{-4}$$
$$x = 2$$

96.
$$5 - 2(x - 3) = 15$$
$$5 - 2x + 6 = 15$$
$$-2x + 11 = 15$$
$$-2x + 11 + (-11) = 15 + (-11)$$
$$-2x = 4$$
$$\frac{-2x}{-2} = \frac{4}{-2}$$
$$x = -2$$

97.
$$7 - 2x = 10 + 4x$$
$$7 - 2x + (-4x) = 10 + 4x + (-4x)$$
$$7 - 6x = 10$$
$$7 + (-7) - 6x = 10 + (-7)$$
$$-6x = 3$$
$$\frac{-6x}{-6} = \frac{3}{-6}$$
$$x = -\frac{3}{6}$$
$$x = -\frac{1}{2} \text{ or } x = -0.5$$

98.
$$-3(x + 4) = 2(x - 5)$$
$$-3x - 12 = 2x - 10$$
$$-3x + 3x - 12 = 2x + 3x - 10$$
$$-12 = 5x - 10$$
$$-12 + 10 = 5x - 10 + 10$$
$$-2 = 5x$$
$$\frac{-2}{5} = \frac{5x}{5}$$
$$-\frac{2}{5} = x \text{ or } x = -0.4$$

99. x = number of students taking math
$x + 12$ = number of students taking history
$2x$ = number of students taking psychology
$$x + (x + 12) + 2x = 452$$
$$4x + 12 = 452$$
$$4x + 12 + (-12) = 452 + (-12)$$
$$4x = 440$$
$$\frac{4x}{4} = \frac{440}{4}$$
$$x = 110$$
math = 110
history = 110 + 12 = 122
psychology = 2(110) = 220

100. x = width
$2x + 5$ = length
$$2(x) + 2(2x + 5) = 106$$
$$2x + 4x + 10 = 106$$
$$6x + 10 = 106$$
$$6x + 10 + (-10) = 106 + (-10)$$
$$6x = 96$$
$$\frac{6x}{6} = \frac{96}{6}$$
$$x = 16$$
width = 16 m
length = 2(16) + 5 = 37 m